Potato production and innovative technologies

Potato production and innovative technologies

edited by:
Anton J. Haverkort
Boris V. Anisimov

Wageningen Academic
Publishers

ISBN: 978-90-8686-042-5

First published, 2007

Wageningen Academic Publishers
The Netherlands, 2007

Recommendation committee

J.E. Godfrey	Chairman International Potato Centre (CIP), United Kingdom
A. Gordeev	Minister of Agriculture of Russian Federation
Dr. C.P. Veerman	Minister of Agriculture, Nature Management and Food Safety, The Netherlands

Scientific committee

Dr. B.V. Anisimov	All-Russian Potato Research Institute, Russian Federation
Dr. S. Banadisev	Belorussian Potato Institute, Belorussia
Dr. A. Bondazehuk	Ukraine Potato Institute, Ukraine
Dr. A. Boronin	Institute of Biochemistry and Physiology of Micro-organisms (IBPM), Russian Federation
Prof. Dr. J. F Guenthner	Idaho Centre for Potato Research and Education, University of Idaho, United States
Prof. Dr. A. Hanafi	Institut Agronomique et Veterinaire Hassan II, Morocco
Dr. A.J. Haverkort	Plant Research International, The Netherlands
Dr. S.D. Kiru	All-Russian Plant Research Institute, Russian Federation
Dr. Ir. I. Mastenbroek	NAK Agro, Inspection and analyses, The Netherlands
Dr. E. Simakov	All-Russian Potato Research Institute, Russian Federation

Advisory committee

Z. Bautina	Director of APK VVC Ltd., Russian Federation
M. Binin	Deputy Director of the Department of scientific and technology policy of the Ministry of Agriculture of Russian Federation, responsible secretary
J.H.J. Haarhuis	Managing Director Europoint b.v., The Netherlands
B. Hamchiev	Director of the department of Plant growing of the Ministry of Agriculture of Russian Federation, Vice-Chairman
M. Musaev	General Director of All-Russian exhibition Centre, Russia
L. Orsik	Director of the Department of scientific and technological policy of the Ministry of Agriculture of Russian Federation
V. Ryabov	Advisor of the Minister of Agriculture of Russian Federation

Organisers

Europoint b.v.
Ministry of Agriculture of the Russian Federation
All-Russian Exhibition Centre (AREC)

Table of contents

Technology

Seed potatoes

Agronomy

Crop protection

Advances in breeding

Advances in biotechnology

Preface

The Potato Russia international conference took place in August 2007 in Moscow. The event gathered hundreds of representatives of the potato producer organizations, breeding, chemical and machinery companies and potato specialist of well know institutions. The conference was a real meeting place of East and West and never before was the state of art of potato technology – with special reference to the Russian federation and surrounding countries – brought forward in such a comprehensive and extensive manner. The urge to improve potato production in central and eastern Europe is even greater than in western Europe. Both parts of the world share similar concerns and challenges such as rapidly evolving consumer wishes and governmental regulations regarding the impact of production on the environment. Beside this the East of Europe has an enormous area under cultivation. The Russian Federation with a population of less than two hundred million grows potato on three million hectares whereas the European Union with a population of about four hundred fifty million only has two million hectares. This shows the importance of potato production in the largest neighbouring country of the European Union. The relatively greater importance of potato production coupled with a transition from a regulated market to a free market system where new roles and responsibilities are being formed give extra dimensions and challenges to potato production in the East. With the rapid opening up of markets the transition that takes place offers enormous possibilities for entrepreneurs and technology in West and East.

The call for papers has lead to an impressive list of over 40 accepted papers from many parts of the world. The book starts with a series of papers that give an excellent overview of consumer behaviour and marketing with examples of Russia, Turkey and Italy. The quality of processing and ware potato and methods of assessing is addressed by papers from the United Kingdom and Japan. The newest developments in technology, mechanization and storage are highlighted in contributions from Belgium and Russia. The importance of having adequately functioning seed potato systems with up to date rapid multiplication systems is shown in chapters from Cyprus, Russian regions and the commercial quality standards of the United Nations Economic Commission for Europe (UNECE) with authors from Italy, the USA and Switzerland. Recent agronomic and crop management developments are from the Czech Republic, Iran and Russia. Crop protection aspects put special emphasis on diagnostics and detection of resistance levels, among others against wart, with contributions from Switzerland and Russia. The extensive Russian breeding programmes - with value for the global potato community are highlighted in the breeding section with additional papers from Japan and the Netherlands. The book ends with a series of papers on molecular aspects of innovative breeding.

The event in Moscow was of international importance and having all presentations gathered in a single relatively comprehensive volume will enhance its value as a reference for the years to come.

Anton J. Haverkort and Boris V. Anisimov (editors)

Consumer behaviour and marketing: Russia and the World

The potato sector in the Russian changing economy

E.A. Simakov and B.V. Anisimov
All-Russian Potato Research Institute, 23, Lorch Street, Kraskovo, 140051, Moscow region, Russia

Abstract

The transition from a government regulated economy to a free market deeply influenced the Russian potato sector. At the same time the Russian Federation remains one of the largest potato producers in the world after China. More than 10% of the world potato production is planted in Russia. Potato is one of the main foodstuff products in the country. In 2006 the total planted area of potato in Russia was more than 2.9 million hectares and the gross yield was 38.7 million tons. In this paper the authors have represented the information regarding potato production, usage, market quality system improvement, international cooperation and other aspects, connected with potato industry in Russia. To know in detail the actual problems of potato sector in Russia is very important for the development of Russian and in general of the international potato sector.

Keywords: production, usage, market, quality, standard

Potato production: Russia and the world

The analysis of the world level of potato production shows that the potato production in the world is increasing continuously. In 2005 328 million tons of potato were produced in the world (FAO, 2005). Potato gross yield in the EU was 67 million tons from 2 million ha, in Russia it was 37 million tons from 3 million ha (Table 1).

The leading position in potato production level belongs to China (75 million tons), Russia is the second potato producer (37 million tons). The following countries are India (25 million tons), USA (20 million tons), Ukraine (19 million tons), Poland (15 million tons), Germany (13 million tons), Belarus (8 million tons), Netherlands (7 million tons), France (6.9 million tons), Great Britain (6 million tons) and Canada (5 million tons). In 2006 in Russia the total planted area of potato was 2,974,000 ha and the gross yield was 38.5 million tons.

Table 1. Potato production in the world, EC and Russia.

	Planted area (million ha)	Production (million tons)
The world*	19.0	328
EC*	2.0	67
Russia**	3.0	37

*FAO, 2005
**Rosstat, 2006

Traditionally potato production in Russia is located practically in all parts of the country. The main regions of potato production are Central (11 million tons), Privolgskij (9.5 million tons) and Siberian (6.3 million tons) regions. Other significant regions where potato is grown are Ural (3.2 million tons), Southern (3.3 million tons), Northwestern (2.5 million tons) and Far-Eastern (1.6 million tons) regions. Table 2 shows the data of the area planted and potato production in the main Russian regions.

According to the management forms during the last 15 years potato producers in Russia generally are divided into 3 main types: agricultural enterprises (AE), peasants farms (PF) and small private farms (SPF). Different types of farms have their own features of areas under crops, gross yield, crop capacity and the technical level and efficiency of potato production. Agricultural enterprises (AE) include the collective farms (former kolkhozes), the state farms (former sovkhozes), the associations, the join-stock companies and the agricultural producing cooperative societies. In 2006 the potato in AE was grown on an area of 155,000 ha, gross production was 2.7 million tons or 7% of the total volume of potato production in all categories of farms. The average crop yield in AE was 18.1 tons/ha.

Peasant farms (PF) have a new form of management in Russia. They were formed about 15 ears ago. In 2006 they had 75,000 ha of planted potato area and produced more than 1.1 million tons of potatoes, that is about 3% from the total production volume in the all categories of farms. Small private farms (SPF) are represented by a great variety of minor forms of management. This category includes the country resident's plots, gardens and country plots of city dwellers, and some others. These minor forms comprise 16 million small farms. In 2006 SPF produced 34.7 million tons of potato that is 90% of the total volume production in all categories of farms.

Thus the analysis of the structure and volume of potato production according to the different categories of the farms shows that in Russia after the transition from a government regulated economy to a free market the major part of potato production is transferred to the private sector of small family farms. Most of them have small potato fields (less than 2 ha). The share of large agricultural enterprises in the gross potato yield was reduced from 34% in 1990 to 10% in 2006 (Figure 1). Minor areas with the small production units of cultivated potato prevail.

Table 2. The potato production in significant Russian regions in 2006.

Region	Planted area (1000 ha)	Production (million tons)
Russia, total	2974	38.5
Central region	930	11.4
North western region	199	2.3
Southern region	319	3.2
Privolgskij region	709	9.8
Ural region	205	3.7
Siberian region	484	6.5
Far-Eastern region	128	1.6

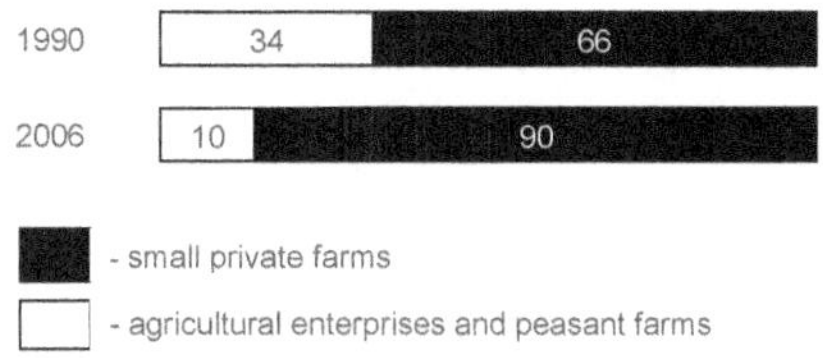

Figure 1. The share of different categories in potato production (%).

Potato production by SPF is to a great extend done manually using the primitive technologies and minimal mechanization of labour. Average crop yield in the most regions is not more than 11-12 t/ha.

We observed also that the present storage system and technologies limit producers' capabilities to prepare potato for the market. For this reason the potato producers have to spend significant efforts to sell the potato surpluses. The sector of larger agricultural producers belonging to agricultural farms and peasant farms are produced only 10% from the total gross potato yield. In the large agricultural enterprises average crop yield in 2006 was 18.1 t/ha. Yield in Central and Ural regions was significant above the average (more than 20 t/ha, Table 3).

Use of potato

The structure of potato use as well as potato production is highly fragmented. For a large proportion of the rural population in the main Russian regions, potatoes are a semi-subsistence crop with production targeted at farm household needs for home consumption and animal feed. Therefore the usage structure is mainly defined by both table potato demands and consumption as animal feed.

During the last years the consumption of table potato was constant and estimated at a level of 120 kg per person. The total annual fresh table potato consumption amounts to 18 million

Table 3. Potato yield in Russia in 2006.

Region	All producers (t/ha)	Agricultural enterprises (t/ha)
Russia, total (in average)	13.0	18.1
Central region	12.3	20.3
North western region	11.6	15.8
Southern region	10.0	14.9
Privolgskij region	13.9	18.2
Ural region	18.0	20.3
Siberian region	13.4	15.6
Far-Eastern region	12.9	13.1

tons, i.e. about 50% of all potato production in Russia is used as table potato. As animal feed about 6-7 millions tons of potato is used. Usually in years of abundant potato crop increased potato use in animal feed is found more than in dry years.

Demand for seed potato for all types of farms is 9 millions and potato suitable for processing is 0.5 millions tons. It is expected that the proportion of potato production for processing will increase as the food processing industry develops. So, taking into account the fact that the storage losses are still significant (about 2 millions tons), the total annual demand for the potato in Russia is about 36-37 millions tons.

Potato market

Nowadays Russia has a comparatively low level of potato market development. The ware/fresh potato market is mainly formed under the influence of the small private farms (SPF). They are dominant suppliers on the ware/fresh potato market and don't depend on the seasonal changes of sales volume, i.e. they determine the price level during of whole sale period.

The SPF share is about 70% of total volume (about 3-3.5 million tons annually) of potato traded. But taking into account that the largest share of the harvest is used for the producers' own consumption the marketable surplus or proportion traded is thus comparatively small and the level of potato traded is not more than 10% of total national production.

The distribution system is not developed either and this is a very important problem of the Russian potato market. Nowadays we can observe many different variations in the potato distribution system including farm sales, street trading, market places, as well as specialized greengroceries and supermarket. Since a few years the supermarkets with increasing market share is an important supplier of potatoes for citizens.

The greengroceries, market places and street trading continue to keep a significant part of their market share, but problems of product quality in this sector are becoming more and more important as consumers are looking for guarantee of quality. Usually quality control is higher in supermarkets. Supermarkets increasingly demand guarantees regarding product quality from their suppliers. In its turn these suppliers lay down these demands to their producers, especially to those ones, which collaborate on the basis of long-term contracts.

In the past the average consumers used to buy potatoes mostly from the market places, street trading and partially in the greengroceries. Nowadays they more and more buy their potatoes in the supermarkets. So, the Russian potato sector is little by little beginning to find the real opportunities to improve the products quality. Thus the production of table potato is oriented mainly to the local market. The Russian potatoes practically are not presented at the international market. Only small amounts of potato were exported to the republics of Central Asia in the last years. As a result today a participation of Russia on the international market is considerable less significant than in the past. Import of potato also is very modest (about 2% of total gross yield) and it can not noticeably influence the potato market.

The search for new opportunities for strengthening of integration processes in this sphere including the import and export is very important, in order to keep track of developments at the domestic and foreign potato market.

Consumer behaviour and potato quality

Nowadays in Russia quality improvement of fresh unpeeled table potato is becoming more and more important. The basic requirements for quality of fresh table potato, which is sold for catering and retail market distribution systems, is defined by the State Standard of the Russian Federation. In the light of international trade development it was necessary to harmonize the Russian national standard of fresh table potato with the international standards of quality requirements. First of all they are FFV-30 and FFV-31 EEC UNO standards, concerning the selling and the goods quality control of early potato and fresh table potato respectively, which go for sale to the international market.

It is expected that harmonization of national standards with international requirements will allow to use produced potato more fully and to meet the needs in potato of the whole population. Furthermore the state price adjustment will stimulate the production of high quality products. It will give an opportunity for producers to improve seed industry, to extend the range of goods, to transfer modern technologies, machinery and equipment for presale products preparation. Also it will allow consumers to use high quality potato products.

According to the new Russian standard the fresh table potato is divided depending on realization term to early table potato (which is sold till September 1) and late table potato (which is sold after September 1). There is the similar classification of fresh table potato in the international practice (FFV-30 and FFV-31 EEC UNO standards). So the classification of table potato is also uniform to the international classification. Depending on quality the fresh table potato is divided into the following classes:

- early table potato to 1st and 2nd classes;
- late table potato to extra-, 1st and 2nd classes.

The tuber appearance and size are the major characteristics of fresh table potato quality. Tubers should be fresh, clean, whole, dry, free from defects and without sprouts. According to standard's requirements there are restrictions for content of tubers with insignificant defects, which don't make worse the marketable state of table potato. They are light-green spots till 2 square centimetres, which can be removed by peeling; not serious tuber damage (cuts, pits, chaps, dents); potato scab, oosporosis and wireworm damages. Standard's requirements concerning the tubers size of fresh table potato for sale have significant importance. The difference between minimum and maximum tubers diameter in the same packing should not exceed 20 mm for 'Extra'-class and 30 mm for the 1st class.

Good quality of fresh table potato is defined by the absence of mechanical damages (cuts, pits, chaps, dents) and other defects such as change of tuber shape, second growth, potato scab, rusty potato tuber blotch. Taking into account differences of Russian agroclimatic zones, where fresh table potato is cultivated, the standard supposes that the contents of tubers with above defects should not be more than 2% from weight for 1st class and 5% for 2nd class of early table potato. For late potato this index is 2.0% for 'extra'-class, 5.0% for 1st class and 10.0% of weight for 2nd class.

In Russia traditionally the main part of table potato is stored, prepared and sold on the farm or enterprise where it was produced. Nets of packing services and wholesale centres are not sufficiently developed and they are grading only small amounts of potatoes. The requirements to grading and packing of fresh table potato for sale are harmonized with international ones also. The 'extra'-class potato should be washed and potato of 1st and 2nd classes should be cleaned with or without water. Extra-class and 1st class potato should be packed into tissue, polymeric or mesh packets. However according to the standard the potato of 1st and 2nd classes meant for catering trade don't need to be packed. The packets with potato should be marked with indication of class ('extra'-, first, second). The washed potato must have the last date of expire indicated.

The Russian consumers are very conservative toward potato. Traditionally they prefer well cooked potato with white flesh. However the preferences are changing lately. For example, a part of the consumers today approves of varieties with more firm flesh and also with yellow pulp and red peel.

A very important development is that in Russia the modern consumers are increasingly concerned about their health and in line with this they developed a new attitude towards food. A part of consumers creates specific demand for potato varieties, types, production methods, etc. For example, nowadays some consumers prefer low starch content potato varieties and potato that have been produced in a so called environment-friendly way on lower inputs fertilizer and chemicals (organic production). With an increase of this specific demand the new varieties, which differ in cooking quality, size, shape and taste appeared at the market. Many Russian potato varieties, which were bred during the last years, became very popular because they met the specific consumer's demand. It is expected this trend will continue.

The size and type of packages have changed also. Thus, many shops often sell potato packed in smaller packs, such as 1.5, 2.5 and 5 kg. For consumer, plastic (polyethylene) foil, coloured for protection against light and perforated for ventilation, is mostly used for the packing of potatoes. Greengrocers also sell the potatoes in loose paper bags. Hopefully these changes will give to Russian consumers a broader assortment of high quality table potato.

Potato products

Since a few years convenience potato products are becoming increasingly popular and nowadays they are already well known by Russian consumers. This category consumer can find a wide range of products: washed potato, peeled uncooked potatoes, peeled and pre-cooled potatoes and deep frozen potato products. These products are being sold in the specialized greengroceries and supermarkets.

Washed potato was well appreciated by many Russian consumers after several years of consumption. At the same time other more simple potato products such as peeled uncooked potatoes is a rather new development. With these products the consumers always have an easy and quick meal. Since these products are in line with the needs and preferences of many groups of consumers, it can be predicted that the volume of this products will continue to grow and they will occupy more and more important place at the Russian consumers market. It is expected that this trend in Russian food processing industry will continue in the future.

The deep frozen potato products as well as crisps (respectively, chips), dry potato and many other complete meal potato, and starch also did not go through a positive development in Russia during recent years. Now these kinds of products are imported into Russia from European countries mainly.

Import of processing potato to Russia is rising continuously. Thus, in 1995 Russia imported 20,000 tons of potato products, but this amount increased to 150,000 tons in 2004-2005. Nowadays the level of potato processing to food products is very low in Russia. Production of processed potato decreased in fivefold in comparison with 1990 (Table 4). In fact all of 40 potato starch factories, which were built in the Russian Federation, don't produce potato starch anymore. Nor do factories, producing dry mashed potato and alcohol from potato. They use only 3-10% of their production capacity.

Taking into account the importance of this branch, a research and development program was proposed and submitted to the present Government of the Russian Federation for consideration. Within the framework of this programme the future prospects and new real ways for the development of potato production in Russia are defined for the period till 2010.

The exceptional value to Russia will be the overall improvement of potato yield on farms to the level of 18-20 tons/ha, reduction of losses by 25-30%, and the improvement of economic parameters of production. That will be enable us to attract investments from non-sanitary sources (venture capital, entrepreneurs, interested foreign planters and other) into the development of infrastructure, renovation and modernization of the storage base, improvement of technical and technological levels of the potato production in Russia.

Table 4. Processed potato volume in Russia 1990-2005 (1000 tons).

Product	**1990**	**1995**	**2000**	**2004-2005**
Total processed potato products	133.3	10.1	12.6	27.8
including:				
starch	98.8	3.2	4.3	2.8
dried potato	7.9	1.5	0.4	0.9
quick-frozen potato	16.5	1.5	0.2	0.1
mashed potato and crackers	4.0	1.3	0.1	0.1
fried potato	6.1	2.6	7.6	10.7

International cooperation

Cooperation with international companies is very important for the development of potato production in Russia. The Ministry of Agriculture and many Russian regions and agricultural enterprises have worked actively in creating connections between Russian and international organizations and companies in the field of potato production and transfer of modern Potato technology.

For example, Dutch Potato companies such as 'Agrico', 'HZPC', 'APH Group' and other successfully work in Russia and have gained a positive experience of cooperation with different regions, Agricultural enterprises and Peasants farms. Some Germany companies such as 'Europlant', 'Norica', 'Solana', Finland companies 'Sadokas', SPK and others are well known in Russia. There are the good contacts with the Canadian company 'McCain', British company 'Higgengroup' and other foreign companies for the cooperation development in the future.

One of the most important parts of international cooperation is the organizing of production of farm machines and implements in Russian industrial enterprises (for example, 'Kolmag' in Kolomna, 'Eurotechnik' in Samara and others) together with leading European companies. Development of international cooperation will help to provide economically efficient potato production and introduce advanced growing, harvesting and storage technologies, and processing industry development in Russia.

Many Russian agricultural enterprises and potato producers are ready to work as the cooperative partners with any foreign company which wants to create connections to Russia on the way to an open market economy. The main directions of national and international cooperation development are the following:

- breeding of new potato varieties which are adopted to the local conditions, resistant to diseases, best use of environmental and technological resources;
- production of virus free material and original seed potato;
- improvement of quality control methods and certification system;
- improvement of potato technology and crop agronomy (tillage seed bed preparation, fertilization, planting, crop maintaining, harvesting);
- improvement of storage technology;
- development and promotion of potato processing;
- improvement of potato crop management;
- training for potato growers.

The Ministry of Agriculture of Russian Federation provides the following essential measures in order to improve potato production and marketing:

- to assist and support specialized potato growing farms as the major producers and suppliers of high quality seed and ware potato for large industrial centres, the northern regions and processing enterprises;
- to increase the efficiency of potato production in agricultural enterprises;
- to introduce the new technologies;
- to pay more attention to providing private sector, farms and private households with high-quality seed material.

Conclusion

It is expected that the exceptional value to Russia will be the overall improvement of potato yields on farms to the level of 18-20 tons/ha and the improvement of economic parameters of potato producers. That will enable us to attract investments from alternative sources (Venture capital, entrepreneurs, interested foreign planters and other) into the development of infrastructure, renovation and modernization of storage and the improvement of technical and technological level in potato production.

Curriculum vitae – Evgeniy A. Simakov

Evgeniy A. Simakov studied at the Agricultural University in Penza where he received a Master of Science degree in 1972. In 1977, after a short period of practical work, he was invited to the All-Russian Potato Research Institute, where in 1980 he received a PhD in Biology. In 1977-1985 Evgeniy Simakov worked in the Potato Research Institute as a researcher. In 1985-1996 he worked as a Head of Breeding Laboratory. In 1997-2004 – Head of Breeding and Seed Potato Centre. Since 2004 until present – Director of All-Russian Potato Research Institute. E. Simakov is a specialist in breeding and seed potato production.

His current research interests include problems of increasing genetic variability, breeding of new potato varieties and rapid multiplication of virus-free original material and basic seed potatoes. At present he is coordinator of the new program of potato research in Russia for the period 2006-2010. E. Simakov took part in national and international potato conferences and congress: Romanian (1997), Finland (1999), Poland (2000), German (2002, 2006).

Curriculum vitae – Boris V. Anisimov

Boris V. Anisimov studied at the Agricultural University in Stavropol where he received a Master of Science degree in 1959. In 1966, after a short period of practical work, he was invited to the All-Russian Potato Research Institute, where in 1970 he received a PhD degree in Biology. In 1970-1980 Boris Anisimov worked in the Potato Research Institute as a Head of the Department of Seed Potato Research. In 1981-1986 he worked as a Director of Scientific Pathological Laboratory (SPL USSR) of the Ministry of Agriculture, Ambo, Ethiopia. In 1987-1992 he worked as a Head of Breeding and Seed Potato Centre. In 1993-2003 – Head of Potato Section Program Development at the Ministry of Agriculture of the Russian Federation. Since 2003 until present – Deputy Director of All-Russian Research Potato Institute.

Boris V. Anisimov was an initiator and coordinator of seed potato research production and seed certification program development in Russia, the breeding and seed potato program development at the Potato Research Institute, was a team-leader responsible for the breeding and seed potato program (1970-2003).

At present he is an initiator of the development of a new concept for seed production in Russia for the period 2005-2010. Boris Anisimov is an active member of Russian Academy of Natural

Science since 2001. He took part in national and international potato congresses as a member of the Organizing Committee in Finland (1999), Netherlands (2005), Germany (2006).

Company profile – All-Russian Potato Research Institute

The All-Russian Potato Research Institute (VNIIKH) was founded in 1930 in order to improve yield and quality of potatoes in Russia. The main objectives of the All-Russian Potato Research Institute are coordination of the national potato research and development program; improvement of genetic variability and breeding of new varieties; production and delivering original seed potato of high biological quality; development and promotion of modern potato technologies (growing, harvesting, storage); national and International cooperation development.

The Institute includes the following research departments: Genetics, Breeding, Biotechnology, Seed potato, Agrotechnics and Fertilizers, Plant protection, Technology and Mechanization, Storage and processing, Marketing and management, Scientific and Technical Information.

For 75 years the All-Russian Potato Research Institute has been responsible for improving the quality of potatoes in Russian Federation. The Institute's core business in the coordination of the national potato research program, the improvement of existing potato varieties and the breeding of new varieties. Other key tasks are the production of high quality starch potatoes and the development and promotion of modern technologies in the field of breeding, harvesting and storing of potatoes.

Main results of the All-Russian Potato Research Institute are following facts: The breeders of the Institute created more than 90 potato varieties, amongst which there are new high yielding disease-resistant and very popular varieties, such as Jukovskiy Ranniy, Udacha, Lukianovskiy and others. The Institute annually produces about 80,000 micro-plants *in vitro*, over 1,000,000 mini-tubers on which basis the Institute's experimental farms produce about 10,000 tons of original seed material (pre-basic seed material of the varieties recommended for growing in different regions). Up to 2,000 diagnostic kits for testing potatoes for viruses and bacterial pathogens are produced annually for seed certification program development. Fifty young specialists and farmers and 120 regional trainers annually attend the practical training course organized by Institute.

Environmental and genotypic influence on sensorial and typical potato characteristics

L. Tedone[1], E. Palchetti[2], M. Manzelli[2], S. Benedettelli[2], S. Romagnoli[2], D. Traversi[3], F. De Cillis[3], V. Vecchio[2] and V. Marzi[1]
[1]University of Bari, Department of Plant Production, Via Amendola 165/a 70125, Bari, Italy
[2]University of Florence, Department of Agronomy and Land Management, P. le delle Cascine 18 50144, Florence, Italy
[3]Centro Ricerche Bonomo, Contrada Castel del Monte 70031 Andria, Bari, Italy

Abstract

Italy has a great culinary and market food tradition, playing a central role in the ambit of typical products with more than 20% of European certified productions. Nowadays, the agriculture and diet issues are becoming crucial and complex, especially due to its direct and indirect reflexes on life quality. Potato plays a primary role in Italian horticulture for the possibility of cultivation in different periods of the year guaranteeing fresh product spaced out in time. However Italian products suffer the market competition of other Mediterranean countries, with special regards to products without origin guarantee. In order to find parameters useful in characterizing territorial origin, a collegial research was implemented on potato, involving typical areas for potato cultivation, located in Tuscany, Apulia and Sicily, and using three typical potato varieties, respectively Rossa di Cetica, Sieglinde and Arinda. Eco-physiological parameters were recorded during the crop cycle in order to evaluate vegetative development. Total yield and commercial tuber size classes, mineral element content, market and sensorial characteristics, volatile compound determination were measured. Preliminary results underlined differences in eco-physiological patterns, with greater vegetative development of Rossa di Cetica, in production trends, with greater productivity of Arinda and Rossa di Cetica and in market characteristics, with dry matter, ACB and consistence at cooking highest values in Rossa di Cetica. Panel sessions underlined differences in terms of typical taste, chestnut taste, grass odour in relation to varieties and cultivation areas. Aromatic analysis highlighted 56 chromatographic compounds some of which discriminating for variety and others for cultivation area. Macro and micro element determination highlighted the importance of some elements in discriminating in function of environment or genotype. Such analysis allowed to correlate the product to the environment of cultivation by means of objectively verifiable indicators confirming the importance of the territory in the characterisation of typical products.

Keywords: typical food, fingerprinting, mineral elements, sensorial analysis

Introduction

Agriculture and diet have always been strongly linked and exercise both the direct and indirect effects on the quality of life in human beings. In particular, agriculture has always played an important role in determining food security, either when referring to food availability in quantitative terms, or, especially in modern societies, as a guarantee of a safe and functional food production. In terms of a maximum integration, today it is possible to talk about a global system aimed at quality, specifically in food sector, and is composed of various steps starting with the environment and involving all production, transformation, preservation and distribution processes.

Undoubtedly quality is a conventional parameter defined by sectorial experts. Quality is not constant in time, is of limited duration and is regulated by specific rules. Moreover, complex and diversified aspects are involved in the definition of quality, including extrinsic and intrinsic food features along all sectors of the agro-food chain. When considering the extrinsic features, it is appropriate to focus on environmental, geographic and agronomic characteristics as well as historical-cultural traditions concerning production and consumption of food modalities, whereas the intrinsic features are prevalently determined by chemical-physical food characteristics. Generally, food quality definition also includes the relationship between the food and consumer, in which choice is affected by subjective parameters linked to specific sensory perceptions. These aspects, despite having received previous attention, are, today, particularly important because scientific and technological progress have allowed a more profound exploration of the chemical and biochemical composition of food. Their importance also resides in the increasing role that nutritional and sensory characteristics are playing in agricultural product market competition.

According to the above mentioned information, the present trend is aimed at valorising typical products is significant, as is well confirmed by the attribution of brands to these products certifying the provenance (PDO, PGI, TSG, etc.). Referring to certified food products, Italy ranks first between the countries in Europe, covering the 22% of whole production (http://ec.europa.eu/agriculture/qual/it/sv_it.htm). Typical food has an interesting added value and, in recent years, the market of these foods was unaffected by the effects of the agricultural sector crisis, caused by a surplus offer and stronger international competition. Typical food has specific sensorial characteristics, particularly appreciated by some categories of consumers. Moreover, the typical conditions of production and transformation can result in a larger content of secondary nutrients, such as mineral salts, vitamins and phyto-complexes, in which effects on human health have recently been recognised.

The cultivation of potatoes typical to Italy

Potato plays an important role in Europe, representing one of the principal sources of carbohydrates along with cereal crops. In Italy, potato consumption is less than in North-European countries because of the prevalence of cereal food. Nevertheless, in some regions of South Italy, extra-seasonal potato production occupies an interesting niche market (Table 1), particularly appreciated in North Europe.

The possibility of exploiting potato cultivation all year round as well as the different pedo-climatic conditions on the growing environment has resulted in a differentiation of potato production areas. Nowadays, it is possible to individuate some areas growing typical potatoes where producers are organised in consortia adopting specific protocols of cultivation. Between them, it is interesting to distinguish between the consortium *Typical Potato of Bologna* for ware potato and the consortium *Typical Potato of Syracuse* for early potato. Moreover, other consortia are promoting their own potato production, such as the consortium of Salento (Apulia) that has

Table 1. Area, yield and production of potato in Italy.

	Area (ha)	Yield (ton/ha)	Production (ton)
Early potato	18,460	19.53	360,573
Ware potato	54,003	27.18	1,468,017
Total potato	72,463	25.23	1,828,590

recently requested the certification of the local potato production mainly represented by an old German variety, namely Sieglinde, particularly appreciated by German consumers.

Cultivated potatoes are mainly represented by recently selected varieties which are commercially diffused. Nevertheless, in recent years, growing attention has been paid to the recovery of old local varieties, such as Quarantina, Cannellina and Brugnoa in Liguria (Angelini, 1999), Rossa di Cetica in Tuscany, Turchesa in Abruzzi and Viola Calabrese in Calabria (Scazziota *et al.*, 2005; Lovatti and Vecchio, 2003; Andrenelli *et al.*, 2002, Ghiselli *et al.*, 2002, 2001 a, 2001 b). Italian potato production is typically associated with the territory of cultivation. Characteristic factors are often represented by local varieties. In order to guarantee typicality of production, it is important to have objective, verifiable markers which are able to discriminate between variety and provenance, either within Italian potato production or between Italian and imported potato products.

Table 2. Early potato import/export in Italy (ISTAT, 2006).

Country	Import 2000	Import 2001	Import 2002	Export 2000	Export 2001	Export 2002
Austria	595	752	1,642	5,555	5,232	5,240
Cyprus	2,740	1,640	4,064			
Egypt	54,575	65,538	86,440			
France	24,565	16,369	25,716	2,118	3,019	4,173
Germany	1,690	8,185	8,438	119,256	125,680	153,809
Israel	350	1,158	298			
Netherlands	855	197	650	9,387	12,469	8,226
Poland	-	-	21	6,210	8,167	8,069
United Kingdom	47	-	13	5,882	7,535	3,545
Slovakia	-	-	-	856	2,104	7,356
Slovenia	-	2	-	1,112	1,597	1,622
Others	1,799	661	197	18,470	31,762	33,117
Quantity (ton)	87,216	94,501	127,478	168,846	197,566	225,158
Value (€)	17,840,080	24,788,437	37,125,284	48,804,897	76,175,196	60,803,389

Techniques for the determination of geographic provenance of food products

Concerning the identification of potato varieties, in addition to the traditional morphological analysis, modern techniques have been adopted in recent years, such as methods based on enzymatic protein screening (Nieto *et al.*, 1990) and DNA analysis by means of molecular markers (Polzerova, 2001; Gorg *et al.*, 1992; Coombs *et al.*, 2004; Feingold *et al.*, 2005). All these techniques have provided interesting results (Cooke, 1999), representing a valid alternative to morphological identification.

Studies on methods which are useful in the identification of geographic provenance are less numerous. Nonetheless, the interest in linking agricultural products to production area by means of reproducible markers is increasing, especially for those products with the higher market value. Of these products, it is interesting to include wines, for which provenance screening methods have been adopted based on the content analysis of some organic compounds, such as phenols (Peña-Neira *et al.*, 2000), amino-acids (Herbert *et al.*, 2000), biogenic amines (Heberger *et al.*, 2003), aromatic alcohols and other volatile substances (El Jazouli, 2000; Guerzoni, 2004), which are either determined chemically or by means of sophisticated spectrometry analysis (Liu *et al.*, 2006; Picque *et al.*, 2005). Similar studies have also been conducted on animal food products. For example, meat products were analysed taking into account lipid acidic composition and isotopic ratio between elements present in the organic matrix (Franke *et al.*, 2005). However, with animal feed, the composition can be varied conveniently and this largely affects these parameters. Analogues results were obtained from analysing milk and cheese (Brescia *et al.*, 2004; Manca *et al.*, 2006; Pillonel *et al.*, 2002). Cheese was also analysed by determining volatile compounds using gas-chromatography, mass spectrometer and electronic nose (Pillonel *et al.*, 2003) and free amino-acids, in particular taking into consideration amino acid ratios (Pillonel *et al.*, 2003).

Some of the above mentioned methods could also be applied to vegetable products such as potatoes. However potatoes are not rich in volatile compounds and the organic compound content largely depends on the duration of preservation period. Thus, great interest is directed towards those techniques useful in determining substances unaffected by quantitative variation. In this regard, it is possible to make mention of the methods for the determination of macro (P, K, Ca, Mg, Na) and microelement (Al, Cu, Fe, Mn, Zn, Ba, Cd, Cr, Co, Li, Mo, Ni, Pb, Rb, Se, Sr, V, etc.) content. These mineral elements, even if present in different chemical compounds, remain quantitatively stable if analysed with methods that eliminate any possible chemical bonds, such as those associated with different types of spectrometry analysis (FAES, FAAS and ICP-AES). Atomic emission spectrometry (ICP-AES) is assuming a central role in this type of analysis, mainly because of the rapidity in sample preparation and the large screening capacity.

Mineral element analysis was applied to different food products (Franke *et al.*, 2005; Brescia *et al.*, 2003; Del Signore *et al.*, 2004; Coetzee *et al.*, 2005; Ariyama *et al.*, 2007; Ariyama *et al.*, 2004). All those studies detected significant differences, often revealing clear divisions between food products grouped in function of their geographic provenance.

Concerning potato, recent studies have provided interesting results. Pulp tuber mineral content was utilised in distinguishing DOC potatoes (Kennebec cultivar) grown in Galicia (Spain) from other potatoes cultivated in the same region as well as in other Spanish regions (Padín *et al.*, 2001; Peña *et al.*, 2001). Differences in the composition of elements permitted the screening both within and between potato provenances.

Research conducted in the Canaries Islands (Rivero *et al.*, 2003) applied a similar methodology on local potato cultivars belonging to *Solanum tuberosum* ssp. *Tuberosum, Solanum tuberosum* ssp. *Andigenum* and the hybrid species *Solanum* × *chaucha*. Even though some differences were detected, genetic screening was incomplete, whereas a better discrimination was observed between potato provenances (North and South of Tenerife Island) confirming the efficacy of this technique in food geographic origin studies.

In the United States (Anderson *et al.*, 1999) mineral element analysis was applied to verify the geographic origin of potatoes grown in Idaho. Differences in metal contents detected in potatoes derived from different regions was highly significant for a large part of the mineral elements analysed. Multivariate statistical methods permitted the correct classification of 97% of the Idaho potatoes and 95% of the potatoes grown in other regions.

The above mentioned results highlight that the mineral element content of potato tubers is strongly correlated to the growing environment, even though in some cases, discrimination between varieties is less clear. Moreover, it is important to emphasise that determination based solely on a mineral element is insufficient in detecting correlations, whereas a multi-element analysis may represent a valid tool in food provenance *fingerprinting*.

Sensorial characteristic determination in potato tubers

Since ancient times, food was evaluated in function of sensorial characteristics. Nowadays, sensorial evaluation, as defined by the *Institute of Food Technology* (IFT), represents a methodology utilized to measure, analyze and explain food characteristics, as perceived by the five senses.

Individual sensorial organs act as analytical tools, with subjective thresholds for each parameter. This explains the preferences and/or refusals for a specific food. Moreover, the brain response to a specific sensorial stimulation is not constant, but is affected by exogenous (temperature, humidity, light, etc.) and endogenous factors (endocrinal, metabolic and psychological).

Perception of sensorial characteristics of a food product is the result of sense stimulation at different intensity levels by food physical-chemical properties. Food sensorial characteristics can be grouped, as defined by ISO standards, in the following categories:

- Aspect: food sensorial characteristics mainly perceived by sight.
- Taste: combination of taste and olfactory sensations.
- Consistence: sensorial characteristics mainly perceived by touch and movement sensation.

Food sensorial quality depends on different levels. For example, it is clearly demonstrated that the packaging can positively affect the purchase of some food products, and that other food products, usually not well appreciated, may be consumed in particular situations, such as during parties (Multon *et al.*, 1985). Another example of the psychological influence on sensorial perception is represented by the brand of food. In the presence of a modification of sensorial characteristics, food is often not appreciated or refused. For example the taste, smell, colour or consistence not corresponding to what is considered typical for a specific food immediately activates a self-defence or refusal mechanism. A food may have an excellent nutrient composition or have indispensable properties for health, but if the latter is not appealing in a sensory manner, it will be not consumed, thus annulling its nutritional value.

Steam cooked potatoes are characterised by a light but typical aroma, caused by the production of different precursors during the cooking process. The greater contribution to potato flavouring is associated with a small number of volatile compounds, such as metionale and different pyrazines.

In general, the cooked potato aroma is weak and strongly different from raw potatoes. This is dependent on the presence of lipoxygenase, that catalyses the oxidation of unsaturated fatty acids (Galliard and Phillips, 1971). The reaction is rapid with cell breakage, as happens during potato pealing (Galliard and Matthew, 1973), and results in the production of a high number of compounds (Hsieh, 1984). Enzymatic reactions can also occur during the boiling process due to the movement of active enzymes from the central uncooked part of tubers towards external parts where enzymatic activity is inactivated (Josephson and Lindsay, 1987).

Aroma tuber evolution during the cooking process is composed of three main reactions:
- Compound degradation during the cooking process and thus a decrease in the aromatic compound content in comparison to raw potatoes.
- Non enzymatic auto-oxidation of linoleic and linolenic acids, and the production of different compounds more concentred in cooked potatoes.
- Strecker degradation of amino-acids in aldeids.

Aromatic compound determination techniques emphatically showed the production of a defined number of aromatic compounds during cooking, which are linked to the above mentioned reaction typologies (Petersen *et al.*, 1998).

Metionale is a degradation product of methionine. Pyrazines are derived from the degradation of amino-acids and sugars during the Maillard reaction. Moreover, more than 140 volatile compounds contributing to the aromatic food profile were identified. Volatile aromatic compounds tend to increase with temperature, in fact in boiled potatoes up to 250 compounds were recognised (Whitfield and Last, 1991).

The presence of volatile aromatic compounds is associated with transformation techniques and cooking methods, with significant differences between raw (Petersen *et al.*, 1998), boiled (Josephson and Lindsay, 1987; Nursten and Sheen, 1974; Oruna-Concha *et al.*, 2002b; Petersen *et al.*, 1998), baked (Butter *et al.*, 1973; Coleman and Ho, 1980; Coleman *et al.*, 1981; Duckham *et al.*, 2001; Duckham *et al.*, 2002; Ho and Coleman, 1980; Oruna-Concha *et al.*, 2001), micro-oven cooked (Oruna-Concha *et al.*, 2002a; Oruna-Concha *et al.*, 2002b), and French fried potatoes (Carlin *et al.*, 1986; Wagner and Grosch, 1997; Wagner and Grosch, 1998). Aromatic compound determination is more sophisticated and sensitive than micro-element determination since it is affected by harvest time, preservation time and modalities, possible tuber damage, and possible facilities conferred to food products such as pre-peeling (Thybo *et al.*, 2006).

There are different techniques for the extraction and concentration of the aromas. These techniques can be summarised as follows: liquid-liquid extraction (Monje *et al.*, 2002), distillation (Horval *et al.*, 1991), supercritical fluid extraction (Stashenko *et al.*, 1996), and Solid Phase Micro Extraction (SPME) (Niesser and Klampil, 2000). It is possible to recognize different determination methods, and better results are obtained with the SPME technique (Sanches-Silva, Lopez-Hernandez, Paseiro-Losada, 2004).

In defining the influence of compounds on cooked potato aroma, it is necessary to link sensorial evaluation to chromatographic determination in order to define those parameters that influence the flavour. Many studies (Hulrich, 2000) highlighted the possibility in evaluating cooked potatoes qualitatively by means of panel tests in order to obtain a quantitative representation which can be well correlated to the aromatic profile obtained by means of gas-chromatography-olfactometry.

Panel test results can be quantified by means of Quantitative Descriptive Analysis (QDA) on steam cooked potato samples, taking into descriptions based the parameters such as aesthetics (tuber aspect, colour, etc.), rheological (firmness, mealyness, humidity and granulation) aromatic (typical and retro nasal odour) (Lovatti and Castagnoli, 1999), which are all correlated to instrumental results (Stone *et al*, 1985).

Recently, sensorial technique utilisation is increasing in the field of vegetables and fruits. In potato, the main objective consists of defining, for each cultivar, culinary destination (salad, steam, fried, purée, small dumpling) and tactile and tasting characteristics, as to orient the consumer towards purchase.

The above mentioned aspects are utilised diffusely in the product evaluation. For example sensorial characteristics, combined to different esthetical descriptors, are utilised to evaluate some typical production such as Galician potato (Montouto-Gran *et al.*, 2002).

Research objective

With regard to the above mentioned information, a biennial research program financed by the Italian Ministry of University and Research (MIUR) entitled *Environment, genotype and typicity in potato and artichoke* was implemented, involving three research units belonging to the Universities of Florence, Bari and Catania. The research is aimed at detecting useful parameters to correlate a typical food product to a geographic production area by means of innovative screening techniques.

Materials and methods

Identification of experimental sites and field trial implementation

Field experiments were conducted in areas where potato is traditionally cultivated: The Tuscany region (province of Arezzo), near the village of Cetica (900 m.a.s.l.), where a local potato variety

(Rossa di Cetica) is traditionally cultivated; the Apulia region (province of Lecce), close to the sea, in an area suitable for an early potato crop, where the Sieglinde variety is predominantly grown; the Sicily region (province of Syracuse), at sea level, in a location suitable for an early potato crop, where the Arinda variety is usually grown. The above mentioned varieties were grown in all experimental sites. Tuber seeds, obtained in each region of provenance, were utilised in each plantation in order to standardize the physiological age of the tubers.

In Apulia and Sicily, the early crop cycle was performed, with planting in January and harvesting in late May-early June, while in Tuscany the common crop cycle was carried out, with planting in May and harvesting in September.

Plant material (tubers) analysis

Macro- and microelement determination

Tuber samples were rinsed and then skinned. The skin and pulp were sampled separately, frozen in liquid nitrogen, stored at -20 °C, and then freeze-dried. Freeze-dried samples were ground and analyzed in the laboratories of the Multidisciplinary Centre of Research on Food Sciences (CeRA) of Florence University. Analysis was performed to determine the following mineral elements: Al, Ba, Ca, Cd, Co, Cr, Cu, Fe, K, Li, Mg, Mn, Mo, Na, Ni, P, Pb, Se, Sr, V and Zn.

Sample digestion was carried out following standard procedures and the mineral elements analysed by ICP optical emission spectrometer.

Aromatic compound determination

The aromatic compound determination procedure was developed by the Department of Crop Sciences of Bari University in collaboration with Research Centre Bonomo of Bari.

Tuber samples were boiled in water for 30 minutes, to ensure the complete cooking of the tuber core. Samples were then dried, cooled, peeled and mashed to obtain an homogeneous sample.

Headspace Solid Phase Micro extraction (SPME) analysis was performed by means of a Supelco fibre. GC/MS analysis was performed using a GC Mass Shimadzu 17/A coupled to a detector MS/QP5050. A Restek (Bellefonte, PA) GC column Rtx-1 (0.32 ID, 60m) was used. Amounts of individual components were expressed in semi-quantitative terms, as GC peak area, according to the formula proposed by Oruna-Concha *et al.* (2001).

Sensorial analysis

Raw material, stored in the cold room for 20 days, was used for the following marketable determinations, according the EAPR (European Association Potato Research), USDA and CNIPT-ITCF indications, respectively: after cooking blackening (ACB), fry colour (USDA scale and colorimetric Munsell scale) and sensorial evaluation of steam cooked potatoes using the Quantitative Descriptors Analysis (QDA).

Results and discussion

Mineral elements

Results of analysis carried out with the ICP-AES spectrometer are reported in Table 3 and Table 4, for both skin and pulp samples. The content of certain elements (Co, Li, Mo, Pb, Se, V) are not reported as they were barely detectable.

The mineral elements detected in the present study are in agreement with those reported in the literature (Anderson *et al.*, 1999; Padín *et al.*, 2001; Peña *et al.*, 2001; Rivero *et al.*, 2003).

In general, mineral concentration is higher in skin tuber samples. This difference is particularly evident for Al, Ca and Fe, while it is low or absent for Mg, Na, P and Zn. The Rossa di Cetica variety showed a lower mineral concentration in skin tuber samples, whereas that trend was shown to be less observable in pulp tuber samples. The reason may be attributable to the tuber shape irregularity, which resulted in the sampling of thicker skin samples containing a certain portion of pulp.

Pulp sample ANOVA detected significant differences between all growing areas for all mineral elements, whilst differences between varieties were evident only for few elements. Determination coefficients indicated that the area component was a principle influence in determining total variance, with the exception of Al, Cr, Mg and Mn in which prevailed the

Table 3. Mineral element content (mg/kg of dry weight) in potato tuber pulp samples.

Area	**Ugento (LE)**			**Cassibile (SR)**			**Cetica (AR)**		
Variety	**Arinda**	**Rossa di Cetica**	**Sieglinde**	**Arinda**	**Rossa di Cetica**	**Sieglinde**	**Arinda**	**Rossa di Cetica**	**Sieglinde**
Al	3.627	4.499	2.326	2.874	3.541	1.475	2.805	3.552	1.295
Ba	0.594	1.042	1.017	0.280	0.485	0.296	1.318	1.883	1.629
Ca	263.7	303.8	320.6	244.4	213.2	234.8	147.6	138.7	178.6
Cd	0.209	0.295	0.237	0.154	0.209	0.187	0.124	0.104	0.228
Cr	0.448	0.266	0.380	0.481	0.280	0.371	0.499	0.528	0.381
Cu	5.345	7.255	6.311	7.323	6.953	7.093	8.772	9.950	9.036
Fe	35.53	34.88	36.02	20.67	22.24	18.08	31.29	30.47	32.00
K	20388	21983	21435	15018	17288	16718	14653	15215	13133
Mg	1124.0	826.1	1051.8	965.5	735.7	1057.1	1045.5	679.2	1022.0
Mn	8.338	4.511	8.975	6.918	4.048	7.712	9.019	4.468	9.964
Na	58.05	104.58	135.10	32.84	42.97	44.37	9.452	24.39	15.23
Ni	0.815	0.580	0.868	0.211	0.143	0.163	1.409	1.292	1.518
P	3730	3344	4196	2736	2388	3006	3013	2445	3216
Sr	0.705	0.859	1.077	1.046	1.093	1.039	0.482	0.648	0.680
Zn	13.61	13.39	14.36	18.97	19.81	19.25	20.48	20.57	25.27

Table 4. Mineral element content (mg/kg of dry weight) in potato tuber skin samples.

Area	Ugento (LE)			Cassibile (SR)			Cetica (AR)		
Variety	Arinda	Rossa di Cetica	Sieglinde	Arinda	Rossa di Cetica	Sieglinde	Arinda	Rossa di Cetica	Sieglinde
Al	645.7	454.3	537.8	305.6	125.6	166.3	220.5	137.3	137.4
Ba	5.557	4.507	5.064	4.374	2.140	2.535	11.175	10.579	8.183
Ca	1494.0	841.0	1665.5	2092.0	844.8	1621.3	746.1	460.7	826.6
Cd	0.411	0.419	0.428	0.330	0.264	0.335	0.213	0.156	0.389
Cr	1.306	1.125	0.914	1.347	0.817	0.519	1.150	0.870	0.807
Cu	9.830	8.950	11.478	10.452	7.868	9.360	10.833	8.768	10.544
Fe	482.6	287.0	391.1	262.5	101.5	161.3	232.5	125.5	174.0
K	35018	28798	32543	35558	23545	31768	25943	20500	25630
Mg	1290.5	752.2	1254.3	1210.3	655.9	1381.8	1187.8	557.9	1346.3
Mn	12.833	7.160	11.160	10.716	4.540	8.365	15.323	5.804	13.365
Na	129.54	85.15	101.60	101.42	62.15	71.96	17.57	21.11	13.40
Ni	1.681	0.929	1.679	0.735	0.100	0.344	2.786	1.880	2.914
P	4247	3217	4761	2599	2175	2722	2671	2324	3020
Sr	6.167	5.014	6.904	12.620	7.104	9.116	3.663	3.387	3.671
Zn	20.45	13.61	18.70	21.79	15.64	16.97	22.16	16.83	23.40

genetic component. The same trend was also observed for skin tuber samples.. In the latter, the genetic component was highly discriminating for the elements Cr, Cu, Mg, Mn and Zn.

The Principal Component Analysis (PCA) was initially performed taking into consideration all the elements reported in previous tables, both for skin and pulp tuber samples. For the pulp samples, PCA clearly discriminated the Tuscany samples from remaining two groups, which were partially overlapping (Figure 1).

In reducing the analysis to the elements discriminating for locality, it was possible to obtain a clear distinction between all three sample groups (Figure 2). Variety distinction within each area group was only possible in the Cetica area. For the skin samples, preliminary PCA did not detect any differences between samples grouped in function of cultivation area. In reducing the number of elements in the analysis, only the Tuscany samples were detected as being distinct. No variety discrimination was highlighted within provenance groups.

Market, sensorial and aromatic compound analysis

Market, sensorial and aromatic compound analysis highlighted significant differences within both varieties and geographic areas. Referring to market parameters (Table. 5), washability was strongly affected by provenance area. Tubers from Tuscany attained the highest value (7.2) in comparison with samples from Sicily and Apulia (respectively, 6.5 and 5.5). On the contrary, fry colour was affected by both variety and locality. The *Arinda* variety and tubers from Sicily and

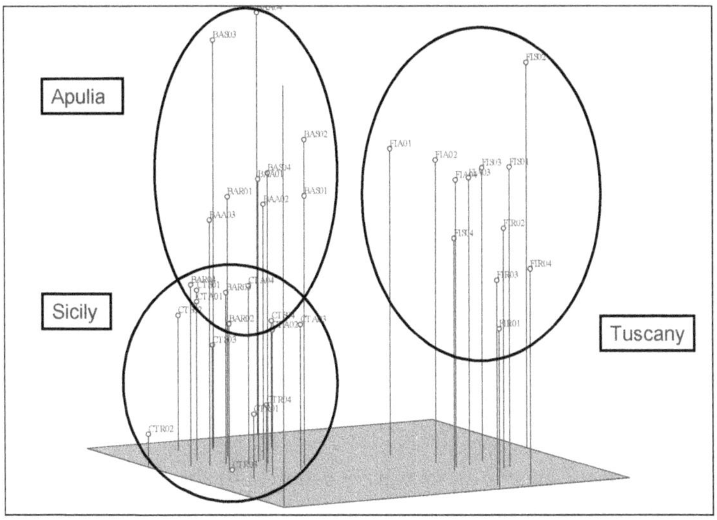

Figure 1. 3D projection of pulp potato samples subjected to PCA based on all 15 measured elements.

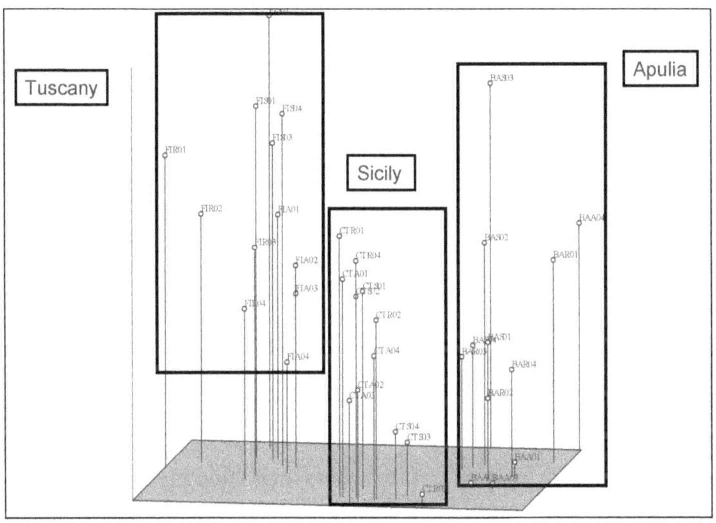

Figure 2. 3D projection of pulp potato samples subjected to PCA based on the 11 elements related mainly to the location.

Apulia showed a lower fry colour. Referring to the ACB parameter, no significant differences were detected, even though *Rossa di Cetica* showed a higher blackening tendency.

Sensorial analysis indicated significant differences both for varieties and growing areas (Table 6). In general, sensorial parameters such as sweetness, granulation and some aromatic

Table 5. Market characteristics revealed in potato tuber samples.

	Washability (CNIPT)	Fry colour (USDA)	ACB
Arinda	6.3	0.6 b	0.0
Rossa di Cetica	6.5	1.7 a	0.6
Sieglinde	6.3	1.2 a	0.3
Sicilia	6.5 ab	0.8 b	0.2
Toscana	7.2 a	1.7 a	0.3
Puglia	5.5 b	1.0 b	0.4

Table 6. Sensorial and aromatic parameters revealed in potato tuber samples (the parameters brought in capital letters are meaningful different for $p \leq 0,01$).

	Parameters affected by			Parameters non significantly affected
	locality	variety	locality and variety	
Sensorial parameters	Sweet-like, grainy	SWEET-LIKE, GRAINY	Sweet-like, grainy	crumbly
Aromatic parameters	Grassy, BURNT, starchy, wheat-like	Earthy, fruity	FODDER, WHEAT-LIKE	Chestnut like

ones (grassy, cooked, starchy and wheaty) were mainly affected by the provenance area. In referring to the variety component, more significant differences were detected for sweetness, granulation, typical taste. The aromatic component (fruity and earthy) was particularly pronounced in Arinda and Sieglinde.

Aromatic analysis performed on cooked potato tubers identified several compounds grouped in the following categories: hydrocarbons (7), aldehydes (13), furans (1), terpenes (4), carboxylic acids (1), esters (1), alcohols (1), phenols (1), others (4). Compound classes were not uniformly distributed neither between varieties, with a higher compound concentration, as in Arinda and Rossa di Cetica, nor between localities, with a greater compound concentration in tubers from Sicily. PCA did not distinguish between samples from different growing areas, whereas it discriminated between varieties within each locality (Figure 3, 4 and 5).

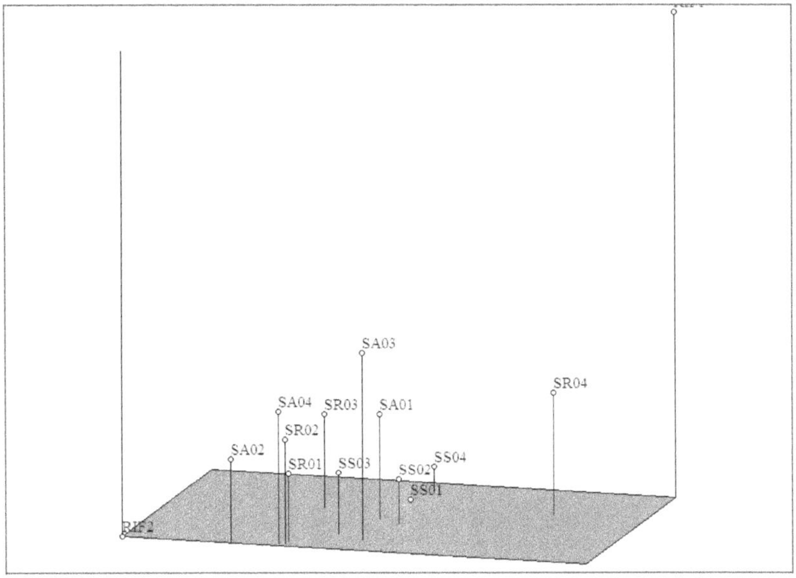

Figure 3. SPME analysis results in potato tubers from Sicily. Second letter reported for each sample indicates the variety: A = Arinda, R = Rossa di Cetica, S = Sieglinde

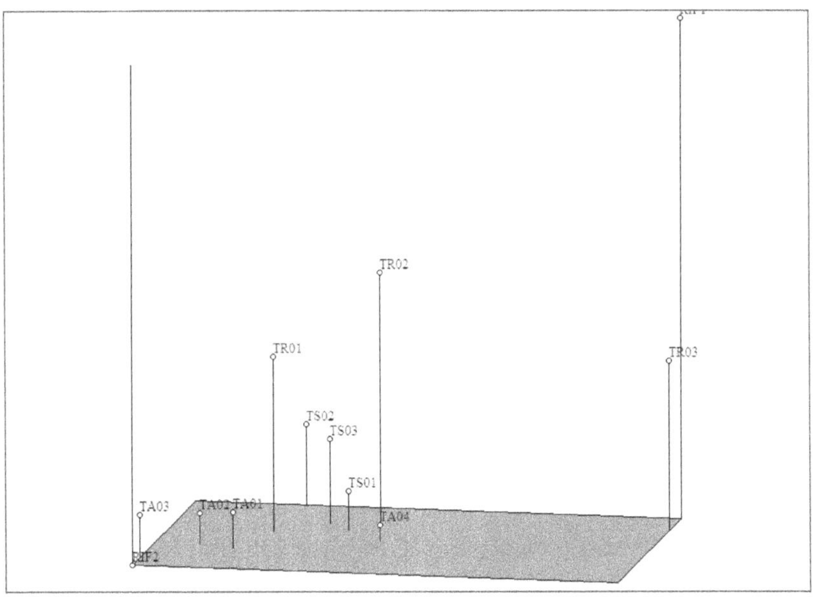

Figure 4. SPME analysis results in potato tubers from Tuscany. Second letter reported for each sample indicates the variety: A = Arinda, R = Rossa di Cetica, S = Sieglinde

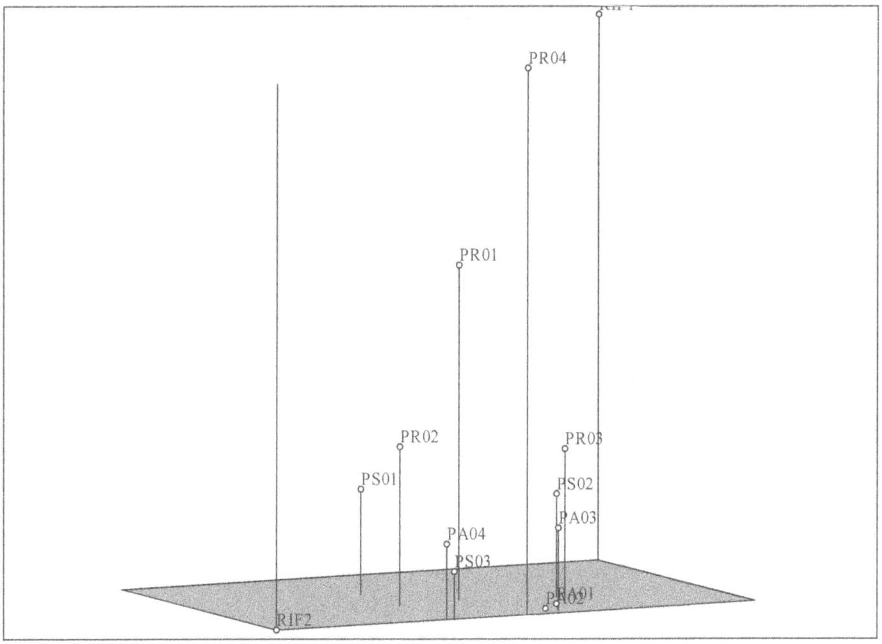

Figure 5. SPME analysis results in potato tubers from Apulia. Second letter reported for each sample indicates the variety: A = Arinda, R = Rossa di Cetica, S = Sieglinde

Conclusions

According to previous studies, preliminary results highlight that the above mentioned analytical techniques are useful in detecting the geographic origin of potato production.

Mineral content analysis permitted the discrimination between different provenance areas. Results on pulp tuber samples resulted particularly stable and reproducible, despite the lower mineral content of the vegetal matrix, especially for certain microelements. Within localities, tuber pulp samples were less variable, especially if compared with tuber skin samples. Geographic origin identification is simple and immediate, even though it is important to take into consideration that the three localities differ in several environmental and agricultural characteristics. Further analysis will be aimed at identifying those environmental parameters that particularly affect mineral content, with a special emphasis on the correlation between mineral soil availability and tuber mineral content.

Market, sensorial and aromatic compound analyses may constitute a valid tool in the discrimination between variety and locality, especially if integrated with mineral analysis. A clear distinction in function of genetic origin was made, while discrimination in function of geographic origin was shown to be less powerful with respect to the mineral analysis. The presence of some aromatic compounds, even if further analysis is required, in only certain varieties grown in specific areas is particularly interesting. In fact those compounds may represent markers that could be useful in the identification of potatoes produced in areas of typical production.

References

Anderson, K.A., B.A. Magnuson, M.L. Tschirgi and B. Smith (1999). Determining the geographic origin of potatoes with trace metal analysis using statistical and neural network classifiers. Journal of Agricultural and Food Chemistry 47: 1568-1575.

Andrenelli, L., E. Palchetti, L. Ghiselli and V. Vecchio (2002). Valorizzazione di germoplasma locale e coltura *in vitro* per la produzione di tubero seme di patata (Solanum tuberosum L.) (Local germplasm evaluation and *in vitro* culture for tuber seed potato (Solanum tuberosum L.) production). Italus Hortus 9 (3): 5-6.

Angelini, M. (1999). Le patate tradizionali della montagna genovese (Traditional potatoes of Genoese Mountains). Quaderni del Co. Re. Pa. (Comitato promotore per il recupero e la valorizzazione del patrimonio varietale di patata della Montagna genovese), Genoa, Italy.

Ariyama, K., Y. Aoyama, A. Mochizuki, Y. Homura, M. Kadokura and A. Yasui (2007). Determination of the geographic origin of onions between three main production areas in Japan and other countries by mineral composition. Journal of Agricultural and Food Chemistry 55: 347-354.

Ariyama, K., H. Horita and A. Yasui (2004). Chemometric techniques on inorganic elements composition for the determination of the geographic origin of Welsh onions. Analytical Sciences 20 (5): 871-878.

Brescia, M. A., V. Caldarola, G. Buccolieri, A. Dell'Atti and A. Sacco (2003). Chemometric determination of the geographical origin of cow milk using ICP-OES data and isotopic ratios: a preliminary study. Italian Journal of Food Science 15 (3): 329-336.

Brescia, M. A., M. Monfreda, A. Buccolieri and C. Carrino (2004). Characterisation of the geographical origin of buffalo milk and mozzarella cheese by means of analytical and spectroscopic determinations. Food Chemistry 89: 139-147.

Coetzee, P.P., F.E. Steffens, R.J. Eiselen, O.P. Augustyn, L. Balcaen and F. Vanhaecke (2005). Multi-element analysis of South African wines by ICP-MS and their classification according to geographical origin. Journal of Agricultural and Food Chemistry 53: 5060-5066.

Cooke, R.J. (1999). New approaches to potato variety identification. Potato Research 42 (3/4): 529-539.

Coombs, J.J., L.M Frank and D.S. Douches (2004). An applied fingerprinting system for cultivated potato using simple sequence repeats. American Journal of Potato Research 81(4): 243-250.

Del Signore, A., F. Di Giacomo and M. Giaccio (2004). Determinating the regional origin of cheeses with trace metal analysis using statistical classifiers. Journal of Commodity Science 43 (3): 133-144.

Duckham, S.C., A.T. Dodson, J. Bakker and J.M. Ames (2002). Effect of cultivar and storage time on the volatile flavour components of baked potato. Journal of Agricultural and Food Chemistry 50: 5640–5648.

El Jazouli, A. (2000). Instrumental and sensory analysis of Greek wines: implementation of Principal Component Analysis (PCA) for classification according to geographical origin. CIHEAM-IAMC, Chania, Greece.

Feingold, S., J. Lloyd, N. Norero, M. Bonierbale and J. Lorenzen (2005). Mapping and characterization of new EST-derived microsatellites for potato (Solanum tuberosum L.). Theoretical and Applied Genetics 111 (3): 456-466.

Franke, B.M., G. Grénaud, R. Hadorn and M. Kreuzer (2005): Geographic origin of meat – elements of an analytical approach to its authentication. European Food Research and Technology 221: 493-503.

Galliard, T. and J.A. Matthew (1973). Lipids of potato tubers. II. Lipid-degrading enzymes in different varieties of potato tuber. Journal of the Science of Food and Agriculture 24: 623-627.

Galliard, T. and D R. Phillips (1971). Lipoxygenase from potato tubers. Biochemical Journal 124: 431-438.

Ghiselli, L., L. Andrenelli, S. Benedettelli and V. Vecchio (2001a). Caratterizzazione e valorizzazione della biodiversità in patata (Solanum tuberosum L.). (Characterisation and valorisation of potato (Solanum tuberosum L.) biodiversity). VI National Congress on Biodiversity, 6-7 September, Bari, Italy: 26.

Ghiselli, L., L. Andrenelli and V. Vecchio (2001b). La Viola Calabrese. (The Viola Calabrese). Gazzettino della Patata 2: 6-11.

Ghiselli, L., S. Benedettelli, V. Vecchio and N.G. Lacerenza (2002). Genetic study of Italian potato landraces. 15th Triennial Conference of the European Association for Potato Research (EAPR), 14-19 July, Hamburg, Germany: 238.

Gorg, R., U. Schachtschabel, E. Ritter, F. Salamini and C. Gebhardt (1992). Discrimination among 136 tetraploid potato varieties by fingerprints using highly polymorphic DNA markers. Crop Science (USA) 32 (3): 815-819.

Guerzoni, M.E. (2004). Prodotti tipici: oggettività dei loro legami col territorio (Typical products: objectivity of their links with the territory). Atti dell' Accademia dell'Agricoltura di Bologna: 139-147.

Heberger, K., E. Csomos and L. Simon-Sarkadi (2003). Principal component and linear discriminant analyses of free amino acids and biogenic amines in Hungarian wines. Journal of Agricultural and Food Chemistry 51: 8055-8060.

Herbert, P., P. Barros and A. Alves (2000). Detection of Port wine imitations by discriminant analysis using free amino acids profiles. American Journal of Enology and Viticulture 51 (3): 262-268.

Hsieh, R. J. (1994). Contribution of lipoxygenase pathway to food flavors. In: C. T. Ho and T. G. Hartmann (Eds.) Lipids in Food Flavors. American Chemical Society, Washington DC, USA: 38-48.

Josephson, D.B. and R.C. Lindsay (1987). c4-Heptenal: an influential volatile compound in boiled potato flavour. Journal of Food Science 52(2): 328-331.

Liu, L., D. Cozzolino, W.U. Cynkar, M. Gishen and C.B. Colby (2006). Geographic classification of Spanish and Australian Tempranillo red wines by visible and near-infrared spectroscopy combined with multivariate analysis. Journal of Agricultural and Food Chemistry 54: 6754-6759.

Lovatti, L. and V. Vecchio (2003). Alla scoperta di un patrimonio antico. (The discover of an ancient patrimony). Il Divulgatore, 6: 26-29.

Lovatti, L. and M. Castagnoli (1999). Sensory analysis of table potato varieties and breeding clones for the Italian retail market. 14th Triennal Conference of the European Association for Potato Research (EAPR), 2-7 May, Sorrento, Italy, pp. 708-709.

Manca, G., M.A. Franco, G. Versini, F. Camin, A. Rossmann and A. Tola (2006). Correlation between multielement stable isotope ratio and geographical origin in Peretta cows' milk cheese. Journal of Dairy Science 89: 831-839.

Monje, M.-C., C. Privat, V. Gastine and F. Nepveu (2002). Determination of ethylphenol compounds in wine by headspace solid-phase microextraction in conjunction with gas chromatography and flame ionization detection. Analytica Chimica Acta 458(1): 111-117.

Montouto-Gran, M., E. Fernandez-Fernandez, M.L. Vazquez-Oderiz and M.A. Romero-Rodriguez (2002). Development of a sensory profile for the specific denomination 'Galician potato'. Food Quality and Preference 13: 99–106.

Multon, J.L. (1985). Che cos'è la qualità di un prodotto alimentare e quali ne sono gli operatori? (What is the food product quality and who are the operators?). In: J. L. Multon and Y. Davenas (1985). Qualità dei prodotti alimentari. Tecniche Nuove, Milan, Italy, 3-27.

Nieto, A.R., A.C. Sancho, M.V. Barros and J.L. Gorge (1990). Peroxidase zymograms at constant and gradient pH electrophoresis as an analytical test in the identification of potato varieties. Journal of Agricultural and Food Chemistry 38: 2148-2153.

Oruna-Concha, M.J., J. Bakker and J.M. Ames (2002a). Comparison of the volatile components of eight cultivars of potato after microwave baking. Lebensmittel Wissenschaft und Technologie 35: 80-86.

Oruna-Concha, M.J., J. Bakker and J.M. Ames (2002b). Comparison of the volatile components of two cultivars of potato cooked by boiling, conventional baking and microwave baking. Journal of the Science of Food and Agriculture 82: 1080–1087.

Oruna-Concha, M.J., S.C. Duckham and J.M. Ames (2001). Comparison of volatile compounds isolated from the skin and flesh of four potato cultivars after baking. Journal of Agricultural and Food Chemistry 49: 2414–2421.

Padín, P.M., R.M. Peña, S. García, R. Iglesias, S. Barro and C. Herrero (2001). Characterization of Galician (N.W. Spain) quality brand potatoes: a comparison study of several pattern recognition techniques. Analyst 126: 97-103.

Peña, R.M., S. García, R. Iglesias, S. Barro and C. Herrero (2001). Authentication of Galician (N.W. Spain) quality brand potatoes using metal analysis. Classical pattern recognition techniques versus a new vector quantization-based classification procedure. Analyst 126: 2186-2193.

Peña-Neira, A., T. Hernández, C. García-Vallejo, I. Estrella and J.A. Suarez (2000). A survey of phenolic compounds in Spanish wines of different geographical origin. European Food Research and Technology 210 (6): 445-448.

Petersen, M.A., L. Poll and L.M. Larsen (1998). Comparison of volatiles in raw and boiled potatoes using a mild extraction technique combined with GC odour profiling and GC–MS. Food Chemistry 61(4): 461–466.

Picque, D., T. Cattenoz, G. Corrieu and J.-L. Berger (2005): Caractérisation des vins par spectroscopie infrarouge. Classification selon l'année de production et l'origine géographique. (Discrimination of red wines according to their geographical origin and vintage year by the use of mid-infrared spectroscopy). Sciences des Aliments 25(3): 207-220.

Pillonel, L., B. Albrecht, R. Badertscher, U. Buetikofer, J.O. Bosset, J.F. Chamba and R. Tabacchi (2003a): Metodi analitici per la determinazione dell' origine geografica del formaggio Emmental. Parametri della proteolisi e reologia (Analytical methods for the determination of the geographic origin of Emmental cheese. Parameters of proteolysis and rheology). Italian Journal of Food Science 15 (1): 49-62.

Pillonel, L., S. Ampuero, R. Tabacchi and J.O. Bosset (2003b): Analytical methods for the determination of the geographic origin of Emmental cheese: volatile compounds by GC/MS-FID and electronic nose. European Food Research and Technology 216: 179-183.

Pillonel, L., M. Collomb, R. Tabacchi and J.O. Bosset (2002): Analytical methods for the determination of the geographic origin of emmental cheese. Free fatty acids, triglycerides and fatty acid composition of cheese fat. Mitteilungen aus Lebensmitteluntersuchung und Hygiene 93 (3): 217-231.

Polzerova, H. (2001). The use of molecular genetic techniques to potato variety identification. Rostlinna-Vyroba-UZPI (Czech Republic) 47 (11): 482-487.

Rivero, R.C., P.S. Hernández, E.M.R. Rodríguez, J.D. Martín, C.D. Romero (2003): Mineral concentrations in cultivars of potatoes. Food Chemistry 83: 247-253.

Scazziota, B., E. Palchetti, V. Vecchio, F. Ieri, L. Andrenelli, L. Ghiselli and M. Manzelli (2005). Agrobiodiversità e qualità integrale - caratterizzazione e valorizzazione di due vecchie varietà di patata italiane: Viola Calabrese e Rossa di Cetica. (Agro-biodiversity and integrated quality – characterisation and valorisation of two old Italian varieties: Viola Calabrese and Rossa di Cetica) In proceedings: VII National Conference on Biodiversity. 31 March – 2 April, Catania, Italy (in press).

Sensory Analysis Methodology. General Guidance, 1985. ISO, Tour Europe, Paris, France.

Web references

http://ec. europa. eu
http://www. cetica. it
http://www. europotato. org
http://www. istat. it

An analysis of the potato production and marketing system and evaluation of development trends in Turkey

M. Bülbül and H. Tanrıvermiş
Ankara University, Agricultural Faculty, Department of Agricultural Economics, 06110, Diskapi, Ankara, Turkey, bulbul@agri.ankara.edu.tr, tanriver@agri.ankara.edu.tr

Abstract

This paper analyzes the economic aspects of production and the marketing system of potato, especially the production volume, trade, and consumption. In Turkey, 4.1 million tons of potatoes were produced on 154,000 ha by 158,487 producers, of which 7-10% was seed and 58-71% was for consumption, while the remainder is accounted for by discarded products, animal feed and losses. The average yield is 25-26 t/ha, which is below that of the major producing countries but above the world average. Potato production techniques and marketing systems vary depending upon the region. The variations in production, yield, and producer and consumer prices was found to be lower than anticipated, and differences in real gross income obtained per unit area were found to be high. Diseases such as potato wart and potato brown rot have caused a 24.2% fall in production area and a 23.8% drop in production quantities after 2000, and the cultivation of potatoes and other tuber crops have been banned on 15,387.7 ha land. The average increase in the rates of potato production and consumption are almost at the same levels. Of the average annual production, 1.9% is exported and the remainder stays in the domestic market. The average per capita consumption of potato is 47.1 kg, which is lower than that of the EU countries. Although the supply elasticity (0.034) and demand elasticity (-0.018) of the potato is considerably low, and it has been identified that an increase in income positively affects potato consumption, although at a low rate. As the fluctuation of market prices negatively influences producers and consumers, there is a requirement for efficient production planning and market regulation.

Keywords: potato farming, production and marketing system, trade, consumption

Introduction

Although at different levels in developed and developing countries, it is well known that the potato is one of the most common staple foods, and that potato production is second only to cereals in many countries. Low-income households consume cheap potatoes more than other vegetables and potato consumption in Turkey is second only to cereals as in other developing countries (Moazzem and Fujita, 2004; Arıoğlu *et al.*, 2006). Due to its high nutritional value and versatility the potato is one of the main foodstuffs that can reduce the problem of an insufficient and unbalanced diet in developing countries, where these problems are commonplace (Arıoğlu *et al.*, 2006). The main advantages in the production and marketing of the potato include low price, high per unit area yield, nutritional value, ease of digestion, wide range of usage, and its ability to be cultivated on irrigated lands in many climatic conditions (Günel *et al.*, 2005). Due to their high nutritional value, potato tubers are used directly by humans and consumed in various forms after processing and are used as additives to wheat flour in bread production (at a rate of 2.5-3%), while discarded products are used as animal feed.

In Turkey, the production of various types of potato cultivars that have been improved in developed countries is common, with no domestic types preferred for production. Potato production and marketing systems vary, based on production objectives (for seed, consumption, industrial use and other purposes), production methods (organic, low input and intensive farming) and by region. Potatoes containing low amounts of protein and with a high amount of starch are divided into two types, as table potatoes and industry potatoes, for which the production models are different. Many different types of potato are produced in Turkey, the number of which changes significantly in time in accordance with market demands. A significant part of the total production falls to the share of table potato production, whereas foreign dependency in potato seeds continues, and the share of processing production is quite low in total production volume. Fluctuation can be found in annual potato production according to the year, based on climatic conditions, market price in the preceding year, and cost and profitability relations. This has a detrimental affect on market prices as well as producers and consumers. Although potato production takes place in nearly all provinces, cultivation in the majority of these provinces is carried out for family requirements, with market-oriented production intensified only in certain provinces (Nevşehir, Çorum, Erzurum, Trabzon, Kayseri, Sivas, Niğde, Konya, Kütahya, Adapazarı and Afyon). While various production models are utilized on farms, differing models in the marketing of products (such as contract farming, direct marketing, various types of market agents) are employed. This paper analyzes the developments in potato production, trade and consumption since 1950. The current situation of production and marketing system of the potato in Turkey is evaluated focusing on the economic relations among farmers, traders, processors and storage owners, and the structural changes in the production and marketing field through a quick assessment survey that has been conducted. The study also discusses the main arrangements required for the potato production and marketing system in Turkey.

Materials and methods

Data from state and private institutions is used in the study to analyze the changes in potato production and marketing systems in the 1950-2006 period with regards to changes in the number of producers, cultivated land, amount of production, exports, imports and consumption levels. Trend equations for potato production, cultivation area and consumption are estimated for the examined period, and the average rates of increase are studied. The elasticity in supply and demand of the potato have been demonstrated and commented upon using the simple models. Major integration models used in potato farming; the contributions of various models to producers and consumers; distribution of the total production in terms of utilization; and main production and marketing issues and future expectations are evaluated using data obtained via a rapid evaluation questionnaire administered to industry and trade firms in the sub-sector, state institutions and cooperatives, and thus fundamental postulations on the sector are developed by using a SWOT analysis.

Analysis of the cultivation area, production and productivity in potato farming

Approximately 19 million ha land is dedicated to potato production around the world, accounting for 0.4% of the farmland and 6.9% of irrigated land. Of this land, 49.3% is in developed countries and the rest is in the developing countries. Total potato production is 330.3 million tons, with the major producer countries being China, Germany, India, Poland, the Russian Federation and the United States. Product yield in the producer countries varies between 10.2 t/ha to 43.5 t/ha, with an average productivity of 17.4 t/ha. The countries with the highest level of product yield are the United States, Germany, Turkey, India, Poland, China and the Russian Federation. Turkey has 0.9% of the world's total potato production area, providing 1.5% of the total production.

Potato production in Turkey dates back to the 1850s, however development has been slow until the 1930s as its nutritional value was not sufficiently understood. The implementation of the Nationwide Potato Project in 1970s, amendments to the Seed Law with issue number 308 in the 1980s, and the incentives provided for the private seed sector after 1984 were factors contributing to the increase in potato production. Turkey has a distinguishing position with regards to potato production thanks to the agro-ecological characteristics. It is possible to carry out potato production in nearly all regions of the country and provinces within the regions in certain periods of the year. The potato production area was 25,550 ha in 1925, reached 67,300 ha in 1950 and 203,593 ha in 2000, after which production area decreased due to the potato wart disease that entered into the country with imported seeds and became rapidly widespread. In the 2000-2005 period, the production area shrank by 24.2% (Table 1), while in the 1925-2005 period the production area expanded six times as opposed to 2.3 times in the 1950-2005 period.

Table 1. Developments in potato production area in Turkey (ha per year) (source: FAO and SIS).

Years	Turkey	İzmir	Afyon	Sakarya	Bolu	Niğde	Nevşehir	Ordu	Erzurum
1950	67,300	1,827	967	4,800	799	2,650	753	3,590	3,350
1960	160,000	3,490	3,705	4,330	1,032	5,100	7,770	30,039	4,400
1970	169,000	9,483	4,063	5,230	2,430	7,625	20,860	11,445	2,993
1980	180,000	8,465	5,039	3,710	4,664	18,882	20,320	12,660	9,113
1990	200,400	9,912	7,275	3,476	11,628	24,582	23,109	10,291	7,735
2000	203,593	11,980	8,944	4,133	10,130	35,241	25,611	9,204	6,849
2001	199,673	12,527	9,578	4,016	10,891	32,709	24,652	9,507	6,336
2002	197,744	13,588	9,631	1,503	9,911	34,609	24,261	10,265	6,167
2003	194,443	13,263	9,905	1,232	10,033	34,448	24,676	8,908	5,500
2004	178,900	13,166	10,220	1,076	10,157	27,312	24,698	6,878	5,411
2005	154,300	13,060	8,738	847	10,118	20,260	13,450	5,928	5,386

After the proclamation of the Republic of Turkey, and especially after the 1970s, potato farming has been concentrated around certain cities. Among the 81 cities of Turkey, potato farming is carried out in 68 cities, and the share of the leading seven cities (Niğde, Nevşehir, Bolu, Sakarya, Ordu, Afyon and İzmir) in the total production area increased from 22.9% in 1950 to approximately 40.0% in 2000, increasing to 50.4% in 2004 (Table 1). There have been significant changes in the share of the total production areas and production quantities of different cities. These changes began to be observed in Sakarya (Adapazarı), while more recently Niğde, Nevşehir and Ordu have been affected due to various diseases. In certain Central Anatolian cities, such as Afyon, Çorum and Konya, the increase in potato production arises from the fact that sugar beet plantation areas have been restricted in these cities, and farmers have been forced to switch to potato production.

The share of the potato farming areas in total farmland is 0.58%. Some 158,487 farmers are producing potatoes, with the average potato farming area being 0.5 ha (SIS, 2004). Among the potato producing farms, 45,000 producers target the commercial market, while the rest produce for their own domestic consumption. The fact that producers are generally numerous, small size, diverse and lack an efficient organizational structure results in low levels of income when compared with their counterparts in industry and commerce. This structure also poses a disadvantage for the producers in the procurement of raw materials and the marketing of products.

Potato production, which was around 73,000 tons in 1925, increased by 74 times to 5.4 million tons in 2000, and by 56.2 times to 4.1 million tons in 2005. Potato production is concentrated in the cities of Niğde, Nevşehir, İzmir, Bolu, Afyon, Erzurum and Ordu in the Central Anatolian, Black Sea, Aegean and Eastern Anatolian Regions. The share in total production of the leading eight cities in terms of plantation areas was 39.6% in 1950, 62.3% in 2000, and 56.9% in 2005 (Table 2). The concentration of potato production in certain areas increases productivity and reduces marketing costs problems. The main reasons for the volatility in potato production can

Table 2. Developments in Potato Production in Turkey (Ton per year) (source: FAO and SIS).

Years	Turkey	İzmir	Afyon	Sakarya	Bolu	Niğde	Nevşehir	Ordu	Erzurum
1950	446,100	9,225	5,624	115,650	6,493	21,014	3,079	15,443	33,500
1960	1,400,000	51,210	52,000	100,000	15,079	51,000	62,140	54,003	65,200
1970	2,100,000	139,130	63,365	59,180	46,950	140,095	177,112	71,835	39,277
1980	3,000,000	206,177	101,110	54,900	75,000	385,800	283,150	128,250	179,520
1990	4,600,000	249,740	174,213	46,899	268,324	899,339	854,798	114,470	97,577
2000	5,370,000	338,261	255,699	79,821	282,601	1,296,020	979,323	114,059	114,218
2001	5,000,000	409,050	278,899	69,778	279,302	1,095,249	784,610	115,546	122,001
2002	5,200,000	399,388	374,418	34,153	270,932	1,206,876	848,730	121,458	118,253
2003	5,300,000	339,875	359,498	24,355	281,308	1,293,694	911,850	130,444	106,055
2004	4,800,000	325,831	332,439	22,957	283,283	929,256	977,999	99,573	102,461
2005	4,090,000	319,983	325,370	21,209	282,940	712,865	475,260	88,123	102,303

be attributed to marketing, market prices and the procurement of high quality seeds. Following years in which farmers experience difficulties in marketing their products, production volumes usually drop as farmers lack sufficient funds to secure the necessary production inputs.

Trend equations ($Y = ab^t$ or $\log Y = \log a + t \log b$) have been estimated using the potato plantation area and production quantity data for the period 1950-2005. According to the plantation area trend equation ($Y = 5.238 + 0.00185X_t$ [$p<0.00$]), the increasing trend in plantation area is continuing and is calculated as 0.19% per annum in this period. The production quantity trend equation has been estimated as ($Y = 6.319 + 0.0125X_t$ [$p<0.00$]) and the average increase in productivity has been calculated 1.3% per annum (Table 3). The reason for the increase in production quantity being higher than the increase in plantation area is due to the improvement in production efficiency. Based on the data for 1950-2005, the variation coefficient (CV) in plantation area (CV = 7.68) is seen to be lower than the variation in production quantity (CV = 29.59) and average yield (CV = 25.10).

Over the last 20 years, average yield has improved considerably due to the use of high quality seeds and advanced production techniques. Thus, total production quantities have increased, while production costs have decreased. The yield was 2,840 kg/ha in the 1920s. In the following years, the use of chemical fertilizers, irrigation water, machinery and quality seeds helped yield to increase to 6,629 kg/ha in the 1950s, then to 12,426 kg/ha in 1970, 22,954 kg/ha in 1990, 26,376 kg/ha in 2000, and finally to 26,786 kg/ha in 2005. Average yield is 1.5 times higher than the world average (17,400 t/ha), but 1.6 times lower than the average of the leading producer countries. There are significant differences in terms of yield among different regions, in addition to the variations in other characteristics. The cities with the highest yield are Nevşehir, Niğde, Afyon and Bolu, where the yield is higher than the world and leading producer country averages. Hakkari, İçel, Adıyaman, Siirt, Hatay, Diyarbakır and Bingöl are the cities with the lowest yield, where average yield varies between 2 and 9 t/ha; Siirt has the lowest yield with 2 t/ha.

Table 3. Coefficients of trend models of production area and quantity.

Production area (ha)		**Production quantity (ton)**	
log a	5.238	log a	6.319
A	172981.63	a	2084490.6
log b	0.00185	log b	0.0125
B	1.0042	b	1.0292
R^2	0.310	R^2	0.88
F	15.736	F	259.719

Analysis and problems of potato production models

The potato has a considerable capacity for adaptation. Potato production systems show variations among different countries and regions within the same country depending on climate and soil conditions and the socio-economic structure (Horton, 1987; Günel *et al.*, 2005). As there are variations in agro-ecologic conditions between different regions in Turkey, different potato production systems are being utilized. Potato production is possible during certain periods of the year depend upon the climate conditions. The early potato production possible in the irrigated parts of the South eastern Anatolia and in the Mediterranean and Aegean Regions during May-July, while production in all the other regions is obtained during August-September period in a year. Harvesting starts in May in the Aegean, Mediterranean and South-eastern Anatolian Regions, while it continues until September in the other regions (Table 4).

It is observed that there is a segregation among the regions in terms of the type of potato produced, i.e. seed, table and industry potato. The best quality seed potato is produced in the higher altitudes, plateaus and mountains of Kars, Erzurum, Bursa and Bolu. Seed production is carried out in areas free from diseases and pests. The leading potato producing provinces are Nevşehir, Erzurum, Niğde, Konya, Kütahya, Adapazarı and Afyon. Potatoes produced for industry is concentrated in the İzmir, Bolu and Afyon provinces. Private firms usually meet their demands through contract farming, and the share of potato processing in total volume varies in the range of 4-5%. The remaining part of production is for the table and other uses. The contract farming accounts for 80-100% of seed and processing potato production. Although still at low levels, organic potato farming is carried out and organic products account for only 0.5% of total production, but there is high potential in the highlands.

As potato is an arable crop, it has a significant use in crop rotation and is a preferred crop of producers due to its relatively higher yield compared to other products (Özçelik *et al.*, 1999; Arıoğlu *et al.*, 2006). In the agriculture of potatoes, pricing and marketing problems, high cost of seed and other inputs, insufficient diversification, import dependency for seeds, insufficiency of integration between agriculture and industry, growth in diseases and pests, insufficiency of

Table 4. Potato Production Periods by Region in Turkey (Source: Arıoğlu et al.*, 2005).*

Geographical regions		Sowing time	Harvest time	Vegetation period
Trace - Marmara		March - April	August - September	5-6 months
Aegean	spring	February - April	June – July	4-5 months
	autumn	August	December - January	4-5 months
Mediterranean		December - January	May – June	4-5 months
Central Anatolia		March - May	August - October	5-7 months
Black Sea		March - May	August - September	5-6 months
East Anatolia		May	September	4-5 months
Southeast Anatolia		February	May	3-4 months

irrigation water and the rising costs of irrigation, storage, training and coordination problems, rising labour costs and the insufficiency of family labour force for production and insufficiency of suitable land for production have adverse effects on yield, costs and profitability.

Although research on potato breeding initiated in the 1950s in Turkey, variety registration efforts gained pace in the 1970s with the National Potato Project (Günel, 2002). In the following years, variety registration operations have been led by the private sector, and the number of varieties registered had increased steadily to 59 by the end of 2006, of which 15 have received a production permit. Even though Turkey is a significant potato producer in the world, it still does not have a local cultivar in commercial production and cultivars derived from Holland, Germany, France, the United Kingdom and the United States are being used. No significant differences in potato cultivars are observed between regions. In their choice for cultivars, knowledge of a particular variety, its yield, the market demand for that particular cultivars, its seed price, its size and the suggestions of the local seed producers are usually effective (Günel *et al.*, 2005; Yulafcı and Duran, 2006).

In general, the potato yield depends significantly on the quality of seed, climate and soil conditions and cultivation practices. The use of high-quality seed has a greater effect on yield and thus a quality seed can increase yields by 15-20%. Private firms and cooperatives supply certified seeds through foreign firms, which ensure a high yield, but the price is rather high. If certified seeds are multiplied, they can also be used as seeds, but only for one year, as multiplication over two years causes a definite yield decline and is not economical. A number of farmers produce seed potatoes on some of their farm plots for their own use in the next season and/or for sale, while others purchase seed potatoes every year from other farmers or sometimes from firms. The reproduction of potatoes by tubers makes it highly vulnerable to the dissemination of diseases (Yıldırım and Yıldırım, 1986). The sale of potatoes that are not certified as seeds creates serious problems in this respect, while the import of mature seeds from abroad has negative effects on production costs. On the other hand, the use of certified seed as table potatoes is another fundamental problem. In order to increase yield in potatoes, it is necessary to change the seed every three years and use seeds free from viruses. It is necessary to apply crop rotations properly in areas dominated by potato farming.

Potato seed production in Turkey is generally achieved through one-time reproduction of imported mature seeds. Certified seed production and procurement through this method satisfied 2-9% of the total seed demand in the 1992-2004 period, which increased to 12.4% in 2005 and 29.7% in 2006 (Table 5). The remaining part of the seed demand is satisfied through the reproduction of second and third generation seeds. Even though the growing risk of disease leads producers to use certified seeds more, the contraction in the plantation areas causes a decline in the quantity of seed used, which is reflected as an increase in the certified seeds used as a percentage of the total requirement. As domestic seed production is not subject to control, seeds are not certified and the quality guarantee lies only with the producer.

It is of utmost priority to establish seed production systems in the potato sector. The production of seed and table potatoes together in the same regions makes the control of diseases and pests difficult. Even though there are no legal requirements, many seed potato producers (reproducers) move their production to new areas. Provinces such as Adana, Kayseri, Bayburt and Sivas have

Table 5. Production, import and distribution of potato seed in Turkey.

Years	Production (ton)	Imports (ton)	Distribution (ton)	Seed requirement (ton)	Distribution coverage of requirements (%)
1992	7,193	4,292	12,632	487,500	2.59
1995	15,312	5,646	14,696	500,000	2.94
2000	23,275	15,524	43,335	512,500	8.46
2001	17,669	3,126	25,941	500,000	5.19
2002	21,375	14,147	26,000	495,000	5.25
2003	27,885	7,570	28,910	487,500	5.93
2004	45,870	8,580	40,406	447,500	9.03
2005	63,901	9,712	47,624	385,750	12.35
2006	93,377	17,893	111,270	375,000	29.67

recently emerged as new seed potato farming areas. Such an approach is aimed at yielding clean and healthy seeds in the short term, but an uncontrolled expansion may lead to greater risks in the longer term. There is a need for an effective and regular state control mechanism in seed production. If seed production continues without the establishment of such a mechanism, there will be no land left for seed production in the near future. First, seed production regions should be determined, then necessary research should be undertaken to specify appropriate seed production techniques and rules, and finally production of table potatoes should be banned in these regions. After the establishment of a system for the control and supervision of seed production, the next step should be to provide tax and credits incentives and foreign trade arrangements to encourage and support high-quality seed production. In this way, the necessary infrastructure for seed production will be established, production will increase and the use of healthy seed in production will become wide-spread.

The timing and method of sowing, fertilization, irrigation, disease and pest control, harvesting and storage are all issues related to the main stages of production and marketing on a farm. In potato farming, the field is ploughed 5-6 times and over-cultivated. In the case of seed potatoes, crop rotations are applied with the planting of potatoes and wheat in consecutive years. However, the same method is not used in the case of processing and table potatoes, and potatoes are planted in consecutive years due to the insufficiency of irrigated land. On the other hand, it has been demonstrated that maize, rape, sunflower, wheat, bean, common vetch and barley have a positive effect on potato yield if planted in the year prior to the potato (Onaran *et al.*, 2006; Kara *et al.*, 2006).

The improper and excessive use of farm chemicals has been seen in potato farming. For example, it has been found out that the average amount of nutritional chemicals used in Nevşehir ranges between 854.2 kg/ha and 957.8 kg/ha, of which 63-76% is nitrogen (Özçelik *et al.*, 1999). On the other hand, experiments in the Nevşehir-Niğde region have shown that the highest tuber yield can be achieved by the administration of 500 kg/ha (Karaca and Demir, 1994). This implies that the farmers in the region are using nitrogen fertilizers excessively or beyond optimum levels.

Instead of bringing any additional benefit to the enterprise, this increases the production cost burden. In several regions, composite fertilizer formulations that are suitable for wheat are commonly used in potato farming. Fertilizing methods based on soil and plant leaf analysis are very seldomly used. The use of foliar fertilizers is expanding rapidly and farmers generally lack the necessary knowledge in making such applications properly. In addition to this, farmers use 2.4-3 kg/ha of pesticide as an active ingredient in the Nevşehir province, where sandy soil is irrigated 10-17 times a year. Irrigation using sprinklers, which is also common in the region in a bid to decrease labour costs, leads to higher electricity costs due to the high frequency of watering. In spite of efforts to achieve a more effective use of the water sources and switch to watering by drop, this has not been so successful at individual small-scale farms. Although irrigation using sprinklers is preferred in some potato production regions (such as Afyon), flooding is still the widely used method on a large portion of the farms. It should be noted that excessive watering is carried out, along with intensive applications of chemical fertilizers and pesticides in the region. Considering the soil characteristics of the region, it is inevitable that such practices will lead to the pollution of ground water sources, which is pumped to potato fields. It is also known that ground water is being used excessively through the use of deep well pumps without a detailed study of the ground water reserves. Therefore, the availability of sufficient water sources for irrigated agriculture and the sustainable management of fertile farmland will be at risk in the future.

As a result of ineffective control and supervision in seed production and distribution, along with improper cultivation techniques, potato wart (*Synchytrium endobioticum)* developed as a serious problem in the 2000s. The main cause is a fungus, which creates warts with a shape similar to a cauliflower on the stalk, stolon and tuber of the plant. Due to its appearance, the disease is referred to as 'cancer' by the laymen and the media, a term that creates fear among the consumers, but it is known that the disease cannot be passed on to humans. The main disease agent can be disseminated through the soil, and affects tubers and farm equipments. The wart spores can stay alive in the soil for 30-40 years (Hooker, 1981; Günel *et al.*, 2005). The Ministry of Agriculture and Rural Affairs issued a decree called 'Fight against potato cancer (*Synchytrium endobioticum* [Schilberszky] Percival)' which regulates the banning of potato cultivation in contaminated soil. The planting of potatoes and other crops of the *Solanaceae* family, or the production of seeds or saplings with other tuber crops that can carry around parts of the soil, is prohibited, with only potato varieties that are known to be immune to *Synchytrium endobioticum* permitted for cultivation. No exact period of the cultivation ban has been laid down, and the Ministry will remove the ban when it is clearly seen that the disease has been fully eradicated. In accordance with the 'Decree on Certified Seed and Sapling Support for Alternative Products to be Planted in Areas Contaminated by Potato Wart,' issued on 15 April 2005, farmers working in contaminated areas have received support payments for the production of alternative products since 2005. Following the contamination seen in Nevşehir, Niğde and Ordu in 2004, Ministry experts' examine all fields in the region that run the risk of contamination. In the cities where contamination is found, plantation bans are initiated.

For each banned field, the farmers receive support of $279.10 to $523.40 per ha for annual alternative crop plantation, and support of $1,395.70 to $2,442.40 per ha for perennial crop plantation. The total payments is $9,051,67.97 to support of 4,236 farmers, covering of 14,139.4 ha of land. In 2006, $9,173,080.77 was paid to farmers whose fields have been quarantined due to potato wart in Nevşehir, Niğde, Ordu, Kayseri and Trabzon within the scope of Decree 2006/18. It is estimated that 20,000 ha of land will be quarantined due to potato wart in the near future.

When it is considered that the disease has generally arrived in Turkey through imported seeds (Arıoğlu *et al.*, 2006), the importance of proper control of imported products is evident. Quarantines within the country and bans on potato production adversely affect the economic and social wellbeing of the farmers concerned. On the other hand, an absence of such bans would alleviate the risk of the disease spreading all over the country. Potato production bans due to potato wart causes a decrease in the production of this high-value product and necessitates imports or the use of new plantation fields not previously used for potato farming. In areas contaminated by potato wart, the cultivation of other products that can cause the spread of the disease is forbidden. Therefore, farmers from Nevşehir and Niğde whose fields have been quarantined rent new fields from Şanlıurfa, Konya, Kayseri and İzmir to continue potato farming. The potato's standing as a traditional product with relatively higher value added compared to alternative products, and the lack of knowledge among the farmers about other products, are the main reasons for this course of action. However, as the disease is disseminated through tubers and contaminated parts of the plant, it is inevitable that the disease will spread to clean areas if contaminated potatoes are planted there. In order to prevent a nation-wide spread of the disease, it is necessary to abide by the quarantine rules and revise the seed import legislation.

Due to the lack of progress made in seed production and variety development, producers fail to produce high-quality and high-yield products. Furthermore, problems with marketing, packaging and transportation weaken their efforts to export their products and their competitive power. Management mistakes and land and labour shortages cause low yields and high production costs and lower operating profitability. High energy costs are the main burden on farmers, especially in Central Anatolia, due to the extensive use of ground water through pumping. Almost all the farmers in the region are in debt to the electricity companies.

The analysis of potato consumption and marketing systems

The total value of potato production is $1,225.22 billion, 80% of which is marketed in Turkey. The share of potato production in total agricultural production of $31,944.36 billion is 3.8%. The share of potatoes supplied to the market varies according to the production systems and the characteristics of the regions. In the case of seed, table and processing potato production, the ratio of marketing amount to total supply is around 80%, while this ratio is very low or even nonexistent in regions where the potato is produced for own-consumption.

Following the harvest, potatoes are divided into three groups according to size. The first and second groups are marketed or consumed at home, while the third group is used as animal feed. In spite of some volatility, potato production exhibited a rising trend up until the 2000s, but this rising trend cannot be said to be continuing. The supply-demand imbalance has an adverse impact on market prices. According to some studies, per capita potato consumption is 60 kg. However, in these studies seed tuber consumption (375-400,000 ton) and storage losses (12-20%) are not taken into account. When taken into account, per capita consumption varies in the range of 39.1-63.6 kg over the last 20 years, and averages around 47.1 kg per capita consumption is lower than the EU average of 81 kg, but higher than the world average of 33 kg. The potato is a basic nutritional element in the EU countries and is consumed in various ways. According to the consumption trend equation ($Y = 6.318 + 0.0122X_t$ [$p<0.00$]), the rate of annual consumption increase is 1.2%, which is almost equal to the annual growth rate in production.

With increased potato production over the past decade, the actual price for potatoes given to growers has changed little, which means that inflation-adjusted price has decreased similar to the previous research results by Pavlisa and Dillon (2005). When the change in potato producers' prices (CV=22.72) and general consumer prices (CV=21.57) are compared for the period 1950-2005, it can be seen that the volatility in potato prices is in line with the level of inflation. The change in gross income per hectare of land calculated with real prices (CV=49.72) is higher than the change in planted land, yield, production and real producer price. It is known that the supply of potatoes, like many other agricultural products, is highly sensitive to price volatility. An increase in prices in the current year may lead to higher production in the following year, which causes prices to decline and a reduction in planted areas in the following year. One of the main theories in agriculture, the Cobweb Theory, is highly applicable to potato farming. If producers are attracted to the high prices of the current year and increase production in the following year, they will be faced by lower prices as demand elasticity is low and increased supply will not be absorbed by a corresponding growth in demand. This will lead to a decline in market prices and the producer will be adversely affected. Under these conditions, producers need to be guided about the market conditions by producer organizations and State institutions, which will enable production planning by the use of effective macroeconomic tools and the establishment of efficient marketing systems.

Based on 56 years of time series data, covering the period 1950-2005, supply and demand elasticity has been calculated taking supply and demand equilibrium ($Q_s=Q_d$). Taking into account the logarithmic distribution of production and consumption, supply and demand equations are stated in logarithmic form:

$$\text{Log } S = a + bP_{t-1} \quad (1)$$

$$\text{Log } D = a + bP_t \quad (2)$$

Equation (1) and (2) direct regression model is solved by the smallest squares method and supply and demand equations are obtained:

$$\text{Log } S = 6.117 + 0.412\ P$$

$$\text{Log } D = 6.211 - 0.225\ P$$

All the coefficients of the equations have statistical significance of 1%. Plugging equilibrium price and quantity into the formula $[(P/Q).(dQ^s/dP)]$, supply elasticity is estimated as 0.034 and demand elasticity is estimated as -0.018 by the formula $[(P/Q).(dQ^d/dP)]$. According to these simple models, there is a positive and statistically significant relationship between potato supply and previous year's prices. The response of potato supply to changes in price is found to be low. Prices have become less responsive to changes in production. Demand elasticity is negative and relatively low, as expected. These results indicate that the potato, as a raw material, is regarded as an inferior product, and changes in price levels do not impact on the demand for potatoes. The demand equation is also estimated as a function of income and

price by using time series data to determine the income elasticity of demand. The following equation is obtained:
Log D = 6.181 – 0.140 P_t + 0.000153 I_t (I= income)

The parameters of this equation have a statistical significance of 1% and indicate that income growth leads to an increase in the demand for the potato, but this increase is rather low.

The potato has a significant market in the world to be consumed as a fresh product or processed food and as seed. While Turkey accounts for 1.5% of global potato production, its share in the total potato exports is lower than 0.5%. Turkey's potato exports were 8,408 tons in 1986 and had risen to 373,969 tons by 1992, but exhibited a volatile trend in this period. The volatility in potato exports became even more pronounced in the 2000s, and exports fell to 42,300 tons in 2006 (Table 6). The share of exports in total production varies between 0.2% and 8.1%, with an average of 1.9%. In this period under review, potato imports are mainly constituted by seed imports (Table 7). A significant volatility in export quantities and values, as well as import quantities and values, is also observed.

Over the last 10 years, the potato has been among the 10 leading export products, with exports in the range of $1.2 to 51.1 million, and imports in the range of $820.120 to $13 million. Due to the volatility in export quantities, significant problems are being experienced in marketing, and there is a significant volatility in seasonal as well as annual prices. International embargoes, bilateral relations with neighbouring countries and export support policies play a role in the volatility of export quantities. In particular, the war in Iraq since the 1990s has adversely affected potato exports to the country. Turkey has an advantageous location in terms of export possibilities. The EU countries, which are also high consumers of potatoes are self-sufficient. However, fresh potato demand reaches a peak in April-May in the EU, bringing imports from Egypt, Morocco, Tunisia, Israel, Cyprus and Italy. Although Turkey's early potato production

Table 6. Developments in potato exports and imports in Turkey (source: SIS).

Years	Imports		Exports	
	Amount (ton)	Value (1000 US $)	Amount (ton)	Value (1000 US $)
1986	9,110.00	3,227.00	8,408.00	1,197.43
1990	14,082.00	5,148.00	22,442.00	4,617.82
1995	9,595.64	3,599.00	103,534.00	23,873.00
2000	10,730.00	4,726.00	140,353.00	24,956.53
2001	2,107.00	820.12	105,886.00	12,950.37
2002	11,822.00	5,766.00	25,179.00	1,908.00
2003	9,254.00	5,221.00	172,301.00	16,070.00
2004	9,075.65	5,858.29	154,907.00	14,535.00
2005	13,701.00	7,693.00	78,042.00	6,399.00
2006	23,221.92	13,003.95	42,324.06	8,278.10

Table 7. Turkey's potato exports and imports by product types (source: SIS).

Products		**Unit**	**1996**	**2000**	**2001**	**2002**	**2003**	**2004**	**2005**	**2006**
Frozen potatoes	Exports	Amount (Ton)	10,002.9	1,062.5	34.6	237.3	814.3	359.3	162.1	1,627.5
		Value (1000 US $)	1,365.8	216.1	12.9	37.3	281.7	160.2	148.7	741.3
	Imports	Amount (Ton)	39.3	1.3	46.0	190.5	-	54.7	70.0	64.3
		Value (1000 US $)	22.5	0.9	18.2	81.7	-	12.7	12.9	52.3
Table potato	Exports	Amount (Ton)	240,232.5	140,353.1	105,885.6	34,398.0	176,647.7	154,828.6	77,890.4	40,674.6
		Value (1000 US $)	29,712.1	24,956.0	12,951.71	2,472.2	16,606.6	14,527.3	6,376.2	7,521.5
	Imports	Amount (Ton)	556.9	239.1	137.9	2,933.4	1,673.5	1,293.3	4,475.7	8,384.4
		Value (1000 US $)	102.1	64.6	23.4	230.3	409.3	87.2	1,120.6	2,466.4
Seed potato	Exports	Amount (Ton)	468.5	-	-	-	103.1	80.0	150.0	22.0
		Value (1000 US $)	145.0	-	-	-	14.4	8.0	22.5	15.3
	Imports	Amount (Ton)	9,872.9	10,490.5	1,968.6	8,888.3	7,581.1	7,777.8	9,225.1	14,773.2
		Value (1000 US $)	7,221.9	4,662.6	797.3	5,534.6	4,811.5	5,770.3	6,572.1	10,485.3
Total	Exports	Amount (Ton)	250,703.9	141,415.6	105,920.2	34,635.2	17,565.1	155,267.9	78,202.5	42,324.1
		Value (1000 US $)	31,222.9	25,172.1	12,964.6	2,509.5	16,902.7	14,695.5	6,547.4	8,278.1
	Imports	Amount (Ton)	10,469.2	10,730.9	2,152.6	12,012.2	9,254.6	9,125.8	13,770.9	23,221.9
		Value (1000 US $)	7,346.4	4,728.1	838.9	5,846.5	5,220.8	5,870.2	7,705.6	13,003.9

is growing in the Mediterranean and Aegean Regions, exports are not increasing at the same rate. Furthermore, Turkey's Middle East neighbours lack suitable climates for potato farming and import large quantities. The annual import amount of nine countries in the region (Iran, Syria, Jordan, Lebanon, Israel, Yemen, Saudi Arabia, Kuwait and the United Arab Emirates) varies between 200,000-300,000 tons, which constitutes a significant potential, while additional potential exists in the Balkan and Central Asian States.

The identification and application of export support programs that are not in conflict with GATT Agreement on Agriculture will be effective to increase exports. Turkey applied an export premium of $19 per ton in the 1 April-31 May 2004 period; $20 per ton in the 1 March-31 May 2005 period; and $68 per ton for only frozen potatoes in 2006. The time limit imposed on premiums and late announcements of such applications prevent a full realization of their expected benefits. For example, the premium to be paid for the 1 Jan-31 May 2004 period was announced towards the end of April. Even if it was announced in a timely manner, a premium period that starts in January increases product storage costs. Similarly, the ending of a premium period on 31 May prevents İzmir from benefiting from it. A longer support period and timely announcement of such policies will be very useful. The support of processed products rather than fresh produce as offered in 2006 is especially significant for the growth of higher value-added production and exports.

In the marketing of potatoes, traders, processing firms, seed firms, cooperatives and producers that market their own products play a significant role. Local wholesalers buy products from producers and sell to distributor wholesalers in other cities and/or directly market to retailers. The type and number of intermediaries in the marketing chain differ according to production targets. When there is over-supply, significant problems are experienced in the marketing chain. Producers have difficulty in collecting receivables. The marketing channel of the potato is illustrated in Figure 1. If we consider only non-seed potatoes (NS), after the harvest of potatoes during May-October period, there are several options for the producers: (i) to sell them to traders immediately, or after drying for two weeks, (ii) to sell them to traders after storage at home for 2–3 months, (iii) to make direct marketing to the consumer, and (iv) to store them in cold storage and sell in the market several months later. The majority of farmers, especially those running large- and medium-size operations, sell harvested potatoes in their homestead and receive cash from traders immediately, or with some part paid later (within 1-2 months). Small farms and landless tenants sell their small amounts of produce to traders in the nearby marketplace or on the field to provide cash for demand. Seed (S) and processing (P) potatoes are directly sold to companies involved in contract farming and the products are delivered from the field. The distribution of products among different physical marketing channels varies from year to year, but the improvement in the selling power of the producer and the development of direct marketing from the producer to the consumer will lead to an increase in the welfare of the producer.

Around 7-10% of potatoes produced are reused in the farms as seed, and there are storage losses of 12-20% due to improper storage facilities. A further 4-5% of production is used as raw material in the industry, 4-5% is consumed to feed animals, 2% of production is exported and the remaining part (58-71%) is used in domestic consumption. There are an excessive number of intermediaries in the chain from producer to consumer. When the real producer prices and

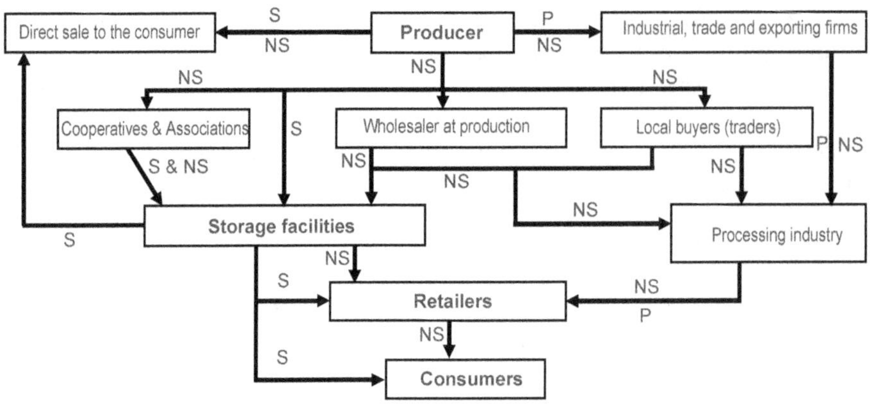

Figure 1. Distribution channels in potato marketing in Turkey.

consumer prices are analyzed for the period 1950-2005, it is seen that 59.1% of the price paid by the consumer is received by the producer. This ratio is lower than 50% in many agricultural products, so it should be noted that this ratio seems relatively high. Still, it is possible to increase this ratio through a more extensive usage of contract farming and direct marketing to the consumer. Products are sold for cash and/or on a maximum two-month term, and producers usually find prices at 50-70% of their expectations.

The lack of efficient organization among the producers in line with market demands, the inability to identify problems in production and marketing and the non-existence of producer organizations to develop policies and applications effectively are the root causes of the problems faced in marketing. There is no producer organization established within the scope of Cooperatives Law No. 1163 to specifically deal with potato farming, processing and marketing. In Niğde, Bolu, Afyon and Bitlis, four potato producers' associations with a combined membership of 200 have been established in accordance with Agricultural Producer Association Law No. 5200.

Except for seed, table or processing products directly sold from the field, products are stored for 2-5 months and various types of storage (soil depot, stone depot, house, hayloft and cold storage) are used for this purpose. High losses arising from improper storage remain as a significant problem. Production reaches a peak in September-October (2.5-3 million tons), and since consumption is spread over a long period, and the fact that potatoes are 70-80% water, potatoes have to be stored in proper facilities. Farmers who do not want to store their products under improper and primitive conditions sell their products early at low prices. The presence of special stone storage units suitable for potatoes in Nevşehir (Ortahisar) and Afyon (Sandıklı) enables potatoes to be stored for long periods without deterioration. Products stored in this way have a higher chance of being sold at good prices when compared to potatoes stored at the farmhouses. In order to enable proper storage of potatoes destined for the domestic and export markets, storage facilities to be built in accordance with the 'Law on Licensed Storage for Agricultural Products' No 5300, issued in 2005, should be encouraged and supported. This will in turn increase the welfare level of the producers.

Potato processing industry and analysis of export possibilities

In the potato industry, there are nine firms active in the production of frozen fingerling potato, leaf chips and flake potatoes. The current production capacities of these facilities range between 160-180 tons per year and the industry has a total raw material processing capacity of 200,000 tons per year. According to data obtained from these firms, distribution of annual production is as follows: French fries (frozen potato) - 48%, flake potato - 3.5% and potato chips - 48.5%. Specific potato types suitable for fried potatoes and chips are provided to the farmers by the processing companies, and the products are later purchased based on contract farming. Seeds for industry potatoes are also imported in the mature stage by the processors and are distributed to the farmers. Production is secured through contracts. Domestic raw material production is not sufficient for the chips and frozen potato processing capacities, but as per capita consumption in Turkey (0.4 kg) is lower than the EU average (7-11 kg) (Aslanoğlu *et al.*, 2006), processing facilities operate at low capacity or secure raw material directly from the market if necessary. The development of the domestic and export market for processed potato products will put production planning on a healthy course and will control excessive price fluctuations. Industrial processing will increase the added value of the products and expand export opportunities.

An overall analysis of production and marketing

The existing problems, opportunities, risks and strengths of the potato sector in Turkey have been analyzed using the SWOT technique to determine ways to develop potato production and marketing. The existing problems, opportunities, strengths and weakness have been determined as follows, based on the data gathered through a quick assessment survey conducted with potato producer associations, industrial establishments, seed producers and other representatives in the sector (Table 8).

Conclusions

Potato production in Turkey displayed an increasing trend until the 2000s, since when a contraction in plantation areas and total production has been experienced due to a spread of disease and pests. Thanks to developments in agricultural techniques employed, relatively higher average yields compared to the world average have been achieved. Efforts to develop local varieties have been neglected and imported foreign cultivars are used in domestic production. Serious problems have been experienced in high quality and yield seed production and distribution, and only 30% of the seed demand can be satisfied. The dependency on foreign inputs causes the loss of genetic sources and deterioration of the land. The quarantines and cultivation bans applied in certain regions due to the spread of disease and pests require a re-evaluation of the existing agricultural policies in general. It is necessary to establish seed production facilities, separate the production of potatoes for table, seed and industry, and readjust the input use such as farm chemicals, irrigation and mechanization.

The progress achieved in potato production in spite of the various drawbacks have not been realized in marketing, and exports have declined to 2% of production since the onset of the Gulf War. Failure to increase foreign demand for fresh and processed potatoes and an inability to find

Table 8. Anclysis of Potato Production and Marketing in Turkey (SWOT Analysis).

Strategy	Strengths	Weaknesses	Opportunities	Threats
The Development of Potato Production and Marketing Systems	4.8 million potatoes are produced on 178,900 ha land. Average yield is higher than world average. The share of potatoes in agricultural production value is high in the producing cities and cash crop is used. There are significant problems in physical inputs and use of seed. Planting bans due to quarantine causes contraction in domestic production and plantation areas. It is seen that potatoes have become a traditional product in many cities after the 1970s.	Producer and consumer prices display volatility on a yearly and seasonal basis. The cost of production inputs is high and most are procured from external sources. Significant problems are experienced in the marketing channels and the producer is unable to deliver products to the consumption areas by himself or with the help of the producer associations. The intermediaries get a significant portion of the final product value. The production and consumption of processed potato products are at a low level. The share of exports in production is quite low.	Yield per planted area is quite high. There is scope for development of the potato processing industry. There is high demand for seed and potential for growth in the seed market. Export opportunities are expected to grow in the future. Proper storage facilities are being established, which will help to increase farmers' income. Improvement in product quality is observed, which implies an opportunity to increase market share.	The necessity of quarantines in new planting areas due to diseases and harmful pests. Spread of disease and harmful pests limit export opportunities. The excessive use of agricultural chemicals, irrigation and other inputs and the resulting economic and environmental problems. Lack of technical preventive measures and the resulting problems. The low level of education, knowledge and organization capability of the producers. Uncertainties in macro policies and regional risks.

new markets has set forth the requirement to limit potato production to a level that can meet the demands of the domestic market only. Elimination of the uncertainties in international markets, enhancement of storage and processing facilities, ensuring stability in production by means of planning, implementation of adequate and efficient export incentives, developing and producing domestic seeds, and establishing a well-organized marketing structure would ensure an enhancement of production and marketing. It is determined that the elasticity in the supply and demand of the potato is much lower than expected. Although potato supply is a function of the previous year's prices, the sensitivity of production quantities to market price is low. Still, some fluctuation is observed in production on a yearly basis, and it is thought that this is related to product price and domestic and external demand levels. Under these conditions, it is necessary to carry out production planning to achieve stability in the product price. In order to protect producers from volatile market conditions, individual producer associations and other institutions should assist in production planning. Also, in the integration of agriculture with industry, the vertical integration method of contract farming should be preferred. An effective organization among the producers, producer associations and cooperatives is necessary for a proper application of the contract farming method, and these organizations should be a party to the contracts signed.

In order to decrease dependency on seed imports, institutions should cooperate to develop high-quality local cultivars. According to the interviews conducted with the stakeholders, there are problems in the procurement of high-quality, sufficient and continuous raw material for seed production in potato farming, processing and trade. The sector's competitive power should be improved by increasing varieties through standardization, quality, packaging and colour. The private sector, producer, consumer and state should cooperate in this respect. Organic and good agricultural techniques should be encouraged, storage problems should be resolved and high storage losses should be brought down to minimum levels. It is necessary to improve storage conditions, to build modern storage facilities, decrease storage losses and thereby increase the added value to the whole economy. Producer associations should be oriented to make production planning through proper support mechanisms, price fluctuation should be minimized, export support premiums should be increased to meaningful levels for fresh and processed potatoes, domestic production of seed potatoes should be emphasized and an effective marketing channel should be established from the point of production to the final consumer to eliminate the current problems in marketing.

References

Arıoğlu, H., 2002. Nişasta ve şeker bitkileri, çukurova üniversitesi genel yayın no: 188, Adana.

Arıoğlu, H., Çalışkan, M.E. and Onaran, H., 2006. Türkiye'de patates üretimi, sorunları ve çözüm önerileri, In: proceedings of IV. ulusal patates kongresi, patates araştırma enstitüsü müdürlüğü, Niğde, pp:1-10.

Aslanoğlu, F., Akay, H. and Sütveren, H., 2006. Türkiye'de cipslik patatesin üretim ve pazarlama durumu ile patates cips üretiminin fabrikasyon aşamaları, In: proceedings of IV. ulusal patates kongresi, patates araştırma enstitüsü müdürlüğü, Niğde, pp.321-324.

FAO, 2006. FAOSTAT citation database results, FAO. http://faostat.fao.org/faostat.

Gül, U. and Tan, S., 2003. Patates, tarımsal ekonomi araştırma enstitüsü, T.E.A.E – BAKIŞ, sayı: 2, nüsha:8, Mart 2003, Ankara.

Günel, E., 2002. Dünden yarına patates yetiştiriciliği, In: proceedings of III. ulusal patates kongresi, İzmir, pp:23-42.

Günel, E., Çalışkan, M.E., Tortopoğlu, A.İ., Kuşman, N., Tuğrul, K.M., Yılmaz, A., Dede, Ö. and Öztürk, M., 2005. Nişasta ve şeker bitkileri üretimi, türkiye ziraat mühendisliği VI. teknik kongresi, Ankara, pp:431-457.

Horton, D., 1987. Potatoes, production, marketing and programs for developing countries, Westview Press, Boulder, USA.

Kara, K., Öztürk, E. and Polat, T., 2006. Patates (Solonum Tuberosum L.) için uygun olan ön bitkilerin belirlenmesi, In: proceedings of II. ulusal patates kongresi, Erzurum, pp.152-155.

Karaca, M. and Demir, Z., 1994. Patateste azotlu gübre miktarı ve uygulama zamanının yumru verimine ve hasat sonrası toprak inorganik azotuna etkisi, TC tarım ve köyişleri bakanlığı tarla bitkileri merkez araştırma enstitüsü, Ankara

Köksal, H., Çengel, V. and Ertunç, Ö., 1999. Niğde bölgesinde patates sulu tarımında yaşanan sorunlar. In: proceedings of II. ulusal patates kongresi, Erzurum, pp.131-141.

Kuşman, N., 2002. Türkiye patates tohumluk endüstrisinin teknolojik, ticari ve hukuksal yapısı, In: proceedings of III. ulusal patates kongresi, İzmir, pp:34-49.

MARA (Ministry of Agriculture and Rural Affairs), 2006. TC tarım ve köyişleri bakanlığı kayıtları (unpublished), Ankara.

Moazzem, K.G. and Fujita, K., 2004. The potato marketing system and its changes in Bangladesh: from the perspective of a village study in comilla district, the developing economies, Vol: XLII-1(March 2004):63–94.

Onaran, H., Çalışkan. M.E., Ünlenen, L.A., Nam, M., Bilgin, M.G. and Arslan, R., 2006. Niğde ve nevşehir koşullarında farklı ekim nöbeti sistemlerinin patates verimi ve kalite özellikleri ile ürünün ekonomik değeri üzerine ekileri, In: proceedings of IV. ulusal patates kongresi, patates araştırma enstitüsü müdürlüğü, Niğde, pp:69-77.

Özçelik, A., Turan, A. and Tanrıvermiş, H., 1999. Türkiye'de tarımın pazara entegrasyonunda sözleşmeli tarım ve bu modelin sürdürülebilir kaynak kullanımı ile üretici geliri üzerine etkileri, tarımsal ekonomi araştırma enstitüsü yayın no: 14, Ankara.

Pavlista, A.D. and Dillon, M.F., 2005. Potato prices as affected by demand and yearly production, american journal of potato research:339-343.

SIS (State Institute of Statistics), 2004. 2001 census of agriculture agricultural holdings (households), publication no: 2924, Ankara.

SIS (State Institute of Statistics), 2006. Agricultural structure (production, price and value) 2004, publication no:3022, Ankara

Yıldırım, M.B. and Yıldırım, Z., 1986. Tohumluk patates yetiştiriciliği, TC tarım ve köyişleri bakanlığı ege tarımsal araştırma enstitüsü müdürlüğü, İzmir.

Yulafcı, A. and Duran, H., 2006. Karadeniz bölgesinde patates üretim tekniği ve pazarlaması, In: Proceedings of IV. ulusal patates kongresi, patates araştırma enstitüsü müdürlüğü, Niğde, pp:167-173.

Curriculum vitae – Mehmet Bülbül

Mehmet Bülbül was born in Samsun province which is situated in the black sea coast in 1939. He graduated form the department of agricultural economics, agricultural faculty Ankara university in 1962. He received a Ph.D. in 1969 at the department of agricultural economics, Kiel University in Germany. He had been worked at the department of agricultural economics as academic staff during 1963-2005 and after the retirement, he has been working at the department of real estate development department at the Ankara University. He is studying the economics agriculture, real estate management, agricultural finance, economics and management of agro-industry and irrigation economics and management. He has experience in the project planning and implementation in public and private institutions.

Curriculum vitae – Harun Tanrıvermiş

Harun Tanrıvermiş was born in Ordu province which is situated in the black sea coast in 1971. He graduated form the department of agricultural economics, agricultural faculty Ankara university in 1991. He received a M.Sc. in 1993, and a Ph.D. in 1996 at the department of agricultural economics, graduate school of natural and applied sciences, Ankara University. Also, he received a second M.Sc. degree on environmental management in 1995. He has been working at the department of agricultural economics as academic staff and also he head and founder of the real estate development department at the university. He is studying the economics and management of natural resources and the environment, real estate management and valuation, and the economics and management of sustainable agriculture. Except academic activity, he has great experiences in the project planning and implementation in public and private institutions.

Profile – Department Agricultural Economics

Agricultural education and researches was started in 1933 with establishment of High Agricultural Institute at the university of Ankara. Department of agricultural economics was established in the body of this institute. Agricultural economics is an applied science branch which is investigated solution ways of problems in agriculture and agricultural industry sector. Agricultural economists should know both technical issues like field crops, horticulture, animal production, processing of agricultural products and issues like planning of farms and agribusiness, management, finance, accounting, marketing, agricultural policy, agricultural law, rural sociology, rural development, agricultural extension and communication. At the same time agricultural economists should deal with plans and programs applied by state aimed at developing agriculture sector, improving farmers living level, protecting consumers from over rising prices and arranging foreign trade of agricultural products. Agriculture is one of the key sector in the economy of Turkey being the suppliers of food as well as raw materials to agro-industries which constitute a large share in manufacturing and food industries and an important source of foreign Exchange earnings. In this context, the subject of agricultural economy always keeps its importance in the education, research activities and also administrative body of the country. Today, the education and research activities of agricultural economics in the agricultural education and research system depend upon a long history and these activities have seen broadly continuing. In the body of agricultural economics education, training and research facilities carry on agricultural management and agricultural policy and extension issues. Agricultural economics researches related especially national and international research results with economic, social and politic developments. There are 17 staff members in the department. The number of periodicals written by the foreign language is 4 and the Turkish periodicals number is 20.

Income as a factor of potato consumption

L. Hadelan, V. Par and M. Njavro
Faculty of Agriculture, Farm Management Department, Svetošimunska cesta 25, 10000 Zagreb, Croatia

Abstract

Engel's Law is the observed phenomenon that when a family's income increases, the proportion of money they spend on food decreases. As an extension of this, the goods are divided into the two groups – inferior and normal goods. In consumer theory, an inferior good is a good that decreases in demand when the consumer's income rises, unlike normal goods, for which the opposite is observed. As an inferior good, lots of sources use potato as an example. According to that, the potato is marked as a cheap, low-valued food article. One of the most accepted methods for life standard determination is Gross Domestic Product per capita. Using this method, a significant increase of life standard in Croatia can be noticed. According to the data from the Yearly Statistics GDP in Croatia has increased by 278 percent during the period 1992-2005. Consumption of potato in the same period has not changed. According to official statistical data about potato consumption, yearly potato consumption per capita in 2005 was 45,0 kg. It was 3,7 percent more than in 2002 when the potato consumption per capita was 43,4 kg. Income elasticity can be easily calculated as ratio of percent change of potato consumption and percent change of income. In this case, income elasticity for potato is 0,01328 what is a clear indicator of potato income inelasticity in Croatia. Therefore, one can conclude that potato in the Croatian example is a normal, inelastic good.

Keywords: potato, consumption, income elasticity

Potato production in Croatia

Potato production in the Republic of Croatia has a very important role in crop production. Concerning arable and harvested area and total production, potato takes the first place in vegetable production of Croatia. In Croatia, potato producers were mainly non-commercial small family farms without any significant subsidies and with low intensity technological level. Technically more equipped and economical stronger farms have used more capital resource production combined with low human labour.

Practicing mostly arable crops farming without livestock production, small farms could not ensure enough manure that is precondition for commercial potato growing. Those farms did not invest in equipment providing and modernization of potato production technology as well.

In most arable crops production and vegetable growing yields were moving up constantly since 40 years. In case of potato, progress has been achieved recently. This delay of potato yields increase can be noticed even in those days. There is a lack of information about the most appropriate potato cultivar for a different agro climate zone as well as information about

Table 1. Production of Potato in Croatia in the period 1994-2003 (Source: Statistical Yearbooks 1995-2004).

	1994	1995	1996	1997	1998	1999	2000	2001	2002	2003
Harvested area, ha	66,356	66,458	65,537	63,189	64,784	66,374	65,232	65,641	64,640	63,097
Production, tonnes	563,285	692,216	666,020	620,032	662,573	728,646	553,712	670,233	736,198	375,050
Yield, tonnes/ha	8.48	10.40	10.15	9.78	10.22	10.96	8.47	10.20	11.37	5.93

modern potato technology. As a consequence, in Croatia there is a strong production and potato price volatility, as well as an undefined market situation. Some small producers have given up on potato production while stronger ones are expecting more information on market needs in future.

In Croatian potato production there was a sharp decrease since the sixties of the 20th century.

Within the period 1966 to 1999 land under potato production has been reduced by 40 percent. In the year 1966 there were 107,000 hectares planted by potato what was 7% of total arable land surface. In 1988, potato has been grown on 78,360 ha or 57.1% of total vegetable production. With average yields of 7.8 t/ha, annual production was 612,958 tonnes. In 1996 potato production was established on 65.5 thousands hectares or 5 percent of the total arable land in Croatia.

The average potato yield has increased to 11.37 tonnes per hectare in 2002. This is almost three times smaller than in EU countries (30 tonnes/ha). Therefore, Croatia is on the bottom of European potato production productivity. However, these figures may not be relevant completely because they include production on intensive commercial farms as well as extensive potato producers that produce for their own consumption. The estimation is that only 15 percent of producers use the modern production technology that ensures yields in range of 25 to 30 tonnes per hectare.

Potato consumption

For a long time, potato has been considered as a food of low-income consumers. Nowadays, the potato is regarded as an 'inferior product' and it evokes memories of hardship. By contrast, the industrial processing of potatoes has increased steadily, even enjoying a boom in the eighties and nineties. The potato is on demand as a renewable raw material. Its starch is increasingly used today for paper and pulp, paste and glue, building materials and packaging, and even washing powder, toothpaste, tablets and many other items.

Estimation of share of potato human food consumption is 44 percent of the total Croatian potato consumption. For the processed industry purpose 1.5% of the total consumption has been used. For the livestock feeding purpose 20 percent has been used and the losses of production are in the same percent. Seed potato consumption is approximately 15 percent of total production. Personal consumption according to the latest results is 45 kilograms per Croatian household member. This is exclusively consumption of raw potato. No doubt that this amount would be larger if the processed potato was included.

According to data from Table 2 it is obvious that potato consumption in Croatia did not change dramatically. Potato consumption per person has increased within the period 1992 to 2005 by 3.7 percent.

Croatian GDP

One of the most accepted methods for life standard determination is using data of Gross Domestic Product per capita. Using this method, significant increase of life standard in Croatia can be noticed. According to the data from Yearly Statistic GDP in Croatia has increased by 278 percent during period 1992-2000.

Table 2. Personal consumption of raw potato per household member in Croatia (Source: Statistical Yearbooks 1995-2004).

Year	**1992**	**1993**	**1994**	**1995**	**1996**	**1997**	**1998**
Quantities consumed, kg	43.4	42.5	43.8	41.8	40.3	42.4	43.5
Year	**1999**	**2000**	**2001**	**2002**	**2003**	**2004**	**2005**
Quantities consumed, kg	40.2	35.8	47.3	44.3	42.4	45.4	45.0

Table 3. Croatian GDP per capita (source: Statistical Yearbooks 1995-2004).

Year	**1992**	**1993**	**1994**	**1995**	**1996**	**1997**	**1998**
GDP, USD $	2,291.0	2,349.2	3,137.2	4,028.9	4,421.9	4,398.2	4,805.1
Year	**1999**	**2000**	**2001**	**2002**	**2003**	**2004**	**2005**
GDP, USD $	4,371.1	4,152.6	4,476.2	5,186.9	6,665.7	7,943.8	8,674.4

Potato consumption elasticity

Engel's Law is the observed phenomenon that when a family's income increases, the proportion of money they spend on food decreases. As an extension of this, the goods are divided into the two groups – inferior and normal goods. In consumer theory, an inferior good is a good that decreases in demand when the consumer's income rises, unlike normal goods, for which the opposite is observed.

As an inferior good, lots of sources (i.e. Wikipedia) use potato as an example. According to that, the potato is marked as a cheap, low-valued food article. However, data on Croatian potato consumption within the period 1992 – 2005 and respective GDP growth show totally opposite situation. In this respect, raw potato consumption did not change although GDP as a life standard indicator has been considerably increased.

The Income Elasticity of Demand measures the rate of response of quantity demand due to a raise (or lowering) in a consumer's income. The formula for the Income Elasticity of Demand (IEoD) is given by:

Income Elasticity = (% Change in Quantity Demanded) / (% Change in Income)

In our case during the period 1992 to 2005 the consumption of potato has raised by 1.6 kg per person or 3.7%. At the same time GDP per capita was increased by 278.6 percent. Hence, potato income elasticity is 0.01328. Using the definition, which explains that good with income elasticity in range (0.1) is inelastic, one can concludes that potato on Croatian example is normal, inelastic good. The same conclusions can emerge from correlation analysis. The standard Pearson product moment correlation coefficient (*r*) measures the extent to which two variables are related.

In examined model r=0.301 what indicates medium positive correlation between potato consumption and GDP growth.

Table 4. Pearson product moment correlation coefficient (Source: author's calculations).

		Potato consumption	**GDP per capita**
potato consumption	Pearson Correlation	1	0.301
	Sig. (2-tailed)	.	0.296
	N	14	14
GDP per capita	Pearson Correlation	0.301	1
	Sig. (2-tailed)	0.296	.
	N	14	14

Future of potato market

The Croatian experts' predictions say that consumption of the 'raw potato' in near future will not change significantly (Kovačić, 2005). A certain small decrease can be expected after life standard change, although our analysis showed that during the last 14 years potato consumption was not correlated with GDP change at all. At the same time, new opportunities for some unique, origin potato brand will emerge. Today, consumers want to consume added product quality and they are ready to pay more if it ensures exactly demanded taste, colour and other quality brand parameters.

Experiences from Austria or Germany showed that consumers can recognize characteristics of 'Pfanni's best Kartoffel' and 'Bramburi' which hold more than 30 percent of raw potato consumption. Croatian producers should assume that the same trend will exist on Croatian market and they should recognize their chance to offer high quality branded potato. In Croatia the most famous potato-region is the mountain region Lika.

Experts from Faculty of Agriculture in Zagreb conducted research in which they have planted the same potato cultivar on different Croatian locations. The potato cultivated in Lika region had a minor yield but at the same time 20 percent additional dry matter. Consequently, it was more enriched with minerals and vitamins than potato from other locations. Some investors from Lika have recognized an added value chance in a form of potato growing properties by establishing a specialized cooperative. The project of the Lika potato branding process has been conducted in cooperation with the Faculty of Agronomy and the Croatian Ministry of Agriculture and Forestry.

Conclusion

Using data on potato consumption and GDP per capita during the last fourteen years in Croatia, potato cannot be considered as an inferior good. Although, potato is a cheap food article, consumers have recognized its nutrition value continuing the level of consumption regardless of changes on its life standard. To the contrary, if processed potato (i.e. frozen French potato) consumption is included in this analysis, we are sure that a lot of arguments would define potato as a normal or superior good. Anyway, potato has its future guaranteed market in Croatia because there are multipurpose uses of it as well as opportunities for some unique original potato brands.

References

Kolega A., Božić M. (2001). Hrvatsko poljodjelsko tržište (Croatian agricultural market), Tržništvo, Zagreb

Kovačić, D. (2005). Izravna prodaja seljačkih proizvoda- teorijska polazišta i praktična primjena (Direct selling of farm product - theoretical background and practical use), Agrarno Savjetovanje, Zagreb

Kovačić,D., Radman, M. Kolega, A. (2002). Behaviour of fruit and vegetable buyers on the city markets in Croatia In: 13th International IFMA Congress of Farm Management, book of abstracts, Arnhem, 2002, 26-27

Uglem, S., Frřlich, W., Holte Stea, T., Wandel, M. (2006). Correlates of vegetable consumption among young men in the Norwegian National Guard, Appetite 48 (2007) 46–53

Wandel, M., Fagerli, R., Kjernes, U.(2001). Changes in potato consumption in different stages of life in Norway, Appetite (2001) 36, 211-223

Curriculum vitae – Lari Hadelan

Lari Hadelan is an agroeconomics PhD student at the Faculty of Agriculture in Zagreb, Croatia, where he graduated in 2000. His first job was Bank Analyst in a middle sized commercial bank in Croatia. In 2003 he was employed by the Farm Management Department at the Faculty of Agriculture in Zagreb as an assistant on modules Farm Management, Microeconomics in Agribusiness, Management and Entrepreneurship in Agribusiness. He associates as a young researcher on the project 'Risk and Economic Efficiency Analysis of vegetable production of family farms' sponsored by Ministry of Science of Republic of Croatia. His field of interest is financial management. Since 2003 he is Member of Croatian Society of Agricultural Economists. As a practitioner, he owns a small family farm specialized in blackberry wine production.

Profile – Zagreb Faculty of Agriculture

The Faculty of Agriculture in Zagreb (founded in 1919) is the leading institution in the Republic of Croatia, with respect to the number of researchers and their activities in the field of agricultural sciences and the profession. It involves over 200 researchers in scientific and research work and comprises 28 departments, 28 laboratories and six experimental farms. The Faculty offers three graduate and eight postgraduate programmes for more than 1700 students. The main research subjects are: preservation of the biodiversity of plants and autochthonous indigenous varieties of animal, pollution of soils and waters, effective plant protection, application of biotechnology in agriculture, application of geoinformatics science in agriculture, new technologies for sustainable and renewable (organic) agricultural production, new food production technologies, evaluation and preservation of landscapes and the heritage of garden-come-park architecture, socio-economic and marketing analyses of Croatian agriculture, and competitiveness of Croatian agriculture in the domestic and world markets.

The comparative economic analysis of potato production on farms and the role of the producers in the marketing system in Turkey

H. Tanrıvermiş, H. Şanlı and M. Bülbül
Ankara University, Agricultural Faculty, Department of Agricultural Economics, 06110, Diskapi, Ankara, Turkey, tanriver@agri.ankara.edu.tr; tanrivermis@gmail.com; sanli@agri.ankara.edu.tr; bulbul@agri.ankara.edu.tr; bulbulmeh@yahoo.com.tr

Abstract

We make a comparative economic analysis of potato farming in the main production areas of Turkey, and examine the role of producers in the marketing phase. The research analyses the data complied 291 farms chosen from the provinces of Niğde, Nevşehir, Ordu, Erzurum, Afyon and Bolu, covering 41.4% of the country's potato cultivation areas and 48.6% of the total production, as well as the firms and merchandisers purchasing products from these farms. Potato producing farms are generally small in size, the level of organization is low and the population and labour force are considerably high. The farms allocate between 15%-63% of their operating lands for potato framing in provinces, where potato farming can be said to be traditionalized and has become the major cash crop. The gross and net profits in potato farming per unit area are considerably high when compared to other crops cultivated on the same farms, which encourages the cultivation of potatoes as a monoculture farming on farms. In areas where land has been quarantined due to potato wart, the sum of the net profit provided by the main alternative crops being cultivated instead of the potato and the state support being provided falls far short of the potato's net economic benefit. It was found that the producers are unsatisfied with the imposed quarantine for economic and social reasons. Between 79%-96% of annual production on farms is offered to the market. The producers' rate of participation in the process of marketing in accordance with provinces ranges between 19% and 36%. It appears that the level of producer participation in the process of marketing is insufficient. In order to increase the small producers' welfare, it is imperative that they participate in the process of product marketing and benefit from vertical integration.

Keywords: production cost, productivity, profitability, competition among activities, producer tendencies

Introduction

Potato farming started in Turkey early in the Adapazarı and Istanbul provinces, followed by the Black Sea, Central Anatolia and Eastern Anatolia regions. Potato farming presently is carried out on 158,487 of the total 3,022,127 farms in the country, and accounts for 0.5% of the total farmland. After cereal crops, the potato is the crop that is cultivated the most, particularly in the central regions and in the higher lands. The potato is preferred by producers for many reasons: it is a crop that can be planted in rotation with other crop, that it is a hoeing plant and yields high income in irrigated areas, demands more labour per unit of the land, and there is an abundance of potential buyers. Production oriented towards the various utilization areas of the potato (for table, processing and seed) prevails in different regions, and the production and marketing systems are observed to vary vastly across the regions.

Potato production became rapidly widespread, particularly in the central regions and higher lands of Turkey after the 1970s; and on irrigated lands and in the regions where precipitation is adequate for potato farming, potato has become the main source of income and employment. The area dedicated to potato production decreased rapidly to 154,300 hectares in the 2000s, and in parallel production fell to 4.1 million tons. Potato farming is intensified in the provinces of Niğde, Nevşehir, Ordu, Bolu, Afyon, Adapazarı, İzmir and Erzurum, where 50.4% of production area and 56.9% of production comes from. In certain provinces, monoculture agriculture stood at the forefront. In order to increase the potato's contribution to the enterprise and the national economy large-scale and high quality production should be carried out and production inputs should be used rationally (certified seeds, labour force, pesticides, chemical fertilizers, etc.), production processes should be improved (planting, irrigation, diseases and pests management, harvest etc.) and the post-harvest product flow should be regulated. In this way an advantageous production and marketing system can be established and operated effectively for the benefit of both the individual producer and the consumer. There are many significant problems in terms of the use of certified seeds, farms chemicals and other inputs in potato farming that can have a negative impact on the economics and environmental sustainability of the production activities.

This paper examines the economics of potato farming in the major production regions (such as Nevşehir, Niğde, Bolu, Ordu, Afyon and Erzurum) and the contributions of potato farming to enterprise economics as compared with other major cash crops. The economic characteristics of potato producing farms are evaluated in terms of production and marketing economy, input use, unit product costs, productivity, gross and net profit per unit area and prices, marketing organization and margins. The economics of potato farming and its advantages and/or disadvantages as compared with other major production activities of potato farming at individual farms have been discussed based on the results of the study. The contributions of potato agriculture to the economies of both the enterprise and the country and factors delimiting production are discussed in general, and the impacts of alternative production and marketing policies on individual farms are set forth.

Materials and methods

The data used in this study to evaluate the economic characteristics of potato farming in Turkey was obtained through a questionnaire, and potato farming was compared on the basis of the farms' physical and monetary data. In order to obtain an economic assessment of the enterprises, data compiled by the questionnaire, applied by the researchers in the 2006 production period, was used. Monetary values were measured in the national currency and then converted into USD using the national average exchange rates. A series of intensive personal interviews were conducted with selected sample farmers in the Ordu, Erzurum, Niğde, Nevşehir, Bolu and Afyon provinces of Turkey, using both qualitative and quantitative survey methods. The area of potato cultivation in farms was used as the basis in the sampling and the suitable size of sample farms was determined through the simple random sampling method. The number of enterprises to which the questionnaire was applied in this way was 291 in total, broken down as 37 in Bolu, 45 in Niğde, 51 in Nevşehir, 56 in Afyon, 41 in Ordu and 61 in Erzurum. The questionnaire was also applied to industrial and commercial firms contracting directly with producers in provinces.

In the economic assessment of potato and other production activities at the enterprises, such indicators as the gross production value, production costs, productivity and gross and net profits were analyzed. The gross production values were found by multiplying the average yields of the farms with the prices received by the farmers. The contribution of potato farming to the living standards of the individual producer is evaluated by using an analysis of the production activity (Lampkin and Padel, 1994; Webster and Bowles, 1996; Turner and Taylor, 1998; Bülbül and Tanrıvermiş 2002). In the analysis of production activities; the gross (margin) profit of the production activity was calculated by deducting the variable costs from the gross production value; and the net profit of the production activity is estimated by deducting the total production cost of the activities from the gross production value. The expected impacts of individual activities such as potato, cereal, sugar beet, vegetables and sunflower were analyzed by comparing gross margin and net profit per unit of cultivated land. A comparative analysis is intended to diagnose problems and provide a basis for remedial action. For comparative purposes such as between enterprises of the same type on farms, between different provinces and farms or change in economic results between production activities, gross margins are a useful tool (Lampkin and Measures, 1999), and in addition to this the net margin is also calculated to avoid misleading results. The net economic benefit of potato farming seed has been calculated by making a comparison between the cases with project (net profit of potato farming) and without project (net profit of other cash crop farming), and a cost-benefit analysis has been used with this aim (Gittinger 1984, Webster and Bowles 1996, Tanrıvermiş 2000). The research results provide an opportunity to design alternative policies for decision-makers.

Main structural characteristics of potato cultivating farms

The average operating land owned by farms cultivating potatoes ranges between 2.1 ha (Bolu) and 15.0 ha (Nevşehir), and the share of owned land in total varies between 63.6% and 89.1%. It has been observed that a structure of self-entrepreneurship prevails on the farms, which have characteristics of small family farms in terms of the existence of the operating land. The amount of operating land of the farms is higher than the national average, except in such Black Sea Region provinces as Bolu and Ordu (6.1 ha). While producers emphasized that their land resources were not sufficient for their livelihood, it was found that the amount of land operated through rental and crop-sharing is considerably high, particularly in Nevşehir and Afyon, where irrigated land is scarce (Table 1). In these two provinces, 32.6% and 36.2% respectively of the total operated land comprises land operated through rental and crop-sharing, with rented lands generally allocated to irrigated and potato farming. Land rental agreements are generally informal, and thus the producer cannot benefit from the support payment on the amount of operated land.

It was observed that the amount of irrigated lands on the farms generally is low, and that the number of crops cultivated on irrigated lands is limited. The share of irrigated lands within the operating land ranges between 10.1% and 86.4%. The fact that the amount of irrigated

Table 1. Land resources and land tenure of farms (ha).

Provinces	Owned land	Land used under rental and/or crop-sharing	Land allocated to rent and/or crop-sharing	Total operating land		
				Irrigated	Dry	Total
Erzurum	11.05	1.62	0.23	5.22	7.22	12.44
Niğde	13.37	1.63	-	11.83	3.17	15.00
Nevşehir	9.56	5.44	-	13.04	2.00	15.04
Bolu	1.89	0.23	-	1.46	0.66	2.12
Afyon	5.57	2.66	0.08	6.80	1.35	8.15
Ordu	5.27	0.59	-	0.59	5.27	5.86

lands, particularly in Ordu, is as low as 10.1% is related with the annual average rainfall in the province, which is higher than 1000 mm. The operating land and irrigated land resources on farms producing potatoes is higher than that of farms not dealing with market-oriented potato agriculture in the same cities. The products to which the land is allocated in farms vary from one province to the other. Some 44.7% of the operating land in Erzurum is allocated to cereal crops, 14.8% to potatoes, 11.8% to sugar beet and 9.7% to sunflowers; while 62.7% of the operating land in Niğde is allocated to potatoes and 29.2% to cereals; 54.1% of the land in Nevşehir is allocated to potatoes and 36.4% to cereals; 40.6% of the land in Bolu is allocated to potatoes and 47.2% to cereals; 45.4% of the land in Afyon is allocated to cereals and 29.5% to potatoes; and 55.5% of the land in Ordu is allocated to cereals and 30.0% to potatoes (Table 2). Considering the share of land allocated for a single production activity within the operating land, one can see that the farms in the Niğde, Nevşehir and Bolu provinces are potato-production oriented and that the diversity of product/activity in farms is at a low level, which increases risks and ambiguities. As the amount of precipitation is insufficient during the vegetation period, the tradition of fallowing the dry land is maintained in the Erzurum province, and cultivating pulses in turn, instead of fallowing, is preferred.

The crop and agricultural product value of potatoes on farms was determined respectively as follows: 29.9% and 17.2% in Afyon; 18.8% and 10.11% in Erzurum; 56.1% and 35.4% in Niğde, 51.1% and 31.2% in Nevşehir; 41.1% and 25.3% in Ordu and 38.4% and 26.7% in Bolu. Clearly, the share of the potato within the households' crop production value and total gross production value is considerably high, particularly in the Niğde, Nevşehir, Bolu and Ordu provinces, and households' income and living conditions are greatly influenced by changes in potato prices and the amount of production. As the household income and means of subsistence comes primarily from potato farming in particular in four cities, it should be remembered that limiting potato production in defence against diseases and pests may have important economic and social consequences.

The average household size ranges from 5.33 to 8.55 people, with the highest household population found in the province of Erzurum and the lowest in the province of Niğde. The amount of labour force, calculated by taking the distribution of household population in terms of age

Table 2. Land use and the crop patterns of farms.

Crops	Erzurum		Niğde		Nevşehir		Bolu		Afyon		Ordu	
	Hectare	Ratio	Hectare	Ratio	Hectare	Ratio	Hectare	Ratio	Hectare	Ratio	Hectare	Ratio
Wheat	2.02	16.24	3.60	24.00	4.22	28.06	0.75	35.38	2.43	29.82	1.11	18.94
Barley	3.54	28.46	0.78	5.20	0.92	6.12	0.25	11.79	1.24	15.21	-	-
Sugar beet	1.47	11.82	-	-	0.62	4.12	-	-	1.03	12.64	-	-
Sunflower	1.20	9.65	-	-	-	-	-	-	0.15	1.84	-	-
Potato	1.84	14.79	9.40	62.67	8.13	54.06	0.86	40.57	2.40	29.45	1.76	30.03
Maize	-	-	0.21	1.40	0.33	2.19	-	-	0.03	0.37	2.14	36.52
Alfalfa	0.41	3.29	0.81	5.40	0.82	5.45	-	-	0.09	1.10	0.11	1.88
Wild vetch	0.26	2.09	-	-	-	-	-	-	0.05	0.61	0.10	1.71
Fruits	-	-	-	-	-	-	-	-	-	-	0.64	10.92
Vegetables	0.11	0.88	0.20	1.33	-	-	0.26	12.26	0.73	8.96	-	-
Fallow	1.59	12.78	-	-	-	-	-	-	-	-	-	-
Total	12.44	100.00	15.00	100.00	15.04	100.00	2.12	100.00	8.15	100.00	5.86	100.00

and sex into consideration, ranges between 4.15 and 6.69 men work units. Between 70.0% and 81.6% of the household populations are within the 15 to 65 age group, accepted as the ages capable of working economically. It was found that 79.3% of the female and 92.7% of the male population aged 7 and above are literate, with an average of five years of education, and that the education received by the household members is generally at the primary school level.

Results of the questionnaire show that 74.1% of the potato producers in Niğde and Nevşehir, 63.4% in Erzurum, 63.3% in Afyon, 47.7% in Bolu and 58.9% in Ordu visited agricultural institutions every time they went to the province/sub-province. Producers carrying out market-oriented production use loans and establish relations with financial institutions. The producers' experience in the field of potato cultivation is at the highest level in Ordu and Erzurum at 45.6 years; followed by Bolu with 41.4 years, Afyon with 38.0 years and the Niğde and Nevşehir provinces with 33.5 years. The producers' level of organization and the current producer organizations' efficiency in the solution of problems concerning potato agriculture and trade are considerably low. Results show that 37.0% of the producers in Niğde-Nevşehir were members of agricultural cooperatives, compared to 64.5% in Afyon, 69.0% in Bolu, 22.3% in Ordu and 21.0% in Erzurum. Across the provinces, between 25% and 63.0% of the producers have become members of agricultural chambers. However, as the current producer organizations do not deal only with the cultivation, processing and marketing of potatoes, their activities in this field limited.

There have been a great deal of improvements in potato farming since the 1970s, when machinery replaced much of the labour force in places where the land size and topographical characteristics are appropriate. There are an average of 0.92 tractors per farm in Nevşehir, 0.88 in Niğde, 0.89 in Afyon, 0.68 in Bolu, 0.32 in Erzurum and 0.11 in Ordu. Farms in Niğde, Nevşehir, Bolu and Afyon have motor pumps and deep well pumps for irrigation and cultivation, as well as ploughs and digging machines. It was observed that in parallel to the introduction of contract farming in the Niğde-Nevşehir Region, modern agricultural tools and machinery and methods have been adopted and put to use by farmers in a short period of time.

Physical input utilization, costs and profitability analysis in potato production

The demand for labour per hectare of processed area in potato agriculture ranges from 547.2 hours to 931.6 hours, and the level of mechanical power utilization ranges from 17.5 hours to 458.0 hours. In the Central Anatolian and Aegean Regions, potato production processes are carried out mostly based on mechanization. The use of labour force and mechanical power is related with the level of mechanization in the production processes. The highest level of labour force demand per hectare is observed as 652.4 hours/ha and 939.9 hours/ha in the Erzurum and Ordu provinces respectively, where the level of mechanization is low (Table 3). The labour demand is at the highest and mechanical power demand at the lowest levels particularly in provinces where the harvesting and ploughing are carried out manually. In provinces where the family labour force is insufficient, this structure leads to an increase in production costs due to the high cost of a large labour force. Such maintenance works as ploughing, fertilizing, irrigation and harvesting in potato production activities usually demand a large labour force, and thus producers employ imported labour as the family labour force is insufficient. The

Table 3. Input utilization, productivity and costs analysis in potato farming.

Indicators	Provinces					
	Erzurum	**Niğde**	**Nevşehir**	**Bolu**	**Afyon**	**Ordu**
Labour Requirements (hour/ha)	939.9	547.2	583.6	692.2	931.6	652.4
Machine Power Demand (hour/ha)	17.5	137.2	143.4	123.2	458.0	25.6
Seed Use (kg/ha))	2,697.0	2,952.5	3,002.0	2,790.1	4,804.6	1,802.8
Fertilizer Use (kg/ha)	810.6	2,510.2	2,880.9	1,459.2	500.2	481.7
Pesticides (kg/ha)	1.9	2.7	3.6	1.4	1.2	0.6
Irrigation Sources and Methods	Surface water and flooding irrigation	Ground water and flooding & sprinkler irrigation	Ground water and flooding & sprinkler irrigation	Surface water and flooding irrigation	Ground water and sprinkler irrigation	Surface water and flooding irrigation
Harvesting Techniques	Labour	Harvesting machine	Harvesting machine	Harvesting machine	Harvesting machine	Labour
Variable Costs (US $/ha)	3,152.9	3,741.7	3,963.3	6,622.7	4,936.2	2,777.9
Production Costs (US $/ha)	3,345.7	4,554.7	5,019.7	7,888.4	5,481.3	3,063.2
Productivity (ton/ha)	17,248.0	31,272.2	34,072.1	36,042.3	30,968.5	18,762.2
Farmers Received Prices (US $ kg^{-1})	0.21	0.15	0.16	2.77	0.21	0.21
Unit Product Costs (US $ kg^{-1})	0.19	0.15	0.15	2.21	0.18	0.16
Net Margin (US $/kg)	0.02	0.00	0.01	0.56	0.03	0.05
Gross Production Value	3,655.7	5,353.6	4,893.5	10,087.4	6,618.5	4,003.3
Gross Margin (US $//ha)	502.8	1,390.3	1,151.8	3,464.7	1,682.2	1,222.4
Net Profit (US $/ha)	310.0	333.9	338.9	2,198.9	1,137.2	937.1

imported labour force either works for 2.5 to 3.0 months continuously during the vegetation period and maintenance and harvesting work, or work for the day. In this case wages are paid monthly or daily, in accordance with the situation.

Seed tubers should be planted in a highly humid, light and loose soil for rapid growth. As the enterprises specialized in potato agriculture know this very well, 4.7% of the labour force demand in Bolu and 19.6% in Ordu; and 8.6% of the machine power demand in Bolu and 17.2% in Ordu per unit area are spent for the soil-preparation process.

The period of changing the seed is generally three years in potato and cereals at the farms which were surveyed. The seeds used in potato farming are obtained from the household, companies and cooperatives; the farms' chemical fertilizers, tools and machinery are obtained from firms and cooperatives. Certified seed use is at the highest in Bolu, with 65%, followed by Afyon (40%), Niğde (39%), Nevşehir (36%), Erzurum (21%) and Ordu (12%). The average rate of seed utilization at farms ranges from 1,802.8 kg/ha to 4,804.6 kg/ha. The level of seed utilization in the Afyon, Niğde and Nevşehir provinces is higher than the suggested amount, whereas it is lower in the province of Ordu. The level of seed utilization per unit area essentially depends on the method of planting and whether the planting is carried out manually or by machine. Planting is carried out in such temperate climates as in the Aegean and Mediterranean Regions in January-March; in the Central Anatolian and Black Sea Regions in April-May; and on altitudes during the spring in order to be protected from late (last) frosts.

Potato has an important role in crop rotation and it is also one of the best plants entering rotation. Sugar beet, sunflower, cereals and pulses can be the other elements of an ideal crop rotation system in the potato producing regions. According to the results of the study, when potatoes are planted after pulses, production increases. However, the system of potato-potato rotation is applied often in table or industry-oriented production in the surveyed regions, where the potato is rarely replaced by other products. The rotation plan suggested and imposed by the companies for potato seed is in the order of 'potato-cereals'. Performing the planting of potato seed on stubble is compulsory. Although the same rotation plan is also suggested in the production of industrial and table types, this plan is generally not implemented. As the land, particularly the irrigated land, resources are scarce in the region, industrial and table types are planted each consecutive year. Particularly in the Ordu, Niğde and Nevşehir provinces, 65% to 71% of producers stated that they inevitably planted potato in consecutive years on the same field (due to a lack of extra fields or different crops with a yield as high as the potato).

The use of fertilizer in potato agriculture depends on many factors, such as the period of cultivation, soil structure, the amount of irrigation and the targeted product values. The use of chemical fertilizer to provide the total plant nutrition in the provinces surveyed was as follows: 183.5 kg/ha in Ordu, 263.8 kg/ha in Erzurum, 530.6 kg/ha in Bolu, 956.7 kg/ha in Afyon, 2,008.6 kg/ha in Nevşehir and 1,879.8 kg/ha in Niğde. Fertilizers are used in the Nevşehir, Niğde and Afyon provinces in amounts that are higher than suggested doses, and the use of fertilizers in potato agriculture is 9- to 15-times higher than that of cereals. Apart from the five farms which were interviewed in Niğde and Nevşehir, there were literally no producers obtaining soil and leaf analyses. It should be noted that the potato is the crop that draws the highest benefit from farmyard manure, which clearly should be taken advantage of. However, as the livestock population of the farms dealing with market-oriented production is low, manure is generally not used in potato farming except in the Ordu and Erzurum provinces, and the farms which do not have animals purchase manure in such provinces as Niğde, Nevşehir, Afyon, and İzmir at relatively higher costs.

The potato plant is exposed to the impact of many diseases and pests during its growth and development. Fighting weeds in potato agriculture is carried out manually or using machines, influencing production costs. Diseases such as potato bug, aphids, late blight and potato root rot cause a great deal of harm to crops. It was found that an average of 0.6 kg and 3.6 kg of pesticides were used per hectare as the total operative substance in potato agriculture in the surveyed provinces. When the fight against diseases and pests on agricultural enterprises is mentioned, aside from unknown weed control and hoeing for the purpose of ridging, generally chemical pest and diseases management comes to one's mind and pesticides are used instinctively, particularly in the central regions. Pesticides used in production are generally purchased from pesticide dealers, but as technical guidance is not provided sufficiently, the incorrect and excessive application of pesticides increases the economic and social costs of production.

Irrigation is carried out annually 10 to 17 times in the Niğde and Nevşehir provinces, 8 to 10 times in Bolu, 6 to 9 times in Afyon and 2 to 3 times in Ordu, using mainly groundwater in the central regions and surface water in other areas. No payment is made for such water resources in general. Methods of flooding and sprinkling are commonly used in potato production. However, such developed methods as drop irrigation have been used commonly until today. Potato is produced on sandy and loamy-sand soil particularly in the Niğde and Nevşehir provinces, however increased groundwater pollution caused by the excessive use of pesticides and chemical fertilizers poses a serious risk. When the fact that the share of irrigation water including labour costs among the variable costs is 22.9% to 23.7% in the Niğde and Nevşehir provinces, 25.6% in Afyon, 5.6% in Bolu and 3.6% in Ordu is considered, the significance of the issue becomes clearer.

The potato harvest time changes within the period of May-October, depending on the required purpose of production (seed, table and out-of-season) and production region. Harvesting is performed manually, using ploughs and semi- and fully-automatic machinery. In the provinces of Ordu and Erzurum, where harvesting is carried out manually, 44.5% and 39.0% of the labour demand per hectare is spent for harvest processes respectively, whereas this ratio is 59.4% in Niğde, 61.1% in Nevşehir and 57.7% in Afyon, where the harvest is carried out using machinery.

As the potato tuber consists for 75% to 80% of water, tubers lose weight and quality following the harvest, and thus the producers try to minimize losses using various methods. Harvested tubers are taken to the market straight away or are stored for a certain period of time. Cellars or food stores, soil grain storages, volcanic rock stores and modern stores are used for storage. Potatoes produced for the main and early seasons and stay in the field for 6 months during the period of May-November and are then supplied to the market. Products produced and collected using various methods during the previous six-month period are supplied to the market in the remaining period of the year. As there is no considerable change in potato demand in these months, 40% to 50% of the potatoes subject to domestic trade are required be stored using various methods (such as cold storage, rock carved-caves, underground, depot) and gradually supplied to the market. The share of storage costs within the operating costs range from 5.4% to 9.2%, depending on the conditions of storage and the period in the provinces.

The variable costs per hectare and total production expenses in potato production activities on enterprises, unit production cost and the profit margins change depending on the purpose of cultivation. Variable costs per hectare for table potato farming range from $2,777.90 to $4,936.20, whereas this amount was estimated as $6,622.70 for seed production. Production expenses per hectare range from $3,063.20 to $5,481.30 in table potato farming and $7,888.40 for seed farming. Production expenses and circulating capital demand for seed potato production is higher than table potato production, and thus the application of rotation becomes inevitable in seed production. Total production expenses in table potato farming and the unit product cost in connection with the average product efficiency ranges from \$0.15 kg^{-1} to \$0.19 kg^{-1}; product prices received by the producer range from \$0.15 kg^{-1} to \$0.21 ^{1}; the profit margin obtained per unit product is zero in the Niğde province, whereas this value is positive in other provinces. Product cost was estimated as \$2.21 kg^{-1} in seed potato production, sales price received by the producer as \$2.77 kg^{-1} and the profit margin obtained per unit product as \$0.56 kg^{-1}. It should also be noted that this margin is higher than table potato farming. It was found that the profit margins in potato production in the farms are at a very low level in terms of the amount and the profit margin's gross product value and its ratio to the sales price. However, a comparative analysis of these indicators with other products that can be cultivated instead of potato in the enterprises will be required.

Gross production value per hectare ranges from $3,655.70 to $6,618.50, and gross profit from $502.80 to $1,682.20 in potato production, equating to a variation of 13.8% to 25.4%. Gross profit per unit area is determined by deducting the variable costs from the gross production values of potato production activity within the enterprises. The net profit per field hectare ranges from $310.00 to $1,137.20 in the farms, meaning a net profit ratio to gross production value of between 8.5% and 17.2%. The gross production value per hectare in seed potato farming is $10,087.40, gross profit is $3,464.70, making a gross profit to gross production value ration of 34.3% and a net profit to gross production value ratio of 21.8% (Table 3). Seed potato farming seems to be more profitable than table potato farming in provinces with similar conditions.

Economic comparison of potato and major crops as alternatives to the potato

The competitive edge of products cultivated on farms depends on such factors as productivity, price, income, supply of production inputs, ease and organization of selling products and producer tendencies. Farmers generally produce more than one product in accordance with the market and/or family needs, and thus these products compete with each other in terms of their relationship to profitability within the enterprise. The production of potatoes and other tuber crops was prohibited on a total of 14,139.4 ha of land in the provinces of Nevşehir, Niğde, Ordu, Kayseri and Trabzon after 2005.

The spread of potato wart is caused by contaminated tubers and soil movements. Products which can carry soil as well as wind, agricultural tools and machinery, animals and humans can all contribute to the communication of the disease in question. According to producers, potato wart derives from imported seeds, instinctive fertilizing and irrigation and the act of year-on-year cultivation of the potato in the same field. According to many researchers, the disease, which had not been encountered previously in the region, was probably transmitted from imported seeds. Based on this claim, certain producers in the region sued the companies from which they purchased the seeds and/or the State, however the courts found in favour of the defendants as the cause and result relationship could not be proven with scientific information and documents. The seed and plant health certificates of imported potato seeds, prepared in compliance with international principles, is an indicator of the fact that the mentioned seed was clean in terms of the mentioned diseases. The planting of potatoes in the same field on consecutive years in places where irrigated land assets are scarce, and using tubers from the previous year's production as seed are important issues. It is also understood that producers do not take the necessary measures to prevent the spread of disease – practicing regular rotation in potato farming, not cultivating crops in contaminated areas, cleaning of agricultural tools and machinery used in the contaminated area, ensuring workers and animals do not transmit the disease from contaminated areas on foot and avoiding the use of diseased tubers as animal feed (as spores are able to survive in the intestines of animals). The practice of quarantine associated with the increased occurrence of diseases, and thus producer losses, should not be the responsibility of the state and/or firms only, as the producer should also be accountable.

It would be useful if net income, land, labour force and capital demands found in potato production were similar to those of alternative crops, which would encourage product diversification and potato production within rotation plans instead of the monoculture agriculture found in the enterprises. In areas where potato wart has been observed and quarantines enforced, such products as wheat, barley, and common vetch are cultivated by the individual producers. However, almost all the producers stressed that these products did not contribute to the family economy as much as the potato. In the six surveyed provinces in the research, the economic performances of the potato and alternative products are examined comparatively. The potato requires a higher labour force per unit area (except sugar beet) on irrigated land in the provinces, and thus field and labour force resources can be put to more advantageous use (Table 4). In all of the six surveyed provinces, the potato is able to produce higher amounts of net profits and positive net economic benefits when compared to the alternatives. In provinces stricken with potato wart, the actual state support has been considered sufficient to compensate for the losses (substitute product's net income + indemnity) suffered by the producers due to the inability to cultivate potato, except in the province of Ordu.

The total losses of the producers due to prohibition on potato cultivation include net income loss deriving from the prohibition of planting on the field, and the provision of additional labour and capital used in potato agriculture compared to other products. The alternative crops that the producer will have to plant instead of the potato, are cereal and fodder crops. If the wheat is watered under the conditions of the Nevşehir and Niğde provinces, 40 to 60 hours of labour force is used per hectare, compared to 800-1,000 hours in potato farming. The labour force demand per thousand square meters increases by 20 times in potato farming, and thus family and imported labour forces can be put to good account at a rate 20 times more than for wheat. Under these circumstances, the amount of idle labour on the farm, which is already high due to the prohibition of potato farming, will make the total income loss increase. Some 12.0% to 23.0% of the labour force used in potato farming in the surveyed enterprises is met through imported labour, with the remaining carried out by the family labour force. It will be necessary to consider the income loss suffered by the producers in terms of family labour due to prohibitions on potato agriculture (as a result of working in agriculture for a shorter period of time and inability to put the increasing labour force into good use in certain fields other than agriculture) in determining the cost of the support payment to be made to the producers. The prohibition on potato cultivation, particularly in the provinces of Niğde and Nevşehir, has had a negative impact on households in the sub-province centres who make a living as farm workers, and the increased unemployment and struggle to make a living in the sub-province centers has encouraged migration to the big cities.

Participation of producers in marketing systems and conditions of success

Potatoes produced in farms are classified and packaged in the field or at the enterprises, and the product is either stored or presented straight to the market. For potatoes, the rate of supply to the market is 93.0% in Niğde, 94.9% in Nevşehir, 78.6% in Erzurum, 80.1% in Ordu, 95.8% in Bolu and 87.8% in Afyon. The rate of supply to the market varies depending on purpose of cultivation and production volume. All of the industry-oriented potato production on the enterprises, 65.0% to 70.0% of the potato seed, and 94.8% to 95.9% of table potatoes are supplied to the market. Although the supply of product to the market in accordance with characteristics of storage is distributed over the entire year, the rate of supply to the market rises in the September-November period, when harvesting becomes intense. More than 50% of the annual production is supplied to the market, particularly in the provinces of Bolu, Afyon, Erzurum and Ordu, during this period, leading to an excess of produce on the market. As the storage is limited and storage conditions inappropriate in the Ordu, Afyon, Bolu and Erzurum provinces, where production levels are low and circulating capital is insufficient, 2/3 of the product is supplied to the market during the period of September-December following the harvest (Table 5). Producers generally store their products under their own conditions following the harvest, with a small number of producers preferring cold storage houses and natural storage. According to the province, between 45.6% and 64.1% of producers store potatoes using their own facilities for a period that generally ranges from three to four months. Product loss during storage is between 13.7% (Nevşehir) and 20.0% (Erzurum).

The potato is a fresh product which should be consumed or processed within one year of production. Potato prices are at their minimum level during the August-October period, when potato demand is high, reaching the maximum level during the April-May period when demand is low. The annual fluctuation in prices also depends on the amount of potatoes supplied to the market, and the marketing channel followed by the potato until it reaches the consumer from the producer. Transaction or exchange activity also generates income for the sellers. However, it should be noted that not all of these income are pure profit. It was determined that an average of 46.0% (Erzurum) and 53% (Bolu and Afyon) of the prices paid by consumers in the provinces surveyed went to the producer, and that 47% and 54.0% corresponded to the marketing margin. The producer and wholesale dealer share 21% to 29% of the marketing margin, and 71% to 99% by the wholesaler and retailer. In order to strike a balance of price in

Table 4. Comparison of production costs, gross production value, and profits of potato and alternative products tha

Indicators	Afyon				Bolu		Ordu		
	Wheat	Potato	Sugar beet	Alfalfa	Potato	Wheat	Potato	Cabbage	Corn
Variable costs ($/ha)	1,043.1	4,936.2	2,859.4	1,520.7	6,622.7	519.8	2,712.3	2,355.9	1,080.8
Seed cost ($/ha)	121.7	2,007.4	49.3	309.7	3,364.6	138.8	411.4	650.1	36.5
Production costs ($/ha)	1,318.6	5,481.3	3,295.8	2,314.2	7,888.4	1,246.6	3,197.1	2,623.8	1,310.5
Gross Production Value (GPV) ($/ha)	1,750.5	6,618.5	3,979.4	2,812.5	10,087.4	1,640.5	3,908.8	3,096.9	1,569.4
Gross profit ($/ha)	707.5	1,682.2	1,120.0	1,291.8	3,464.7	1,120.7	1,196.5	741.0	488.6
Net profit ($/ha)	431.9	1,137.2	683.6	498.3	2,198.9	393.9	917.9	473.1	259.0
Unit cost ($/kg)	0.13	0.18	0.07	0.25	2.21	1.71	0.17	0.24	0.42
Farmer's received prices ($/kg)	0.21	0.21	0.08	0.30	2.77	2.49	0.21	0.28	0.38
Gross profit/GPV (%)	40.41	25.42	28.15	45.93	34.35	68.32	30.61	23.93	31.13
Gross margin as % of total costs	53.65	30.69	33.98	55.82	43.92	89.90	37.43	28.24	37.28
Net profit/GPV (%)	24.67	17.18	17.18	17.72	21.80	24.01	23.48	15.28	16.50
Net Profit/Man Labour Hour ($)	1.00	1.22	1.59	1.16	3.18	0.94	1.41	0.94	0.73
Net economic benefit of potato farming ($/ha)	**453.60-705.3**				**1,805.00**		**433.40-658.90**		

Table 5. Distribution of marketing ratios of potatoes according to months.

Months	Provinces					
	Erzurum	Niğde	Nevşehir	Bolu	Afyon	Ordu
September-October	51.07	40.26	30.00	62.38	51.20	55.60
November-December	21.23	10.77	8.40	20.26	20.20	26.30
January-February	20.11	23.82	26.60	6.45	24.30	17.10
March-April	17.59	25.15	35.00	10.91	4.30	1.00

the product market, production should be directed in accordance with domestic and foreign demands, industry-oriented production should be increased, importance should be attached to cultivars purity (types which are demanded in the domestic and foreign markets should be prioritized) and to standardization (size of tubers desired by the market), packaging should be suitable, potatoes should be supplied to the market for table use in the months of July and *can be cultivated instead of potato.*

	Nevşehir			Niğde			Erzurum					
Bean	**Potato**	**Wheat**	**Barley**	**Potato**	**Wheat**	**Barley**	**Potato**	**Wheat**	**Barley**	**Sunflower**	**Sugar beet**	**Common vetch**
989.9	3,963.3	770.8	700.0	3,741.7	812.5	686.3	3,078.5	638.0	653.8	1,165.1	1,987.8	772.4
68.5	659.2	104.9	75.0	635.6	120.8	79.9	799.1	69.8	81.7	32.1	30.0	140.5
1,216.9	5,019.7	1,071.8	998.8	4,554.7	1,079.7	949.9	3,266.7	753.0	769.2	1,295.9	2,194.7	891.4
1,701.4	5,353.6	1,118.5	1,014.6	4,893.5	1,122.4	987.7	3,569.4	943.4	865.0	1,437.9	2,445.6	962.4
711.5	1,390.3	347.6	314.6	1,151.8	310.1	301.4	490.9	305.4	211.3	272.8	457.8	190.0
484.5	333.9	46.7	15.8	338.9	42.7	37.8	302.7	190.5	95.8	141.9	250.9	71.0
1.08	0.15	0.28	0.27	0.15	0.28	0.27	0.19	0.37	0.31	0.63	0.06	0.15
1.39	0.16	0.24	0.22	0.15	0.25	0.22	0.21	0.26	0.22	0.70	0.06	0.17
41.82	25.97	31.08	31.01	23.54	27.63	30.51	13.75	32.37	24.42	18.97	18.72	19.74
58.47	27.70	32.43	31.50	25.29	28.73	31.73	15.03	40.56	27.46	21.05	20.86	21.32
28.48	6.24	4.17	1.56	6.93	3.81	3.82	8.48	20.19	11.07	9.87	10.26	7.37
1.75	0.57	1.41	0.50	0.62	1.40	1.33	0.32	2.71	1.23	0.23	0.29	0.55
	287.20-318.10			**296.20-301.10**			**51.60-231.70**					

August, which are the dead seasons for the potato, 1/3 of the production should be moved to early types and 1/3 reserved for seed production.

Marketing channels facilitate the flow of potatoes from producer to consumer. A variety of well-established, though informal, marketing channels exist in Turkey for the distribution and sale of potatoes in both the domestic and export markets. Aside from the growers themselves, the participants in potato marketing activities are trade and industry firms, merchants, exporters, wholesalers and retailers. The effective way of reducing the difference between the prices paid by consumers and the prices received by the producers should be through direct marketing by the producers. Direct marketing involves by-passing intermediaries in the marketing channel during the process of bringing the product from the producer to the consumer. A great deal of improvement has been achieved in the flow of potato from the producer to the consumer (such as road-side sales, in province/sub-province centres, in villages and cities) in recent years, and the share of products presented to the market in this way ranges from 25.2% to 32.3% in the different provinces. The flow of product from the producer to the consumer is carried out through sales made by family members, particularly in big cities, and the rental of market space at major fresh fruit and vegetable markets by larger producers, or direct sales from mobile markets. To a lesser extent, marketing channels in the form of producer-wholesaler-consumer, producer-commission merchants-consumer and producer-wholesaler-retailer-consumer have an important share in the flow of the physical product. The share of product flow of the first four marketing channels in the total product supplied to the market ranges from 71.1% to 84.4%. As contracted potato production is common in fields of industry and seed potatoes in Bolu, the share of marketing in the form of producer-industrialist/commercial firm-consumer is at the highest level of 14.4%. As the rate of contracted producers cultivating the potato is 3.7% in Nevşehir and 5.6% in Niğde, the share of this model in these provinces has remained at a lower level. The marketing channels in which marketing agents participate are the paths the potatoes follow from their source of original production to their ultimate destination. A closer look at these marketing agents shows that there are basically four channels that account for the dominant share of potato supply in terms of its distribution. From firms and market data, we derived estimates of the proportion of potato supply absorbed by each of these five types of marketing agents.

Firms deal with seeds, and processing and table potato production and trade. Firms carry out contract farming in the provinces of Bolu, Nevşehir, Kayseri, Adana, Eskişehir, Afyon, İzmir and Konya. These firms provide such inputs as seed, fertilizers, pesticides and cash credits to the producers, which are used during the course of production. Firms are keen to continue the model of contract farming during the coming years. According to the firms managers surveyed, the benefits of contract farming to producers include market, sales and price guarantees, putting the products to good account collectively, optimizing input use and the companies' ability to meet the credit needs of the producers. In the provinces, 45.4% to 81.2% of producers have a tendency towards becoming partners with the companies that purchase their products, and with industrial plants that process the products. However, according to 70% to 95.5% of the producers it would be more beneficial if potato processing and marketing companies were established by producer organizations. It is also thought that if the producers established unions which would go on to become partners of these firms it would be also more beneficial then if they create partnerships individually.

Major problems of potato producing farms and approaches to solutions

Problems that may occur in the farming and marketing system take a number of forms: a) producer characteristics (production scale, marketing ratio, etc), b) crop characteristics (perishability), c) demand patterns (elasticity, seasonality), and d) marketing system complexity (inefficient flow of information, insufficient physical facilities and low financial capacity of the market participants) (Adiyoga *et al.*, 2001). The increase in production and the fact that production is intensified in certain provinces has brought about significant problems, in particular the storing of excess produce. These problems in potato production are experienced particularly in marketing as a result of excessive production. The inability to export potatoes causes the product to be left in the hands of farmers and the prices fall rapidly. Storing potatoes is one of the most important links in the production and marketing chain. The fluctuation in market prices influences the producer significantly, with between 19% and 36% of producers among the regions mentioning the low level and/or instability of product prices as the major problem (Table 6). The lack of storage possibilities, uncertainty in foreign markets and fluctuations in prices, the inability to produce in accordance with demand and the insufficient number and

*Table 6. Producers problems in potato agriculture and marketing.**

Problems	Provinces (%)					
	Erzurum	Niğde	Nevşehir	Bolu	Afyon	Ordu
Instability & Low Level of Product Prices	32.1	27.3	30.5	35.7	34.1	18.9
High Costs of Seeds and Other Inputs	10.1	11.4	10.0	11.4	20.3	26.6
Lack of Confidence in Companies	5.5	12.2	10.9	3.8	2.1	3.5
Payment of Product Costs in Instalments	18.7	15.2	14.8	14.4	11.3	7.9
Storage Problems	10.4	11.1	11.1	10.1	2.2	8.8
Insufficient State Incentives	3.5	5.7	7.4	9.1	9.5	2.2
Excessive Number of Market Agents and Insufficient Producers' Organizations	10.1	3.0	1.0	5.6	10.6	2.6
Increased Diseases and Pests	8.8	13.1	13.3	5.7	9.0	28.5
Others	0.8	1.0	1.0	4.2	0.9	1.0

*Survey participants were allowed to give more than one reason.

capacity of strong marketing institutions make it impossible to put the product to high value, these are the underlying reasons behind price instability.

The low level and instability of market prices are common risks for the producer, while serious problems are also encountered in the collection of payment for produce sold. The issuance of bad cheques to producers and the inability to collect payments for products from purchasers were highlighted as a problem by 7.9% to 18.7% of producers according to the region. It is likely that the most important development that has emerged to remedy this problem is that the producers have been encouraged to make initiatives to market their own products, and thus take a considerable share in the market, particularly in the provinces of Erzurum, Niğde and Nevşehir. In terms of this problem, a serious problem of trust has emerged against the companies purchasing potatoes in the provinces. The excessive number of intermediaries between the producer and consumer, and the producers' inability to be organized and to control the flow of physical products in the marketing channel are also considered as significant problems.

In cultivation, such production inputs as seed, fuel, farm chemicals, and irrigation water are purchased in instalments, as the producers lack the sufficient cash flow, and the subsequent interest payments increase the costs. On the other hand, producers who in debt to input suppliers are obliged to submit their potato produce to the market immediately after the harvest, during the period of September-October. As a result of the excessive amount of produce on the market during this period and the lack of demand causes prices to fall. Producers who have to sell their produce under these circumstances do so at a loss. Another problem is that potato cultivation is carried out on the same field year-on-year. This means that productivity weakens rapidly as the soil becomes less productive and the quality of the cultivated product is reduced. The most important problem of the producers is related to the seed used in production, which is imported from Europe and the US. If breeder seeds are brought instead of the registered seed, the national economy would be provided with more contribution. If the imported seeds were enhanced under controlled circumstances and then distributed to farmers, rather than being distributed immediately, it would be a considerably positive approach for the long-term use of seeds. Producers begin production with the seed types found in the market. A variety of types can be found in the provinces, and farmers are inclined towards foreign seeds. Therefore, it becomes impossible to meet the demand by bringing unilateral types together in the desired amount for potential purchaser countries, and thus the inability to put the product to good account in the domestic market has a negative impact on the sale of the product at the remunerative price. High amounts of inputs, and particularly seed costs, are further important problem in the provinces, which was particularly emphasized by 10.1% to 26.6% of the producers according to region.

The inadequacy of adequate storage capacity on farms and the insufficiency of product storage facilities in appropriate conditions in the production areas are considered important problems that influence the fields of production and marketing negatively. Besides the high cost of potato storage warehouses, depots for storing potato seed and table potato should be considered separately. Possible support and incentive measures for storage investments have had an important impact on the planning of potato production. Although product supply to the market occurs throughout the year, it intensifies during the September-November period when the product is mostly harvested. Although problems of storage are not considered important for the provinces of Ordu and Afyon, where the produce can be marketed easily, storage is a key problem in the other provinces. The other significant problems in potato farming and the marketing systems are the insufficiency of state support and an increase in diseases and pests.

Conclusions

It was identified that farms producing market-oriented potatoes are generally small in size, that the level of organization is low and their population and labour force are considerably high. Across the regions, 15% to 63% of the farms' average operating lands are allocated to potato farming, and the potato is traditional in certain provinces and has become the major cash crop. After the 1970s, potatoes were planted in consecutive years in the central regions, where the amount of irrigated lands is low. The use of inputs has increased in order to obtain high productivity from the scarce lands, but potato wart disease, which is carried with seeds particularly in the provinces of Niğde, Nevşehir and Ordu, and which has become widespread due to bad agricultural practices, has become the major factor threatening production. The level of gross and net profit per unit area in potato agriculture on the farms is considerably higher than other market-oriented products on the same farms. Under these circumstances, it is natural that the potato's competitive power is very high in the allocation of irrigated lands within the enterprises. In regions where quarantines have been enforced due to potato wart, the sum of the net profit provided by alternative crops and the state support is barely close to the potato's net benefit. It was also found that the producers were not happy with the enforcement of quarantines due to certain economic and social reasons. Production processes of the potato and other market-oriented products should be diligently registered and monitored by producer unions, cooperatives and other agricultural institutions to be established. Costs are considerably high in potato farming as production is generally carried out manually, product losses are high, and seed, farm chemicals and irrigation water are used excessively. Product loss should be decreased, more importance should be attached to mechanization and training of the producers, while the produce should be insured against disease and pests, and integrated pest management should be emphasized in training.

According to the region, 79% to 96% of the produce is submitted to the market, products are generally stored under inappropriate conditions following the harvest and product losses are considerably high. Potatoes are classified based on their sizes, and are categorized, packaged, stored or supplied to the market directly by the producers. The level of participation of the producer in marketing the produce in the provinces ranges from 19% to 36%, which is clearly insufficient, however problems encountered during the collection of payments for the produce make it difficult for the producers to participate in the marketing phase. Essentially, the producers' share of the product price paid by the consumer will increase and it will be possible to increase small producers' level of welfare should the producers become more involved in these phases. There is a requirement to establish effective producer unions focused on potato production to ensure that producers can be directed throughout the year to produce certain types and meet market demands under suitable conditions. In addition, enterprises have to increase the importance they place in the storage of potatoes to minimize loss, and to make investments into warehousing and processing following the harvest to maintain a regular supply of produce to the market. As the benefit has been proven in the seed and industry-oriented potato farming, producers should continue to work towards contracted production in the vertical integration model, while integrating the producers into the industrial and commercial sector. On the other hand, major producers and producer organizations should seek to become partners with industrial and commercial enterprises, through which the problem of low-quality raw materials and idle capacity should be solved.

References

Açıl, A.F., 1976. Tarımsal ürün maliyetlerinin hesaplanması ve memleketimizde tarımsal ürün maliyetlerindeki gelişmeler, ankara üniversitesi ziraat fakültesi yayınları: 665, ikinci baskı, Ankara.

Adiyoga, W., Fuglie, K.O. and Suherman, R., 2001. Potato marketing in north sumatra and an assessment of indonesian potato trade, UPWARD working paper series No:7, Los Baños, Laguna, Philippines.

Arıoğlu, H., 1997. Nişasta ve şeker bitkileri, çukurova üniversitesi ziraat fakültesi genel yayın no:188, Adana.

Bülbül, M. und Tanrıvermiş, H. (2002). Vergleichende wirtschaftliche Analyse des konventionellen und ökologischen Haselnussanbaus in der Türkei, Berichte über Landwirtschaft, Band: 80, Heft:2 (Juli 2002), pp.304-320, Bonn, Deutschland.

Erkuş, A., Bülbül, M., Kıral, T., Açıl, A.F. ve Demirci, R., 1995. Tarım ekonomisi, Ankara üniversitesi ziraat fakültesi eğitim, araştırma ve geliştirme vakfı yayınları No:5, Ankara.

Gittinger, J.P., 1984. Economic Analysis of Agricultural Projects, John Hopkins University Press, USA.

Güneş, T. ve Arıkan, R., 1988. Tarım ekonomisi istatistiği, ankara üniversitesi ziraat fakültesi yayınları: 1049, Ankara.

İlisulu, K., 1986. Nişasta-şeker bitkileri ve ıslahı, ankara üniversitesi ziraat fakültesi yayınları: 1960, 279, Ankara.

Karaca, M. ve Demir, Z., 1994. Patateste azotlu gübre miktarı ve uygulama zamanının yumru verimine ve hasat sonrası toprak inorganik azotuna etkisi, tarım ve köyişleri bakanlığı tarımsal araştırmalar genel müdürlüğü, Ankara.
Lampkin, N.H. and Padel, S. (1994). The economics of organic farming: an international perspective, CAB international, UK.
Lampkin, N.H. and Measures, M. (1999). 1999 Organic farm management handbook, third edition, UK.
MARA (Ministry of Agriculture and Rural Affairs), 2006. TC tarım ve köyişleri bakanlığı kayıtları (unpublished), Ankara.
Onaran, H., Ünlenen, L.A., and Doğan, A., 2000. Patates tarımı sorunları ve çözüm yolları, tarım ve köyişleri bakanlığı tarımsal araştırmalar genel müdürlüğü Niğde patates araştırma enstitüsü müdürlüğü, Niğde.
Özçelik, A., Turan, A. ve Tanrıvermiş, H., 1999.Türkiye'de tarımın pazara entegrasyonunda sözleşmeli tarım ve bu modelin sürdürülebilir kaynak kullanımı ile üretici geliri üzerine etkileri, tarımsal ekonomi araştırma enstitüsü yayın no:14, Ankara.
SIS (State Institute of Statistics), 2004. 2001 census of agriculture agricultural holdings (households), publication no: 2924, Ankara.
SIS (State Institute of Statistics), 2006. Agricultural structure (production, price and value) 2004, publication no:3022, Ankara
SIS (State Institute of Statistics), 2006. Tarım istatistikleri özeti 2005, yayın no:3089, Ankara
Turner, J. and Taylor, M. (1998). Applied Farm Management, Second Edition, Blackwell Science, UK.
Tomek, W.G. and Robinson, K.L., 1981. Agricultural product prices, second edition, cornell university press, Ithaca, USA.
Webster, J.P.G. and Bowles, R.G. (1996). Estimating the economic costs and benefits of pesticides use in apples, brighton crop protection conference 1996 pests & diseases, british crop protection council, brighton, UK, pp. 325-330.

Curriculum vitae – Harun Tanrıvermiş

Harun Tanrıvermiş was born in the Ordu province on the Black Sea coast in 1971. He graduated from the Department of Agricultural Economics, Agricultural Faculty of Ankara University in 1991. He received an M.Sc. in 1993 and a Ph.D. in 1996 from the Department of Agricultural Economics, Graduate School of Natural and Applied Sciences, Ankara University. Also, he received a second M.Sc. degree on Environmental Management in 1995. He has been working on the academic staff of the Department of Agricultural Economics and is the head and founder of the Real Estate Development Department at the university. He is studying the economics and management of natural resources and the environment, real estate management and valuation and the economics and management of sustainable agriculture. Aside from his academic activities, he has much experience in project planning and implementation in public and private institutions.

Curriculum vitae – Hasan Şanlı

Hasan Şanlı was born in Nevşehir province in the Central Anatolian region of Turkey in 1977. He graduated from the Department of Agricultural Economics, Agricultural Faculty Ankara University in 2000. He received a M.Sc. in 2004 at the Department of Agricultural Economics, Graduate School of Natural and Applied Sciences, Ankara University and is a Ph.D. candidate in the same department. He had been working on the academic staff of the Department of Agricultural Economics since 2002. He is currently studying economics agriculture, real estate valuation and management.

Curriculum vitae – Mehmet Bülbül

Mehmet Bülbül was born in Samsun province on the Black Sea coast in 1939. He graduated form the Department of Agricultural Economics, Agricultural Faculty Ankara University in 1962. He received a Ph.D. in 1969 at the Department of Agricultural Economics of Kiel University in Germany. He worked at the Department of Agricultural Economics as academic staff between 1963-2005, and since retirement he has been working at the Department of Real Estate Development at Ankara University. He is currently studying economics agriculture, real estate management, agricultural finance, economics and management of agro-industry and irrigation economics and management. He has experience in project planning and implementation in public and private institutions.

Profile – Department Agricultural Economics

Agricultural education and researches was started in 1933 with establishment of High Agricultural Institute at the university of Ankara. Department of agricultural economics was established in the body of this institute. Agricultural economics is an applied science branch which is investigated solution ways of problems in agriculture and agricultural industry sector. Agricultural economists should know both technical issues like field crops, horticulture, animal production, processing of agricultural products and issues like planning of farms and agri-business, management, finance, accounting, marketing, agricultural policy, agricultural law, rural sociology, rural development, agricultural extension and communication. At the same time agricultural economists should deal with plans and programs applied by state aimed at developing agriculture sector, improving farmers living level, protecting consumers from over rising prices and arranging foreign trade of agricultural products. Agriculture is one of the key sector in the economy of Turkey being the suppliers of food as well as raw materials to agro-industries which constitute a large share in manufacturing and food industries and an important source of foreign Exchange earnings. In this context, the subject of agricultural economy always keeps its importance in the education, research activities and also administrative body of the country. Today, the education and research activities of agricultural economics in the agricultural education and research system depend upon a long history and these activities have seen broadly continuing. In the body of agricultural economics education, training and research facilities carry on agricultural management and agricultural policy and extension issues. Agricultural economics researches related especially national and international research results with economic, social and politic developments. There are 17 staff members in the department. The number of periodicals written by the foreign language is 4 and the Turkish periodicals number is 20.

Quality of processing and ware potatoes

Effect of physical properties of skin and underlying tissues of potatoes on bruising

K. Hironaka*, K. Ishibashi, H. Koaze, K. Kumada, K. Matsuda, T. Sato and M. Kojima
Obihiro University of Agriculture and Veterinary Medicine, Hokkaido 080-8555, Japan

Abstract

Physical properties (rupture stress, strain and energy) of skin and underlying tissues of Japanese potato varieties were measured in order to explain the varietal differences in blackspot susceptibility. Bruise–resistant potatoes had more rupture stress, deformation and energy of skin, and higher rupture stress, strain and energy of perimeduallary tissues than susceptible ones. Rupture deformation and energy of skin, and rupture stress and energy of perimeduallary tissues correlated significantly ($p<0.01$ or 0.05) with blackspot index, implying that skin may also play an important role in the protection of blackspot bruise. This information may be useful for potato breeders to develop new resistant varieties.

Keywords: bruising, skin, perimeduallary tissues

Introduction

Potato tubers are largely parenchymatous, lacking specialized secondary thickened tissues (Peterson *et al.*, 1985). As a result, they are susceptible to various forms of damage during commercial production including external and internal defects. Internal damage (blackspot bruise) resulting from the effects of impact on tubers during harvesting operations alone may cause losses in excess of 20% (Storey and Davies 1992). Although it may be a simple matter to grade out tubers showing external damage, blackspot bruise is not visible until after peeling (McGarry *et al.*, 1996). Thus, blackspot bruise is a major quality problem that causes substantial economic damage to potato breeders and the potato processing industry worldwide (Stevens and Davelaar 1997).

Potatoes were first brought into Japan in the early 17th century, but soon after Japan closed its borders to the outer world for approximately two hundred years. In mid-19th century when it opened its doors again, North American and European cultivars were successively introduced. The present leading cultivars in Japan, i.e. Irish Cobbler and May Queen, are the result of early introductions. Potato breeding programs have been publicly organized in Japan since 1902. Until the present, approximately 40 cultivars have been officially released (Hosaka *et al.*, 1994).

Three million tons of potatoes are produced annually in Japan and potato bruising is also a serious problem because the loss of bruising is 10% approximately. Thus, Japanese potato processors and breeders want to know which cultivars are resistant to bruising, and what physical properties of skin and underlying tissues bruise-resistant potatoes have in order to develop new bruise-resistant varieties. However, no information is available on effect of physical properties of skin and underlying tissues on bruising of potatoes. Thus, this study was initiated to examine varietal differences in bruising, and to explicate physical properties of skin and perimeduallary tissues of bruise-resistant potatoes.

Materials and methods

Samples

Tubers of processing variety (Norin 1, Toyoshiro, Hokkai-kogane and White Fryer) and raw consumption one (Sayaka) were harvested at the National Agricultural Centre, Hokkaido in Japan, in September 2002. After harvesting, these potatoes were cured for two weeks at 12C and 90-95% RH for 80 days.
Each cultivar sample used in this experiment was approximately 150 g.

Bruising method

Blackspot bruising was produced (Kunkel *et al.*, 1986) by dropping a round headed plug (100 g) through a cylindrical tube from a height of 50 cm onto the middle of each tuber. The bruised tubers were left for 48 hours at 15-20C. Twenty-five bruised tubers were then assessed using blackspot index described by Dwelle *et al.*, 1974. The bruised areas were peeled until the discoloration was visible. The size and colour intensity of the blackspots were considered and rated according to the following scale:
0 = no discoloration;
1 = very small or vaguely defined spots;
2 = diameter 3 to 5 mm, colour gray or brownish;
3 = diameter 5 to 10 mm, depth less than 5 mm, colour gray to black;
4 = diameter 5 to 10 mm, depth greater than 5 mm, colour black;
5 = diameter greater than 10 mm, depth greater than 5 mm, colour black;
A mean was calculated for 20 tubers.

Physical properties

A 2-mm diameter plunger was pressed into the middle of a potato at a rate of 50 cm per minute. The skin physical properties (rupture stress, deformation and energy) were recorded. Also, potatoes were punched with a cork borer (i.d. 8 mm) to obtain a cylindrical sample. A transverse cut of the zone below the skin was made with sharp knife in 5 mm lengths (cylindrical sample size: diameter 8 mm×length 5 mm). Physical properties (rupture stress, strain and energy of cylindrical samples were measured using a Texture meter.

Data analysis

Data were analyzed using one-way analysis of variance (ANOVA) and the Duncan's multiple range test (SPSS 1992) to determine significant differences.

Results and discussion

Figure 1 shows differences in bruise susceptibility of potatoes stored at 12 °C. As shown in this figure, the most resistant variety (lowest score) was Hokkai-kogane, followed by White Fryer. In addition, Norin 1 variety was quite susceptible to bruising while Toyoshiro and Sayaka were intermediate.

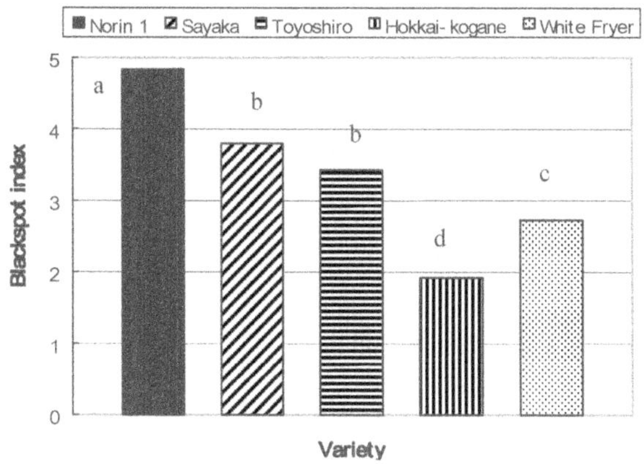

Figure 1. Blackspot susceptibility of potato varieties (means of 0, 40 and 80 days of storage).

Figure 2 shows differences in the skin rupture stress of five varieties. As shown in this figure, Norin 1 potatoes, the most bruise-susceptible variety (See Figure 1), had a lower score than the other four varieties.

Figure 3 shows rupture deformation of the five varieties. As shown in this figure, Norin 1 had the lowest value of the five varieties. Differences in rupture energy of skin for potato varieties are presented in Figure 4. As shown in this figure, Norin 1 variety had also the lowest figure of the five varieties. Figure 5 shows differences in the rupture stress of perimeduallary tissues of potatoes. Norin 1, which was bruise-susceptible variety in Figure 1, had the low score compared with other potatoes.

Figures 6 and 7 show rupture strain and energy of perimeduallary tissues of potatoes, respectively. As shown in these figures, Norin 1 variety had the lowest values of rupture strain and energy.

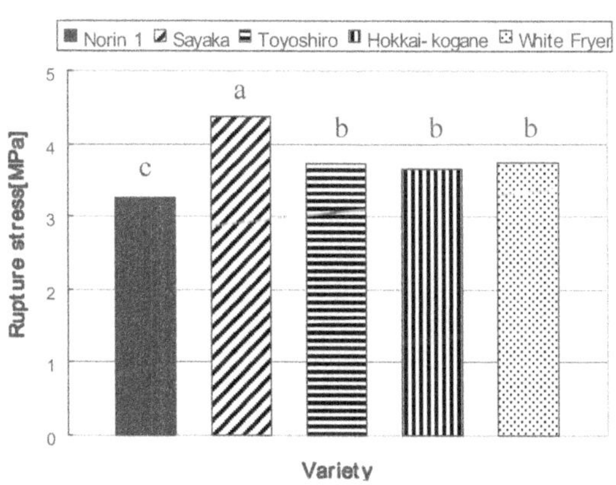

Figure 2. Rupture stress of skin of potatoes (means of 0, 40 and 80 days of storage).

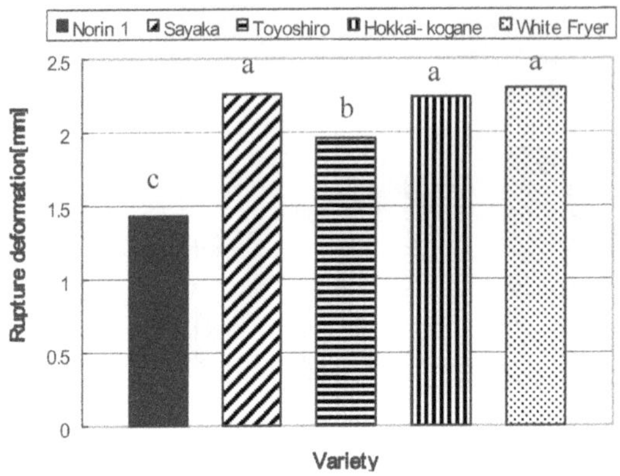

Figure 3. Rupture deformation of skin of potatoes (means of 0, 40 and 80 days of storage).

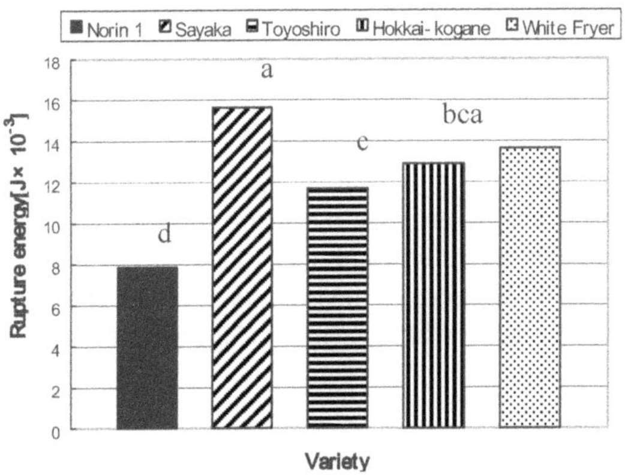

Figure 4. Rupture energy of skin of potatoes (means of 0, 40 and 80 days of storage).

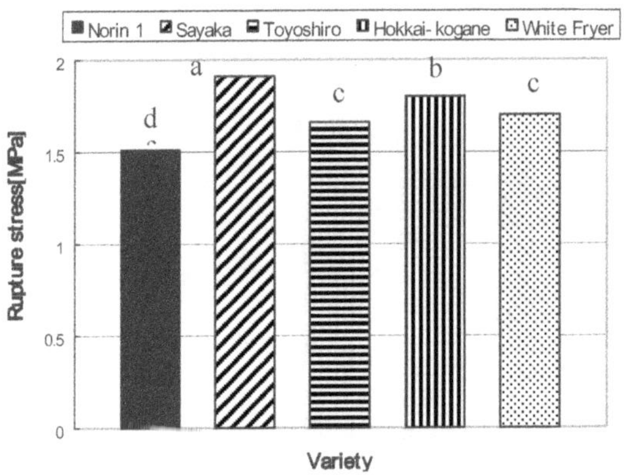

Figure 5. Rupture stress of perimeduallary tissues of potatoes (means of 0, 40 and 80 days of storage).

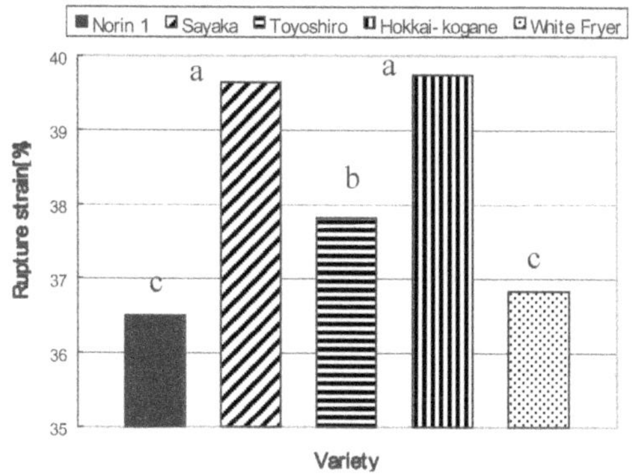

Figure 6. Rupture strain of perimeduallary tissues of potatoes (means of 0, 40 and 80 days of storage).

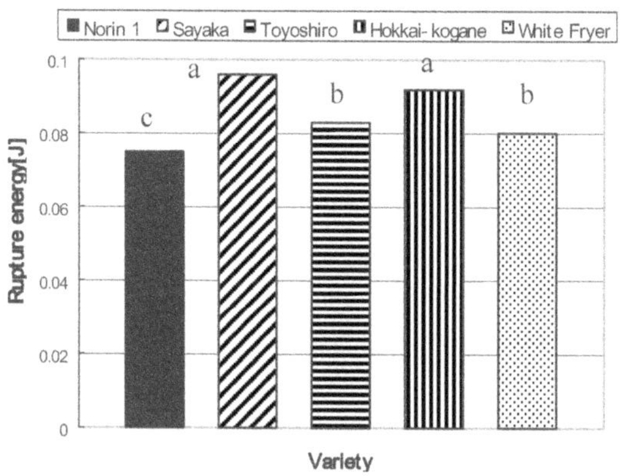

Figure 7. Rupture energy of perimeduallary tissues of potatoes (means of 0, 40 and 80 days of storage)

In addition, in order to investigate which physical parameters of skin and underlying tissues contribute to blackspot index, we attempted to examine correlations between physical parameters and blackspot index. Table 1 gives the results, in which rupture deformation and energy of skin, and rupture stress and energy of perimeduallary tissues correlated significantly ($p<0.01$ or 0.05) with blackspot index, implying that these physical parameters play an important role in the protection of potato bruise. This information may be useful for potato breeders to develop new resistant varieties.

Concerning the physical properties of potatoes, Garry *et al.* (1996) pointed out that cell size, cell wall thickness and skin thickness of potatoes would influence the physical properties of tissue and skin. In addition, Gray and Hughes (1978) speculated on the possible role of high content starch grains in a cell in causing membrane damage if the cell were rapidly deformed as a result of impact. Therefore, further study is required at above cell levels.

Table 1. Correlations between blackspot index and physical properties of skin and perimeduallary tissues of potatoes (n=60).

Rupture stress of skin	-0.175
Rupture deformation of skin	-0.242**
Rupture energy of skin	-0.226*
Rupture stress of perimeduallary tissues	-0.297**
Rupture strain of perimeduallary tissues	-0.133
Rupture energy of perimeduallary tissues	-0.218*

**significant (p<0.01)
*significant (p<0.05)

Conclusion

The results of the present study demonstrated that potato skin may play an important role in the protection of potato bruise.

References

Dwelle, R.B. (1976). Effect of soil, potato treatment and storage temperature on blackspot bruise development in tuber of four *solanum tuberosum* cultivars, American Potato Journal 54: 137-146.

Gray, D. and J.C. Hughes (1978). Tuber quality. In: M. H. Paul (Eds.) The Potato Crop. Chapman & Hall, London, 504-544.

Hosaka, K., M. Mori and K. Ogawa (1994). Genetic relationships of Japanese potato cultivars assessed by rapid analysis. American Potato Journal 71: 535-546.

Kunkel, R., W.H. Gardner and N. M. Holstad (1986). Improvement of techniques for potato blackspot evaluation and some errors associated with measurements, American Potato Journal 73: 436-444.

McGarry, A., C.C. Hole, R.L.K. Drew and N. Parsons (1996). Internal damage in potato tubers: a critical review. Postharvest Biology and Technology 8: 239-258.

Peterson, R.L., W.G. Baker and M.J. Howarth (1985). Development and structure of tubers. In: P. H. Li (Eds.) Potato Physiology. Academic Press, Orlando, 124-152.

SPSS (1992). SPSS for Windows Base System User's Guide Release 5.0. SPSS Incorporation, Chicago, 269-270.

Stevens, L.H. and E. Davelaar (1997). Biochemical potential of potato tubers to synthesize blackspot pigments in relation to their actual blackspot susceptibility. Journal of Agriculture and Food Chemistry 45: 4221-4226.

Storey, R.M.J. and H.V. Davies (1992). Tuber quality. In: M.H. Paul (Ed.) The Potato Crop. Chapman & Hall, London, 507-569.

Curriculum vitae – K. Hironaka

K. Hironaka – born in 1952 –studied food engineering at Hiroshima University (Japan) from which he received a Master degree of Agriculture in 1978. He then worked at Obihiro University and at first studied on storage of Chinese yam. He published papers of visco-elasiticity of Chinese yams stored in a warehouse or at fields. He received a PhD degree from Hokkaido

University (Japan) in 1989. After then, he advanced to the research of potato storage. He worked on the mechanisms of potato bruising or sugar accumulation at low temperatures. He published the paper (2001, Potato research) 'Effect of static loading on sugar contents and activities of invertase, UDP-glucose pyrophosphorylase and sucrose 6-phosphate synthase in potatoes during storage'. He is the author of over 40 scientific papers. He is a member of European Association of Potato Research or Potato Association of America. Hw liked potatoes, especially potato salads made by his wife.

Profile – Obihiro University

Obihiro University is the only national university of agriculture and veterinary medicine in Japan. It was founded in 1941 as the Obihiro Technical School of Veterinary Medicine. In the succeeding years the University added more areas of agricultural sciences and animal husbandry. In 1967, a master's degree program in animal husbandry was set up. In 1990, almost three decades later, a doctoral program in veterinary medicine was established in cooperation with the United Graduate School of Veterinary Sciences, Gifu University. In 1994, a doctoral program in agricultural sciences was established in cooperation with the United School of Agricultural Sciences, Iwate University.

Obihiro is enclosed by the field, and agriculture is active. There are especially much production of potatoes, wheat, and beets. Our Food Engineering Laboratory works on various research projects related to potatoes, e.g. potato bruising, sweetening and potato starch.

The role of plant growth regulators in potato tuber quality assurance

J.P. Maude
Chemtura Ltd., Research and Development Centre, Crop Protection Division, Brooklands Farm, Cheltenham Road, Evesham, Worcestershire, WR12 2LS, United Kingdom

Abstract

High quality potato tubers are achieved through good crop management from the seed selected for planting through to the end use. Good management can only take place with effective tools to enable the grower, store keeper, and processor to apply best practice. This paper focuses on the role plant growth regulators play in tuber quality assurance and how the tools Fazor® (maleic hydrazide) and Aceto Spud-Nic 50™ (chlorpropham) can be employed to achieve best quality particularly in crops used for processing. Fazor is a well established plant growth regulator (PGR) formulated as a soluble granule containing 600 g maleic hydrazide/kg, used world-wide on a range of crops including root and bulb vegetables, tobacco and grass. Maleic hydrazide was first reported to have plant growth regulating properties by D.L. Schoene and O.L. Hoffmann in 1949. Introduction as an agricultural chemical was by U.S. Rubber Co. (now Chemtura) in 1979. Aceto Spud-Nic 50 based on chlorpropham (CIPC) has been used world-wide for potato sprout control for more than 30 years. Chlorpropham was reported as a herbicide in 1951 by E.D. Witman and W.F. Newton and introduced as a herbicide in the same year. Both maleic hydrazide and chlorpropham have modern registration dossiers with toxicology, environmental, efficacy and residue data that has been used successfully to gain Annex 1 listing in Europe under directive 91/414. Aceto Spud-Nic 50 is registered for use in Russia by Aceto and marketed and distributed by Chemtura Limited. Fazor has been submitted for registration by Chemtura Limited in Russia, Ukraine and Belarus. Due to the cropping cycle of potatoes, storage is necessary to meet consumer demands and in particular to facilitate the potato processing industry. Mid to long term storage, up to 10 months is common for processing potatoes and necessary for specific varieties. Fazor is applied in the field through standard hydraulic spray equipment. The active substance, maleic hydrazide, translocates through the plant to the tubers where it inhibits mitotic cell division. With correct timing of application Fazor will assure good tuber development by preventing miss-shapen tubers, secondary growth, and premature sprouting subsequently protecting marketable yield. Fazor will also enable proper tuber curing and provide short term, in-store, sprout control. Spud-Nic 50 is applied in-store through specialized fogging equipment. The active substance, chlorpropham inhibits cell division and will ensure sprout control through long term storage. The combination of the plant growth regulators Fazor and Spud-Nic 50 provides a complete sprout control and tuber quality assurance programme.

Keywords: sprout control, processor, plant growth regulator, quality

Introduction

The potato crop is long established and grown in 180+ countries world-wide. Because potatoes offer a plentiful source of starch in a very versatile form they feature in the major crops grown in the world today. Common use is as a fresh table vegetable with strong and increasing trends in the processing sector for chips and crisps, particularly in western countries. In the majority of climates, particularly temperate, potatoes are grown as a single crop/year. Due to non-continuous crop production coupled to a continuous demand, potato storage is necessary to fill the out-of-harvest periods. Much of the stored crop is used specifically in the processing industry although a small proportion is used for table consumption.

One common aspect of all potatoes grown for human consumption is the need for quality. Quality can be influenced by many factors through the life of the potato tuber, including but not limited to; seed quality, soil preparation/condition, agronomic inputs including; fertilizer, crop protection, irrigation, pre-harvest plant growth regulation, haulm desiccation, harvest and curing, storage and final processing. Effective management through the whole cycle is necessary to assure high quality of the final product.

Pre-harvest growth regulation

Fazor is a soluble granule formulation (containing 600 g/kg maleic hydrazide). The mechanism of action of maleic hydrazide is the inhibition of mitotic cell division (Greulach and Haesloop, 1954, Nooden 1969). The product readily dissolves in the spray tank to produce an effective foliar application through hydraulic spray systems. Once applied to the living foliage, due to its systemic property and formulation optimisation, maleic hydrazide will translocate down through the plant stem and root system into the tubers. Application rate for potatoes is globally standardised at 5 kg/ha product (3 kg active ingredient). Application timing is critical to obtain optimum efficacy. Once tubers have reached 25mm in length, cell division is complete and the remainder of tuber growth is due to cell expansion. This is the critical application timing for Fazor as the tubers need no further cell division to produce high yields and quality tubers. The presence of maleic hydrazide in the tubers will prevent further cell division.

Assuming best practice has been carried out through the initial field phase the next crucial period to secure high tuber quality is tuber development.

Tuber development

In the latter growth stages the potato plant initiates tuber development which takes place rapidly to form the underground seed for the next generation. Normally tuber formation will be complete before flowering is complete and the tubers will lie dormant in the soil, whilst skin set takes place, until desiccation and harvest. Tuber formation is a critical period for high final quality. In many cases crops are irrigated through tuber development in order to achieve consistent growing and tuber bulking conditions to achieve desirable yields. In cases of no irrigation and particular erratic weather conditions tuber formation can be disrupted resulting in miss-shapes and lowering marketable yield. More recently, climate changes through Europe have increased the incidence of re-growth and sprouting prior to harvest, again resulting in a shortage of marketable yield. This has a significant impact on both the value of the crop and the annual harvest volume.

Fazor application takes place within the tuber development phase. Once tubers have reached 25mm in length, cell division is complete and the remainder of growth is due to cell expansion. This is the critical application timing for Fazor as the tubers need no further cell division to produce high yields and quality tubers. If application timing is correct the presence of maleic hydrazide in the tubers will prevent further cell division including; secondary growth, deformation of tubers and early dormancy break and sprouting, without compromising cell expansion. Marketable yield in most cases will be increased over an untreated crop, particularly when conditions promote irregular tuber development.

Well managed crops having received the correct agronomic inputs can yield between 40 – 60 tonnes per hectare. One of the greatest challenges to maintaining the potato crop quality is the act of harvesting.

Harvest

Crop haulm is desiccated to aid harvest and the tubers are mechanically lifted from the soil and transported to packing or storage facilities. Harvest provides the greatest opportunity for tuber damage and subsequent reduction of quality by human intervention. As the operation of harvest involves moving large quantities of tubers from the field through conveyors and into trailers and then storage containers, great care must be taken to minimise skin abrasion that can provide sites for infection on the tuber surface. Minimising damage is the best scenario as complete control over harvest damage is near impossible. In his PhD thesis, *The effects and residues of maleic hydrazide within the potato crop* McKenzie (1989) describes results of wound healing of Fazor treated potatoes and concludes that treated tubers heal equally compared to untreated tubers and in many cases provide improved healing properties. This improved healing is related to the phenolic properties of maleic hydrazide (Miller, 1955) and their ability to stimulate formation of a thicker suberin layer in potatoes (Simonds *et al*, 1953). Wound healing also known as curing is an extremely important component in reducing disease spread and minimising shrinkage loss during long term storage.

A further consequence of potato harvesting is the inevitable loss of berries, chats (small tubers) and un-harvested tubers, each of which remain in the field and will germinate in the following crop if not controlled. The number of volunteers can be from 125-350,000 per hectare and tubers can remain viable for up to 8 years. Problems caused by volunteers include disease and insect carry over to the next crop, competition for light, lodging of the following crop and harvesting difficulties. Due to Fazors' anti-germinative sprout control up to 75% or more of volunteers can be controlled (Buckley D *et al* 2006) removing the need for further herbicide applications in the following crop.

Storage

As already described potato storage is necessary to provide potatoes to supply for table consumption and the largest proportion being stored specifically for the processing industry. The main objective of storage is to maintain the quality of the harvested tubers for the duration required by the market which for processing can be up to 10 months. In this period the natural

Figure 1A Fazor (5kg/ha) treated tubers harvested and then replanted.

Figure 1B. Untreated tubers harvested and replanted.

cycle of un-stored potatoes is to break dormancy and initiate the development of a new plant. The tuber is a biologically active organism that in normal conditions will respire and produce the necessary sugars and subsequently sprouts/shoots to undergo growth. Storage conditions are optimised in order to stop these natural processes.

Sugar development in the tuber has wider implications in the processing industry as reducing sugars have been identified as a cause of dark fry colour in chips and crisps. The reducing sugars, glucose and fructose, were identified as the limiting factor in development of dark fry colour (Roe and Faulks, 1991). The major aim of storage for processing potatoes therefore is to control the development of reducing sugars.

Accumulation of reducing sugars during storage is governed by, but not limited to, four processes; (i) storage at temperatures below 10 °C (ii) dormancy break and sprouting (iii) tuber senescence/dehydration after long term storage and (iv) storage atmosphere (Kumar *et al.*, 2004). The paradox of storing potatoes for processing is that whilst sprout inhibition can be achieved at relatively low temperatures (4 °C) reducing sugars will accumulate in these conditions. The ideal temperature to maintain reducing sugars at acceptable levels is 10 - 14 °C at which temperature in-store sprouting will occur. Plant growth regulators offer the solution.

In-store growth regulation

Fazors' growth regulation starts when maleic hydrazide is first available in the tuber, approximately 2 days following application. The length of persistence of growth regulation is dependant on 4 main factors (i) application timing (ii) weather conditions between tuber maturation and harvest (iii) storage conditions (particularly temperature) and (iv) variety. Because the factors that influence duration of activity are multidimensional the variation can be relatively large. However, factors i and iii are controllable and factor iv is a known quantity. Therefore the opportunity to control the variation is good. Typically duration of optimum sprout suppression will be less than 3 months (see Figure 2), with good confidence at 8 – 10 weeks.

In order to assure high tuber quality is maintained through mid – long term storage sprout control must be maintained passed the effective duration provided by Fazor.

Aceto Spud-Nic 50 is a fogging solution containing 500 g/l chlorpropham. Chlorpropham is approved for use in Russia under the brand name Aceto Spud-Nic 50 and is marketed and distributed by Chemtura Russia. Chlorpropham has been used world-wide for more than 30 years

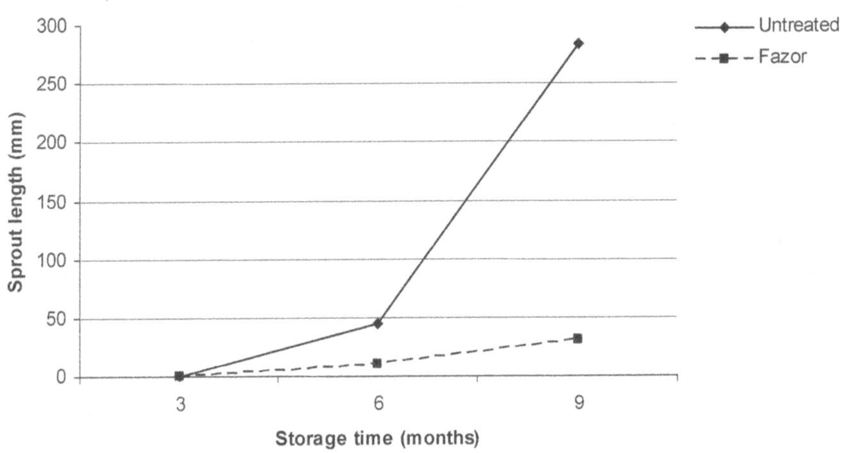

Figure 2. Sprout control resulting from Fazor applied at 5 kg/ha, at tuber size 25 mm, with tubers store at 10 °C. Results averaged from 4 trials carried out in the UK, 2002 on varieties Maris Piper, Cara, Romano, Nadine.

as an in-store potato sprout suppressant. Whilst chlorpropham is an old active substance it has a modern registration dossier with toxicological, environmental, efficacy and residue data. This data has achieved successful registration reviews in Europe and the USA, and chlorpropham was Annex 1 listed in Europe under directive 91/414 in November 2004. Similar to maleic hydrazide, chlorprophams' mechanism of action is mitotic cell division inhibition. Dissimilarly, chlorpropham is not systemic but acts by contact. Application is achieved through specialised fogging equipment carried out by trained operators. The active ingredient is dispersed into the store through fog that caries small solid particles of the active ingredient that distribute through the store. Store design and management is important for even distribution and requires low pressure ventilation through under-floor or duct vents.

Store design can be specified to work optimally with Aceto Spud-Nic 50. Store size should be limited to limited to a size that can be loaded quickly (within 5 days - so that the crop condition is uniform). Smaller or sub-divided stores should be considered - a maximum size of 1000 tonnes has shown positive results. The stored height should not exceed 5m for bulk - up to 8 stacked boxes. The segregation of tuber varieties is desirable to avoid the dosing of dormant crop just because it is stored alongside an earlier-breaking variety. The application port for CIPC treatment should be positioned in consultation with your approved applicator for optimum distribution and to by-pass refrigeration coils and humidification cells. Ventilation systems should be adequate to ensure good airflow before, during and after treatment and to avoid condensation build-up during storage.

Achieving the optimum balance between humidity, condensation and even CIPC distribution is a complex task and an approved specialist should be consulted. External access to fan switches is required to allow control without entering the store during treatment.

Application rate and timing are critical to maintain optimum sprout control. The first application of Aceto Spud-Nic 50 should be made when potatoes are dry and correctly cured, approximately 2 to 6 weeks after loading the store. If application is made earlier there is risk of skin damage

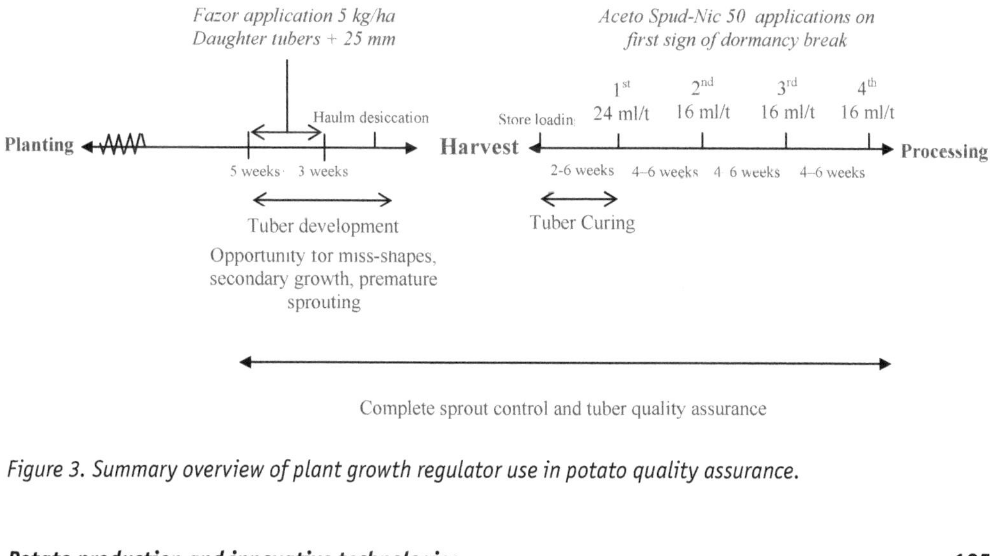

Figure 3. Summary overview of plant growth regulator use in potato quality assurance.

and quality reduction. The curing period is the key interface between the two plant growth regulators Fazor and Aceto Spud-Nic 50 for a complete sprout control and quality assurance programme. Fazor will assure good tuber development and sprout control with proper curing in the initial 2 – 6 weeks in-store which also enables the first application of CIPC to be made safely. The first application of Aceto Spud-Nic 50 should be applied at 24 ml product/tonne (12 g active ingredient). Following application and once the dust has completely settled (8-10 hrs following application) it is beneficial to vent the store for approximately 15 minutes. Duration of activity is expected to last between 4 and 6 weeks depending on storage conditions and potato variety. Three further applications of Aceto Spud-Nic 50 can be applied at a rate of 16 ml/tonne (8 g active ingredient). Stored tubers must be closely monitored to correctly time re-applications which should be made on the first sign of sprout growth. Weekly checks of stores is advised to ensure complete sprout control can be maintained.

Conclusions

Fazor in combination with Aceto Spud-Nic 50 offers a solution to complete potato sprout control and tuber quality assurance, from in the field (pre-harvest) throughout storage to final processor use.

References

Buckley, D. Duncan H., and Anderson E. (2006) Research Review. Maleic Hydrazide in potato volunteer control. British Potato Council 275, 6.

Greulach, V.A. and Haesloop, J.G. (1954) Effects of MH on internode elongation, cell enlargement and stem anatomy. American Journal of Botany 41, 44-50.

Kumar D., Singh B.P and Kumar P. (2004) An overview of the factors affecting sugar content of potatoes. Annals of Applied Biology. 145, 247-256.

McKenzie, J (1989) The effects and residues of maleic hydrazide within the potato crop. Agricultural Chemistry Dept., University of Glasgow.

Miller, D.M. (1995) The polarography of MH. Canadian Journal of Chemistry, 33, 1806-1813.

Nooden, L.D. (1969) The mode of action of Maleic Hydrazide: Inhibition of growth. Plant Physiology 22, 260-270.

Roe, M.A. and Faulks R.M. (1991) colour development in a model system during frying: Role of individual amino acids and sugars. Journal of Food Sciences 56, 1711-1713.

Simonds, A.O. Johnson, G. and Schaal, L.A. (1953) Comparative effects of catechol, some related compounds and other chemicals on suberisation of cut potato tubers. Botany Gazette, 115, 190-195.

Curriculum Vitae – James Patrick Maude

James Patrick Maude – born in the Midlands of the United Kingdom – pursued part time education alongside working, completing a BSc. in Applied Biology at the University of the West of England in 2000 followed by a post graduate certificate in Crop Protection at Harper Adams University College in 2002. He started work for the former crop protection company Uniroyal Chemical Ltd. in 1992 at their UK R&D facility. The majority of former years were spent field testing seed treatments on small grain cereals for global development. Development broadened into root crops including evaluation of maleic hydrazide as a growth regulator in parsnips and carrots, potato seed treatment development, and method design and development for disease inoculation and field testing of potato fungi. More latterly career progression has led into managerial role to technically support plant growth regulators in the Europe Middle East and Africa region including maleic hydrazide used on potato, onion and tobacco globally. Uniroyal Chemical Ltd. has been superseded through various mergers and acquisitions and now is a legacy company of Chemtura Ltd.

Profile – Chemtura Ltd.

Chemtura Ltd. is a multinational specialty chemical company. The company is a supplier of plastic additives and pool and spa products, a marketer of niche agricultural products, active in the petroleum additives market and a marketer and producer of urethane polymers. The Crop Protection business is represented globally and is particularly active in seed treatment and miticides. Crop protection focus is on high value niche crops where many products in the their portfolio are well placed. Crops such as nuts, citrus, grapes, cotton, small grain cereals, vegetables are important targets for Chemtura's crop protection strategies. Chemtura has in-house R&D resource and capabilities to support the business and a strong distribution network to deliver product efficiently and in good condition.

Changes in sugar contents and acid invertase in potato tubers during low-temperature storage

C. Matsuura-Endo[1], A. Kobayashi[1], A. Ohara-Takada[2], S. Takigawa[1], T. Noda[1], N. Hashimoto[1], H. Yamauchi[1], I.S.M. Zaidul[1] and M. Mori[1]
[1]National Agricultural Research Centre for Hokkaido Region, Memuro, Hokkaido 082-0071, Japan
[2]National Institute of Crop Science, 2-1-18, Tukuba, Ibaraki 305-8518, Japan

Abstract

Low-temperature storage of potato tubers causes an increase in sugars; this phenomenon is known as 'low-temperature sweetening' (Isherwood 1973, Burton 1989). The sugar content and the activity of acid invertase in the potato tubers of several Japanese cultivars during storage were investigated. When potatoes were stored at 4 °C, three types of changes in the sugar content were observed among cultivars, which defined as type-1, type-2, and type-3. During storage, the reducing sugars were increased much higher in type-1, but the increasing trend also was observed in type-2 and it was 4- to 6-fold lower than that of type-1. However, in type-3, the sucrose was increased rather than increase of reducing sugars. On the other hand, the activity of vacuolar acid invertase was increased in the type-1, in contrast, in the type-2 and type-3 cultivars, the activities were very low during storage at 4 °C. Reverse transcription-polymerase chain reaction (RT-PCR) analysis of acid invertase showed that the transcripts of the enzyme were accumulated in the type-1 whereas the changes were not occurred in the type-3 during storage at 4 °C. These results suggest that the activity of vacuolar acid invertase is related to the type of changes in the sugar content during low-temperature storage among potato cultivars. The sugars and amino acids contents in the potato tubers and their effects on the acrylamide formation during frying were also investigated. A strong correlation between the reducing sugar content in tubers and acrylamide level in chips was found. Storage temperatures lower than 8 °C increased the reducing sugar content markedly in potato tubers, resulting in an increase in acrylamide level in chips after frying.

Keywords: Low-temperature storage, sugar, acid invertase, chips, acrylamide

Introduction

In Japan, approximately 76% of the total potato production comes from Hokkaido, the northernmost island that is having suitable and cool climate for potato cultivation. Potatoes are harvested after they entered to a rest period, and even after the period ends, their dormancy can be extended forcibly by low-temperature storage. Because of this, potatoes produced in Hokkaido can be supplied for a long period in Japan. The low-temperature storage of potato tubers, however, causes increase in reducing sugars, that phenomenon is known as 'low-temperature sweetening' (Isherwood 1973, Burton 1989). In frying, increased reducing sugars induce a non-enzymatic Maillard-type reaction with free amino acids, forming unacceptable brown- to black-pigmented products (Shallenberger *et al.*, 1959). Furthermore, recent studies have revealed that the Maillard reaction involving asparagines can produce acrylamide, which classified as probably carcinogen (Mottram *et al.*, 2002, Stadler *et al.*, 2002), and potato-based food products cooked at high-temperature contain high level of acrylamide (Tareke *et al.*, 2002). Chuda *et al.* (2003) reported that chips made from potato tubers stored at 2 °C contain high acrylamide. Thus, the increase in reducing sugars due to the low-temperature storage has negative effects on the quality of fried products (Burton 1989, Sowokinos 1990, Chuda *et al.*, 2003).

We have been attempting to clarify the patterns of changes in sugar contents during storage, and the relationship between the changes in the sugar and enzyme activities in each Japanese potato cultivar. The contents of sugars and amino acids in tubers, and their effects on the acrylamide formation during frying have also been evaluated. This paper introduce the outcome of these our recent research.

Changes in sugar contents and activity of vacuolar acid invertase in potato tubers during the growing season and storage period

The cultivars that used in this study are as follows: Irish Cobbler, May Queen, and Kita-akari, which are commonly used as table potatoes; Toyoshiro, which is used for potato chips; White Fryer, an advanced breeding cultivar from Toyoshiro; and, Inca-no-mezame, which is bred for extensive use and has yellowish orange flesh with a very high content of carotenoids. The sugar contents and activity of vacuolar acid invertase (EC 3. 2. 1. 26) in the tubers of these cultivars during the growing season and storage periods at 4 °C and 20 °C were measured (Matsuura-Endo *et al.*, 2004). Figure 1 shows the changes in the sugar content of tubers from three representative cultivars of May Queen, White Fryer, and Inca-no-mezame during the growth and the storage.

Growth period: In young developing tubers (approx. 30 days after emergence, approx. 10 g fresh weight), the content of glucose was high. The glucose content then decreased upon tuber growth and maturation in all the cultivars. On the other hand, the fructose content was found

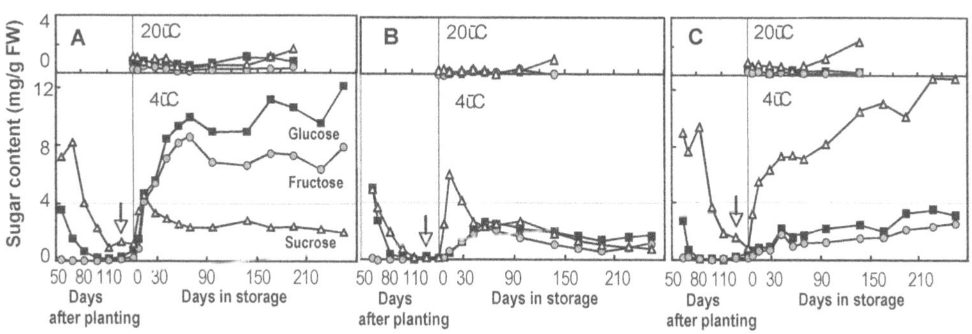

Figure 1. Changes in the sugar contents of potato tubers during the growing season and storage period (from Matsuura-Endo et al., *2004). Seed tubers of potato were planted in late April, and mature tubers were harvested in early September, indicated by arrows. After curing of tubers for 2 weeks, storage was started on the date indicated as 'day 0' on the x-axis. Upper panels, storage at 20 °C; lower panels, storage at 4 °C. A, May Queen; B, White Fryer; C, Inca-no-mezame.*

to be low throughout the growing period. The sucrose content was high in young tubers, which is higher as compare to that of glucose, but decreased to a level approximately equivalent to the glucose level up to the harvest date. Acid invertase showed considerable activities only in the youngest tubers. Pressey (1969) and Davies and Viola (1994) reported that in the growing season the activity of sucrose synthase (in the direction of sucrose breakdown) is predominant over that of acid invertase in potato tubers. Determining the activities of these related enzymes in the tubers of the experimental cultivars are now in progress.

Storage period: At 20 °C, a little change in the sugar contents was found during the dormant periods. This suggests a very low metabolic activity during the dormant period, as reported by Espen *et al.* (1999). A slight increase in the sucrose content was observed at the time of the sprouting in all cultivars, which was in agreement with the results reported by Burton (1989). The activity of acid invertase was extremely low during storage at 20 °C in all these three cultivars.

When stored at 4 °C, three-types of changes in the sugar content were observed: (1) an increase of reducing sugars content occurred at higher level during the storage (May Queen, Figure 1A); (2) an increase of reducing sugar content occurred at low levels throughout storage (White Fryer, Figure 1B); and (3) an increase of sucrose content occurred without increasing the reducing sugars (Inca-no-mezame, Figure 1C).

In the type-1 cultivar May Queen (Figure 1A), firstly, the content of sucrose increased rapidly then reached to the maximum within 2 weeks. Glucose and fructose increased about 1 week later than the sucrose, and reached to the maximum within approximately 60 days, which then remained either steady or increased slightly. Irish Cobbler, Kita-akari, Toyoshiro, and other major cultivars showed almost similar changing pattern as that of May Queen. It is assumed that the type-1 cultivars are 'high sugar-accumulating, cold-sensitive cultivars,' which are in agreement with the reports of Hammond *et al.* (1990) and Trenner *et al.* (1996). The type-2 cultivar White Fryer (Figure 1B) showed a pattern that is also similar to that of type-1 cultivars. The increasing levels of reducing sugars, however, were 4 to 6 times lower than those of the type-1 cultivars throughout the storage period. The White Fryer is the progeny of ND 860-2, which is a 'low sugar-accumulating, cold-tolerant' breeding line. The data obtained in White Fryer were similar to those observed in studies of cold-tolerant lines, ND 860-2 (Blenkinsop *et al.*, 2002), Brodick and 13676ab 1 (Richardson *et al.*, 1990). In the type-3 cultivar Inca-no-mezame (Figure 1C), the sucrose continued to rise after 2 week and reached 13 mg/g fresh weight after 250 days of storage, which about 6-fold higher than that of the other two types. This unique pattern may originate in its genetic background. We are trying to obtain information about changes of sugars during storage using breeding lines related to this cultivar.

The activity of acid invertase in May Queen (type-1) increased rapidly up to 2 weeks of storage, reached a maximum value and then maintained a similar level as storage continued, reflecting the accumulation of reducing sugars. On the other hand, the activities of the enzyme in White Fryer (type 2) and Inca-no-mezame (type 3) were very low during the storage. In case of Inca-no-mezame, the low activity of acid invertase seems to reflect the continuous accumulation of sucrose after 2 week of storage. White Fryer showed low activity of the enzyme during the storage, although the change in sucrose content had a pattern similar to that observed in the type-1 cultivars; i.e. the content increased rapidly and peaked to the maximum within 2 weeks, there after decreases.

Estimation of the relative transcript levels of acid invertase with RT-PCR

We also analyzed whether the activity of acid invertase was controlled by its gene expression using RT-PCR method (Figure 2). For May Queen (type-1), the transcripts of the acid invertase gene might be accumulated in the tubers stored at 4 °C, while only very weak signals of transcripts could be detected in tubers stored at 20 °C. The results are in agreement with a study as reported by Zhou *et al.* (1994) using northern blot analysis in cv. Russet Burbank, suggesting that the activity of vacuolar acid invertase during storage is regulated at a transcriptional level in the type-1 cultivar, May Queen. On the other hand, in tubers of Inca-no-mezame (type-3), stored either at 20 °C or at 4 °C, the transcripts could not be detected clearly. In this cultivar, these results seem to be convincing since the enzyme activity was very low. From these results it is suggested that the acid invertase activity reflects the accumulation patterns of sugars in these two types in low-temperature storage. Interestingly, in the case of White Fryer (type-2), although the enzyme activity was very low during storage, the transcripts accumulated in the tubers stored at 4 °C same as those of May Queen. The reason is not clearly known at this moment, we are now investigating whether any other factors such as an enzyme inhibitor, which is responsible for the low activity of acid invertase during storage in this cultivar, or not.

Relationship between components in potato tubers and acrylamide in chips

We have been evaluating the contents of sugars and amino acids in tubers of Japanese cultivars, and their effects on the acrylamide formation during frying. Using the cultivar Toyoshiro, which is a leading cultivar for potato chips in Japan, it was found that chips made from tubers stored at 2 °C for 2 weeks showed an extremely dark-brown colour (Figure 3) and contained a high acrylamide content (Chuda *et al.*, 2003). Furthermore the acrylamide content in chips began to increase after 3 days of storage at 2 °C, which is due to the increase of glucose and fructose contents in the tubers (Ohara-Takada *et al.*, 2005). To understand which component in tubers had the most effect on the acrylamide level, we have analyzed the correlation between components in tubers and acrylamide in chips. The contents of free amino acids; asparatic acid, asparagine (Figure 4), glutamic acid and glutamine in tubers showed no significant correlation

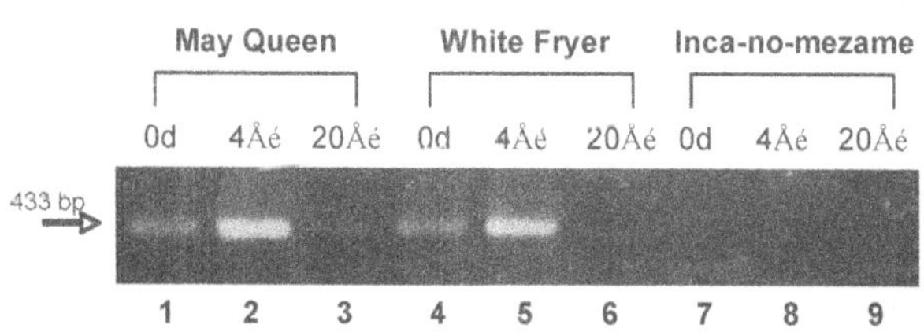

Figure 2. The differences in expression patterns of the acid invertase gene among cultivars (Matsuura-Endo et al.*, 2004). RT-PCR was performed on RNA isolated from the tubers stored for 1 week at 4 °C or 20 °C to detect the expression of the acid invertase gene. A 433-bp fragment of the acid invertase gene was amplified from the RT-PCR.*

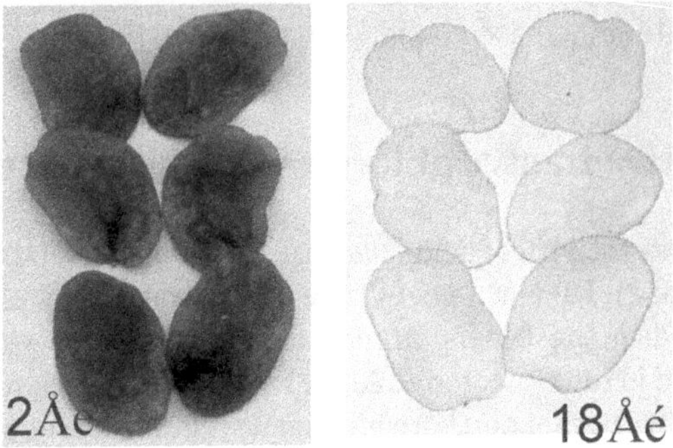

Figure 3. Colour of chips made from tubers stored at different temperature. The colours of chips from tubers (Toyoshiro) stored at 2 °C (left) and 18 °C (right) for two weeks.

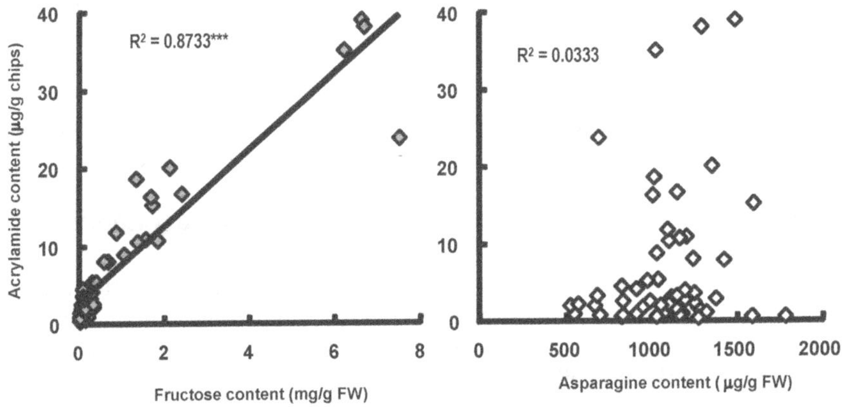

Figure 4. Correlations between fructose (left) or asparagine (right) in tubers and acrylamide in chips (Ohara-Takada et al.*, 2005).*

with the acrylamide level in chips. On the other hand, there were strong correlation observed between the contents of reducing sugars and the acrylamide level (in the case of fructose in Figure 4). The asparagine content is generally steadier and substantially higher than those of the reducing sugars in potato tubers (Lisińska and Leszczyński 1989). From these results, it is concluded that the contents of reducing sugars in material tubers may determine the degree of acrylamide formation in potato chips.

Effects of storage temperature on the contents of sugars in tubers and acrylamide in chips

To clarify the effects of storage temperature on potato components and acrylamide in chips, the tubers from several cultivars were stored at various temperatures (2, 6, 8, 10, and 18 °C), and analyzed the contents of sugars in tubers and acrylamide in chips after frying (Matsuura-Endo *et al.*, 2006). Figure 5 shows the effects of storage temperature on the contents of sugars

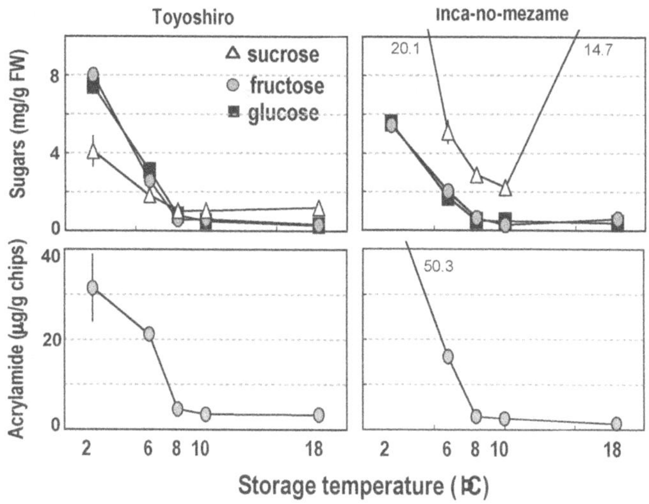

Figure 5. Effects of storage temperature on the contents of sugars in potato tubers stored for 18 weeks and acrylamide in chips (from Matsuura-Endo et al.*, 2006). left, Toyoshiro; right, Inca-no-mezame.*

in tubers stored for 18 weeks and acrylamide in their chips (left, Toyoshiro; right, Inca-no-mezame). At temperatures lower than 8 °C, the contents of reducing sugars increased markedly in all cultivars, with similar trend showed in the acrylamide content and dark brown chip colour. In case of Inca-no-mezame, the sucrose content in the tubers stored at 18 °C was very high after 18 weeks, but the acrylamide level was as low as those in the other cultivars. Sucrose is known to generate acrylamide in the presence of asparagine at 180 °C or higher in model studies (Stadler *et al.*, 2002). The low acrylamide level in chips from Inca-no-mezame stored for 18 weeks at 18 °C (Figure 5) suggests that sucrose in potato tubers hardly contributes to acrylamide formation during chip processing.

Potatoes for chip production are generally not stored at low temperatures (Blenkinsop *et al.*, 2002), and companies processing chips usually measure chip colour and discard discoloured or burned chips before packing (Gould 1999). Chips with such a high acrylamide level, therefore, do not reach the market. However, low-temperature storage (2 to 4 °C) is advantageous in inhibiting sprouting and storage rot, especially in long-term storage. For this reason, recently there has been great interest in the development of potato cultivars whose tubers can be processed into chips with acceptable colour even after low-temperature storage (Blenkinsop *et al.*, 2002). Such cultivars have not been identified in our study. The diverse response of the cultivars to storage temperature indicates that monitoring of asparagine content together with sugar contents in various potato genotypes can provide useful information in the development of new cultivars for chip production. To obtain more information we are currently involved in analytical studies of about 50 cultivars and breeding lines.

Conclusion

We have investigated that the patterns of sugar changes during low-temperature storage are classified into three types among the potato cultivars, and those patterns seem to be closely related to the activities of vacuolar acid invertase. The sugar change as found in the type-3 cultivars indicated mainly the increasing of sucrose but not reducing sugars, that is a new pattern. Therefore, the study of sugar changes in these cultivars would provide useful information for understanding the mechanism(s) of low-temperature sweetening in potato tubers. It has also been found that storage temperatures lower than 8 °C increase the reducing sugar content in the tubers, that might be responsible for an increase in acrylamide level in the chips after frying. We are expecting to utilize the knowledge for developing new cultivars that will be use for chip production as well as a control technology for storage quality.

Acknowledgements

This study was financially supported by the 'Integrated Research Program for Functionality and Safety of Food toward an Establishment of Healthy Diet' of the Ministry of Agriculture, Forestry, and Fisheries of Japan.

References

Blenkinsop, R.W., L.J. Copp, R.Y. Yada and A.G. Marangoni (2002). Changes in compositional parameters of tubers of potato (*Solanum tuberosum*) during low-temperature storage and their relationship to chip processing quality. Journal of Agricultural and Food Chemistry 50: 4545–4553.

Burton, W.G. (1989). The Potato, 3rd edition. Longman Scientific & Technical, New York, 286–522.

Chuda, Y., H. Ono, H. Yada, A. Ohara-Takada, C, Matsuura-Endo and M. Mori (2003). Effect of physiological change in potato tuber (*Solanum tuberosum* L.) after low temperature storage on the level of acrylamide formed in potato chips. Bioscience, Biotechnology, and Biochemistry 67: 1188–1190.

Davies, H.V. and R. Viola (1994). Control of sugar balance in potato tubers. In: W.R. Belknap, M.E. Vayda and W.D. Park (Eds.) The Molecular and Cellular Biology of the Potato. 2nd edn. CAB International, Wallingford, UK, 67–80.

Espen, L., S. Morgutti, A. Abruzzese, N. Negrini, A. Rivetta, M.M. Quattrini, M. Cocucci and S.M. Cocucci (1999). Changes in the potato (*Solanum tuberosum* L.) tuber at the onset of dormancy and during storage at 23 °C and 3 °C. I. Biochemical and physiological parameters. Potato Research 42: 189–201.

Gould, W.A. (1999). Potato Production, Processing and Technology. CTI Publications, Timonium, 85–104.

Hammond, J.B.W., M.M. Burrell and N.J. Kruger (1990). Effect of low temperature on the activity of phosphofructokinase from potato tubers. Planta 180: 613–616.

Isherwood, F.A. (1973). Starch-sugar interconversion in *Solanum tuberosum*. Phytochemistry 12: 2579–2591.

Lisińska, G. and W. Leszczyński (1989). Potato Science and Technology. Elsevier Applied Science, London, 165–232.

Matsuura-Endo, C., A. Kobayashi, T. Noda, S. Takigawa, H. Yamauchi and M. Mori (2004). Changes in sugar content and activity of vacuolar acid invertase during low-temperature storage of potato tubers from six Japanese cultivars. Journal of Plant Research 117: 131–137.

Matsuura-Endo, C., A. Ohara-Takada, Y. Chuda, H. Ono, H. Yada, M. Yoshida, A. Kobayashi, S. Tsuda, S. Takigawa, T. Noda, H. Yamauchi and M. Mori (2006). Effect of the storage temperature on the contents of sugars and free amino acids in tubers from different potato cultivars and acrylamide in chips. Bioscience, Biotechnology, and Biochemistry 70: 1173–1180.

Pressey, R. (1969). Potato sucrose synthetase: Purification, properties, and changes in activity associated with maturation. Plant Physiology 44: 759–764.

Richardson, D.L., H.V. Davies, H.A. Ross and G.R. Mackay (1990). Invertase activity and its relation to hexose accumulation in potato tubers. Journal of Experimental Botany 41: 95–99.

Shallenberger, R.S., O. Smith and R.H. Treadway (1959). Role of sugars in the browning reaction in potato chips. Journal of Agricultural Food Chemistry 7: 274–277.

Stadler, R.H., I. Blank, N. Varga, F. Robert, J. Hau, P.A. Guy, M-C. Robert and S. Riediker (2002). Acrylamide from Maillard reaction products. Nature 419: 449–450.

Tareke, E., P. Rydberg, P. Karlsson, S. Eriksson and M. Törnqvist (2002). Analysis of acrylamide, a carcinogen formed in heated foodstuffs. Journal of Agricultural Food Chemistry 50: 4998–5006.

Zrenner, R., K. Schuler and U. Sonnewald (1996). Soluble acid invertase determines the hexose-to-sucrose ratio in cold-stored potato tubers. Planta 198: 246–252.

Curriculum Vitae – C. Matsuura-Endo

C. Matsuura-Endo – Born in Shizuoka prefecture, Japan – studied Plant Biology at Saitama University from which she received a MSc in 1987. She then studied Plant Physiology at Hokkaido University from which she received a PhD degree in 1992 based upon her work titled 'Low temperature-induced inactivation of vacuolar H+-ATPase with reference to chilling injury in plants.' She revealed subunit composition of vacuolar membrane H+-ATPase from mung bean for the first time, and also investigated the relationship between inactivation of the vacuolar H+-ATPase by low temperature and chilling injury in plants. She then joined National Agricultural Research Centre for Hokkaido Region (NARCH) as a researcher (1993-1996) and a senior researcher (1996-) specializing in potato quality. She is leading research programs on the potato quality, especially during that of storage. She is the author of many scientific papers and conference papers.

Profile – National Agricultural Research Centre for Hokkaido Region (NARCH)

In Hokkaido, the northernmost island of Japan, the agriculture is managed under climatic conditions of cool summers and cold winters. In summer, Hokkaido has favourable agricultural conditions, with long day-length, plenty of sunshine and large day-night temperature differences. Large-scale dairy, paddy rice and upland crop farming has developed utilizing vast land resources, and Hokkaido has become the most important food-supply base in Japan. The National Agricultural Research Centre for Hokkaido Region (NARCH), a part of the National Agriculture and Food Research Organization (NARO), carries out research for the development of agricultural techniques suited to Hokkaido such as: breeding of highly adaptable cultivars in Hokkaido; management of agro-environments; control of plant diseases and weeds, physiological research of crop quality; establishment of the utilization and storage of upland crops; investigation of regulatory mechanisms for chilling and cold tolerance. Memuro Upland Farming Research station, one of the departments of the NARCH, has the Potato Breeding Laboratory. We have been developing new potato varieties using hundreds of varieties and lines as the genetic resources. Recently we have bred unique varieties including potatoes with colourful skin/fresh, potatoes that create little glycoalkaloid, and potatoes that are early maturation and can be harvested before late blight damages.

Technology

Latest developments in 'open mouth' bag closing technology

V. Mpinga and R.A. Bontemps
Fischbein SA, Paepsem Business Park, Boulevard Paepsem 18B, 1070 Brussels, Belgium

Abstract

The importance of bag closing technology for transport, handling, preservation and safe sales of potatoes today is evident for all actors in the sector. Key performances requested to close 'open mouth' bags are certainly:

- cost reduction made possible by an automatic process from simple, operated machines up to machines integrated in high speed lines;
- very light daily maintenance for the operators on the line with machines driven by computers with immediate diagnosis of problems
- reliability for a continuous production 24 hours a day
- flexibility to different bag sizes, bag qualities or bag materials and easy adjustments to any change
- immediate availability of high quality spare parts.

Which are the key factors expected for the bags themselves and their closure?

- Evolutions of the bags themselves in weight, coatings, thicknesses, materials, number of plies.
- A 'mechanical' performance: secure closure of the bags to avoid leakage, costly losses of products.
- An 'esthetical' performance: nice looking, attractive, 'marketing' closures without any single defect.

On the basis of these requirements and performances, the manufacturers of open mouth bag closing machines recently developed various closures:

- by sewing. There are several possibilities:
- plain sewing
  - sewing with folding the top of the bag
  - sewing with crepe paper
  - sewing with easy open tape
  - tape over sewing

Owing to new engineering efforts, very high speeds (up to 2600 stitches/minute) can now be reached, allowing large outputs in terms of bags.

- by sealing: some of the principles of heat transfer are used in industrial continuous sealers:
  - convection with hot air sealers
  - conduction with band sealers
  - radiation

These sealers take benefit of several options improving their performances.

Keywords: bag closing, sewing, sealers, heat transfer, speed

Introduction

Like most dry bulk agricultural products including grains, sugar, coffee and rice, potatoes must be put in a bag once or even several times during their transport, handling, preservation and sale process. A lot of papers will describe multiple and severe efforts to improve:

- the production of potatoes per surface area;
- the protection against external agents;
- the right treatment to increase the size of the potatoes;
- the choices for better quality of the product itself.

All these efforts and research are, of course, necessary and totally justified. But nobody may forget the last steps in the whole process, from harvesting up to eating: potatoes must be transported, weighed, and for some of them, put in a bag to be sold in the retail stores.

Why do we spend so much effort and money to increase the production by X% if these products are then lost or damaged during transport? When these potatoes are put into a bag (a sack), this bag must be closed very carefully. The importance of bag closing today has become evident for all actors in the sector, especially to avoid losses of product, to avoid contamination or deterioration of the product itself and even to help to sell with an 'attractive' bag, easy to lift or to handle. Our paper will be divided in four main sections:

1. The general requirements for the bag closing section in the packaging process.
2. The expected performances for the bags and their closures.
3. The large choice of solutions to close 'open mouth' bags by sewing, by sealing and by gluing, from manual up to automatic operations.
4. The important commitments of the key manufacturers of machines in the sector of 'open mouth' bag closing.

The 'open mouth' bag is by far the most flexible possibility for bags as the spout of the packaging line will accept various sizes of bags of different material (jute, burlap, cotton, paper, woven polypropylene, polyethylene, laminated or coated bags...). Therefore this paper will focus on 'open mouth' bags: from 1-2-3-5-10-20-25 up to 50kg.

Which are the key performances requested during the packaging process to close 'open mouth' bags?

- Cost reduction made possible by an automatic process from simple, manually operated machines up to machines integrated in high speed lines.
- Very light daily maintenance for the operators on the line with machines driven by computers with immediate diagnosis of problems.
- Reliability for a continuous production 24 hours a day with reduced down times.
- Flexibility to different bag sizes, bag qualities or bag materials and easy adjustments to any change.

- Immediate availability of high quality spare parts produced according to the latest technology developments ensuring continuous production shifts even in peak periods.

Which are now the key factors expected for the bags themselves and for their closure?
- A mechanical performance: A strong request for a secure closure, without breakage during the transport and for a sift proof closure to avoid leakage, costly losses, contamination or even theft.
- An 'esthetical' or 'cosmetic' performance: Indeed, marketing purposes often generate new demands for products exhibited on the shelves in supermarkets. Today, marketeers require bags with pictures and/or attractive texts, with nice looking, attractive closures without any single defect.
- The evolutions of the bags themselves in size, thicknesses, quality, materials, coatings, weight, number of plies, and the possible presence of gussets... entail the need of new techniques and new methods.
  Consequently these evolutions necessitate numerous, very careful contacts and exchanges between the marketing departments of the potatoes producers, the bag manufacturers and of course the bag closing equipment manufacturers.

The next developments of this paper will illustrate two key points: 1) Sewing, sealing and gluing offer several appearances to the closed bags, procuring multiple esthetical and mechanical opportunities and 2) Sewing and sealing machines have various technical characteristics procuring solutions for manual, semi-automatic, automatic and even high speed productions. All types of producers from small farmers up to large companies, in different countries, can find an adequate piece of equipment adapted to their bag closing problems with various appearances of closure, for various types of operations in terms of labour force, skill of the operators, number and types of bags to be closed.

'Open mouth' bag closing systems by sewing, sealing and gluing

Sewing remains ideal to close bags of all kinds of material: cotton, jute, burlap, polyethylene, paper, woven polypropylene. Sewing offers today several possibilities giving various appearances of closures:
- Plain sewing: the very simple and well known closure with 1 or 2 threads.
- Sewing with folding the top of the bag, ensuring already less leakage.
- Sewing with crepe paper, possibly with filler cord to fill the small holes created by the needle.
- Sewing with easy open tape, allowing an easy open of the closed bag by the final customer.
- And the new principle: tape over sewing, using a 60 mm wide pre-coated paper tape with either a heat activated hot melt adhesive or a PE coating depending on applications. The tape is here applied after sewing, over the top and both sides of the sewn line of the bag, resulting in a sealed closure that is always strong, clean, durable and sift-resistant.

For manual operations, the Portable Sewing machine remains a very useful tool sold worldwide. This durable, compact, extremely robust machine features a special non-corrosive finish, a reinforced feed mechanism, a presser foot lifter and a lubrication system with oil pump and

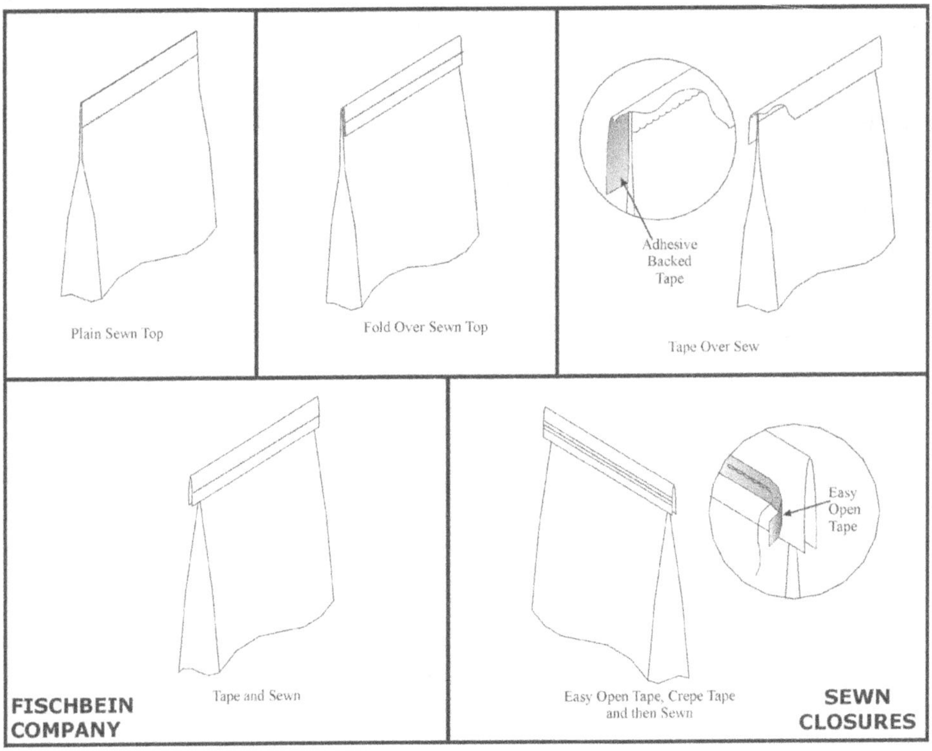

Figure 1. Closures by sewing.

reservoir guaranteeing long life time and performances to 250-300 bags/hour. The body of the portable, with its moving parts, can also be installed on a pedestal to design a complete range of mini-systems for manual up to automatic operations (with an infeed mechanism), allowing an output up to 300 bags/hour. This can certainly be sufficient for some slow packaging processes.

For automatic sewing operations, the most important element of the bag closing line is the Sewing Head itself. Basically, the key characteristics of the Sewing Heads are:

- to be completely sealed in oil, avoiding dust and dirt, allowing a longer life time of the parts, consequently avoiding losses of money;
- a compact design and a robust 'body';
- a very low weight;
- a variable speed, adjustable by pulley;
- a variable stitch length;
- easy adjustments of thread tension;
- a high performance rotary knife: realising a better, one shot, really close to the needle, cut of the thread chain: a great improvement!

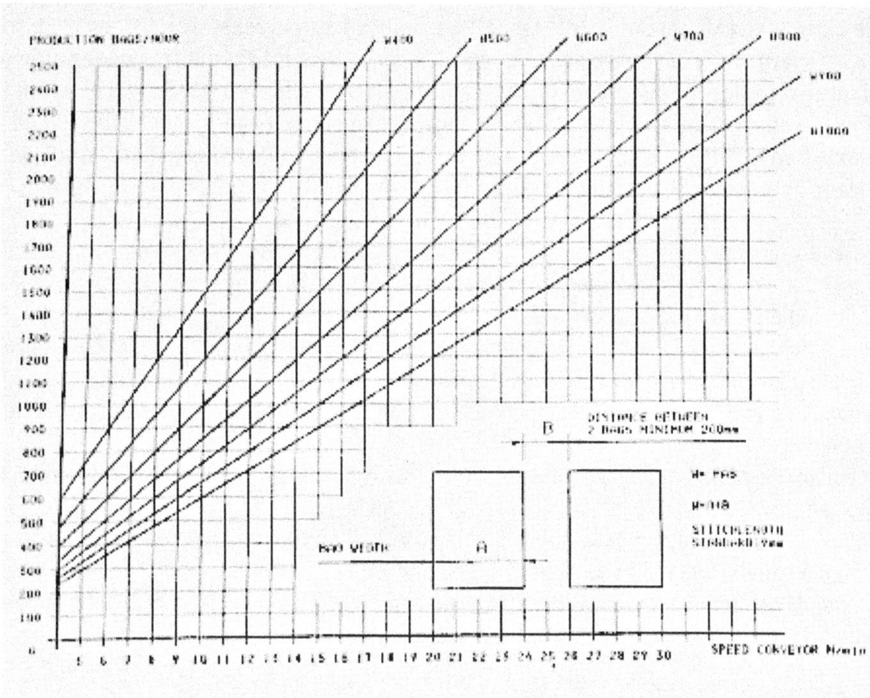

Figure 2. Speed diagram for sewing heads.

Faster speeds can now be reached up to 2600 RPM or 2600 stitches/minute. These high speeds allow consequently large outputs in terms of bags.

To consider a bag width of 400mm, a space between two successive bags of 400mm (total = 800mm); if the requirement is 1500 bags/hour or 25 bags/min, the speed of the conveyor will be 20m/min. It will necessitate, for a standard stitch length of 9 mm, a theoretical speed of the sewing head of 2222 rpm (easily achievable). In practice, 2-3% more, around 2300 rpm. This clearly demonstrates that the maximal output of the packaging line is not at all limited by the bag closing section, but in fact by other components like the weighing scales. An output of 1600 bags/hour can be reached very easily. Other stitch lengths will necessitate other speeds following the calculation:

$$v = \frac{s \times t}{1000}$$

s = stitch length (mm)
v = speed (m/min.)
t = revolutions/min.

These new possibilities of speeds have been obtained by strong engineering efforts.

First, the course of the needle lever stroke was reduced to decrease the forces that cause vibration (primary vibration caused by the needle lever). The stroke reduction does not impact typical applications for the sewing head. Vibration reduction achieved was 25%. This reduction allows the needle arm to be more stable at high speed. Data from in-house tests and the field indicates an excellent reliability at 2600 RPM. The change in stroke was made to address casting inconsistencies (affecting its mass).

Secondly, the revision made to 7 parts provides new geometry for the timing of the needle, looper, feed-dog and proper distance from the throat plate. They were achieved by following/ complementing engineering approaches:

- mathematical analysis of all the functions (to ensure proper timing of all the events);
- kinematic analysis (to establish forces in action);
- materials analysis (to ensure proper choice and behaviour of the alloys selected for each component);
- FEA (Finite Element Analysis, to establish intrinsic properties and reliable behaviour at component level);
- actual on-system vibration analysis to quantify vibration reduction (an accelerometer was used to measure vibration in X, Y and Z axes); and
- high speed motion video to validate interaction among all the significant components (needle, looper, feed-dog, throat plate, presser foot).

As a summary:

- new in depth studies of mass movements;
- redesign of some parts;
- new metallurgic composition;
- redesign of internal oil flow;
- special sophisticated surface treatments like in the car racing.

Consequently, the temperature of the Head is drastically reduced, with the immediate consequence of reduced vibrations and reduced wear of parts. Naturally, sewing systems offer various features depending on the number of bags to be closed, the number of operators, the degree of automation and the type of closure.

Five great categories of sewing systems must be distinguished:

1. A 100% manual operation when the sewing head is suspended to a cable or to the ceiling in difficult environments: here the operator is doing himself all the operations
2. A basic sewing system where the sewing head is on a single leg pedestal: here the already reshaped bag is introduced into the sewing head by hand or automatically. The sewing process starts when the feeler switch is activated. At the end of the process, the operator has to push the thread chain into the knives to cut it.
3. Systems with 'infeeds' for automatic operations where the already reshaped bag is introduced automatically into the infeed.
   What is the role of an infeed mechanism? This is a motorised system designed to bring automatically the top of the bag into the sewing head without any operator. The motor of the infeed is equipped with an adjustable pulley allowing an easy synchronisation with the sewing head and the conveyor belt speeds. A very important characteristic of the infeed is

that the front section is hinged and spring loaded for the passage of various thicknesses of bags, for an adjustable belt tension as well as for the operator's safety. Modern infeeds use 19 mm wide flat belts for a good grip of the reformed bag top instead of dangerous heavy metallic chains. In these systems, the sewing process starts when a photo sensor detects the bag. At the end the rotary knife cuts the thread chain. No operator is needed!

4. The fourth category is represented by high speed sewing systems on a double leg pedestal to decrease vibrations: you have the same sequence of fully automatic operations with a high speed sewing head for performances up to 1600 closed bags/hour.
5. The last possibility is the dual head sewing systems which are a combination of 2 systems, either identical or different. The dual head system is efficient and very flexible. It is the essential element for productions in 3 shifts of 8 hours/day, without any interruption.

In conclusion, sewing remains a very good – and the least expensive! – technique to close 'open mouth' bags, offering various types of closures and various pieces of equipment adapted to various types of operation.

'Open mouth' bag closing by 'Sealing'

Sealing applies to polyethylene bags, or coated, laminated bags or even complex bags with a polyethylene inner liner.

For manual or low speed operations some light, small manual versions can be delivered for applications to close up to 250-300 bags/hour. Among a large choice, some models must be distinguished for their interesting specifications or performances:

- Model Vertipac: double and/or single impulse vertical heat sealer for the heat sealing of large bags made of polyethylene as well as paper with a polyethylene inner liner.
- Model SVV: double impulse vertical heat sealer for the vacuum closing of large bags with or without neutral gas re-injection. The air evacuation is made by two or three (depending on the air volume to suck up) stainless steel retractable side probes. Several cycles to reduce the oxygen level are possible.

All these models require the presence of an operator to close the bags.

For larger outputs, industrial continuous sealers must be considered. Their action is based on some of the principles of heat transfer by convection, by conduction or by radiation:

- Convection relates to hot air sealers where the ambient air is heated by heating capsules and blown to the PE bags to assure the sealing process.
- Band sealers apply the conduction principle where fixed heating elements heat the bag with a contact only through a pair of bands. This is used for coated polypropylene bags, co-extrusions, polylaminates, polyolefin and PVC shrink films.
- Radiation is used when the bag moves through 2 heating elements for special applications like paper bags with a PE coating or a PE inner liner.

For industrial operations, these continuous sealers respond to the most sophisticated requests and offer very interesting features to seal a large variety of bags:

- a temperature strictly controlled within very narrow tolerances, within +/- 2 °C;
- all working parts are fully accessible, reducing maintenance time;

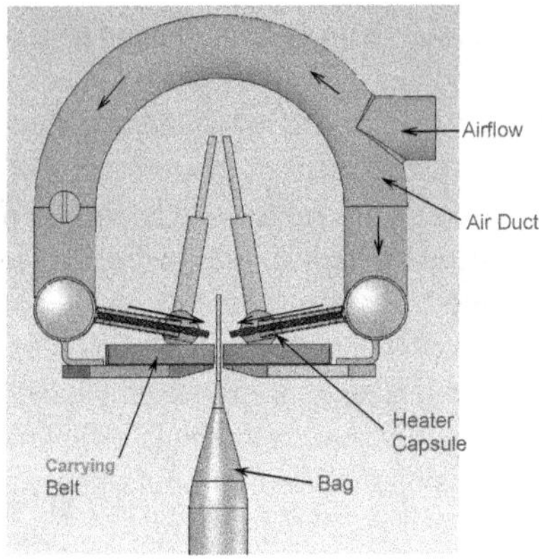

Figure 3. Hot air principle.

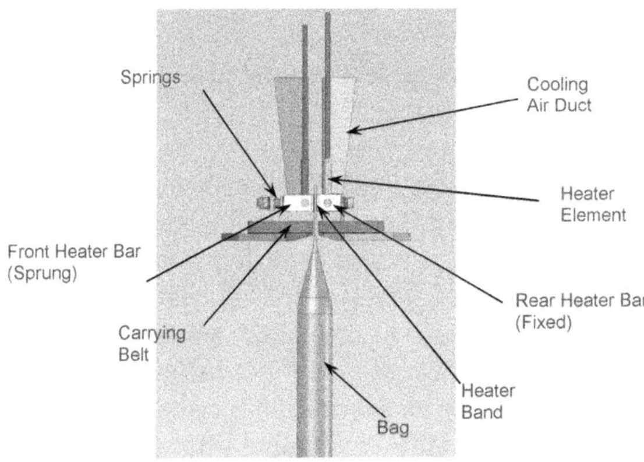

Figure 4. Band principle.

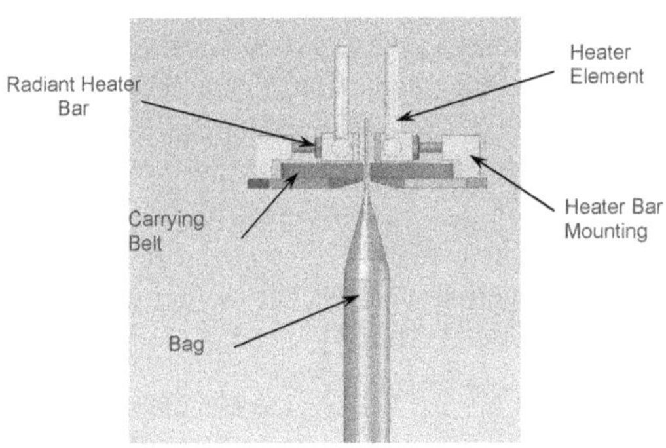

Figure 5. Radiant principle.

- a frame in stainless steel, offering the ability to be used even in environments requesting 'wash-down' procedures
- production rates go up to 1600 bags/hour, with fast change over of seal allowing immediate bag handling and palletization;
- the addition of an air wash bag top cleaner allows to clean the 'mouth' of the bag to avoid the inclusion of dust in the seal and to guarantee sift proof closures;
- the bag top cooler based on the vortex principle allows to decrease very fast the temperature after sealing
- the crimp wheels produce various sealing lines aspects for marketing purposes
- Atex machines are intended to be used in explosive atmospheres, and thus complying with the European Directives for hazardous environments;
- models with a higher power up to 4kW for heavy bags and higher output.

With this variety of industrial sealers:
- the difficult problems of ballooning are minimized;
- the palletising process is facilitated;
- the ability to pack various bag sizes and have quick change over of formats on the same lines is guaranteed.

This enables to respond to almost all applications or requests for sealers in terms of speed (output), automated or manual sealing, aesthetic appearance of the closure, temperature, thickness of bags, polyethylene and/or complex material.

'Gluing' represents a great alternative to close 'open mouth' bags
For industrial operations, when the choice is to close an 'open mouth' paper bag by gluing, two new superb concepts facilitate the operator's life and ensure a sift proof and nice-looking closure for multiwall paper bags.

'Pinch Bag' closer
For pinch bags, the glue is pre-applied during the production of the bag itself, which is a stepped end multiwall open mouth bag. After the filling of the bag, this glue is reactivated by hot air during the closing process.

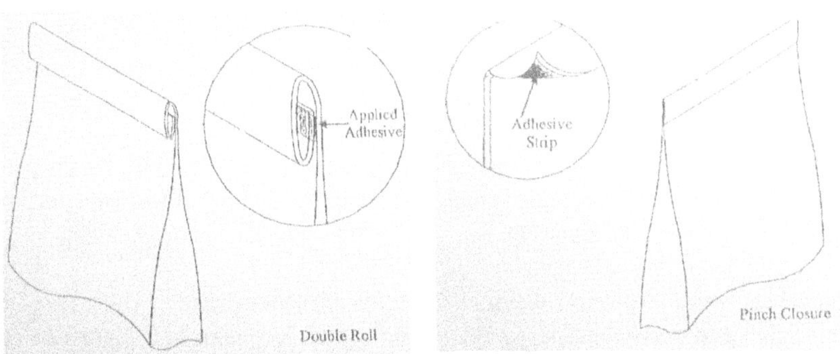

Figure 6. Closure by gluing.

'Double Roll Closer':
The bag top is first trimmed level; all plies are folded a first time and the 'hot melt' glue is applied. All plies are then folded for a second operation on the glue and the bag is compressed through coated belts.

Conclusion

This paper can certainly conclude that the manufacturers of 'open mouth' bag closing machines have recently developed new features and products to meet the new requirements of the potato industry: efficiency, quality, reduced downtime, flexibility, appearance of the bags and to avoid environmental contamination, losses or theft.
Reliable manufacturers certainly commit themselves to continue providing their customers with: Solutions to specific bag closing requirements from an extensive range of top quality equipment; Development using the latest proven technology; A world wide service, owing to a large distribution network; Just in time delivery of spare parts and new machines and Quality products.

Curriculum Vitae – Viviane Mpinga

Viviane Mpinga (38) holds a degree in Languages and Linguistics from the Université Libre de Bruxelles – ULB and a postgraduate in Management of International Trade from the Haute Ecole Francisco Ferrer - Institut Cooremans (Brussels).

She has worked in several international companies (European and American) and joined Fischbein Company in 2001. Since July 1, 2003 she has been promoted to the function of Marketing Manager to help reinforcing operational supports, including a new updated web site and strategic analysis combined with a balanced scorecard implementation. She is in charge of new marketing developments for manual and industrial sealers. She is also conducting marketing studies in the sector of activity of Fischbein, in terms of market segments and new customers.

She has given speeches on 'bag closing technology' in The Netherlands, France and Ghana and has also published, as joint author, articles on the same topic in international magazines.

Curriculum Vitae – René A. Bontemps

René A. Bontemps (52) is a Master of Sciences in Engineering (Ingénieur Civil - Université Libre de Bruxelles – Belgium) and holds postgraduates in Management and in International Trade from the Ecole de Commerce Solvay (U.L.B.).

He has a broad background in international trade (timber, steel) and in the packaging sector with various sales and management functions in several European and American companies active in Europe, Africa and the Middle East.

He is now Vice President – General Manager Eastern Hemisphere Business Unit of Fischbein Company, a world class manufacturer of bag closing equipment by sewing, sealing, gluing and combined means, with manufacturing plants in the USA, the United Kingdom and overseas sales Headquarters in Brussels (Belgium). His sales and market development responsibilities are covering now Europe, Africa, Asia and Oceania.

Since July 1, 2003 he has also been promoted to the function of General Manager of Fischbein-Saxon Ltd in London (UK) to lead the manufacturing and engineering site of all industrial sealers, responsible also for the design and successful launch of a new generation of upgraded sealers.

Profile – Fischbein S.A.

Fischbein S.A. belongs to Fischbein LLC, an American company with its main manufacturing site in North Carolina. The Brussels Headquarters, which were created in 1956, are dealing with the assembling, distribution and marketing of a large range of machines and equipment for Europe, Africa and the Middle East. Open mouth bag closing machines are fabricated, to close jute, burlap, cotton, paper, multi wall paper, polyethylene, woven or non woven polypropylene, net bag, lined/coated papers and foils or laminates. Depending on the material of the bag, different closing techniques are used, such as sewing, hot melt gluing, sealing and combined means. Fischbein S.A. also produces accessories to complete the bagging lines such as bag top formers, tag dispenser, conveyors, bag kickers, bag pushers and bag turner.

Fischbein S.A. deals with customers in all industries that need open mouth bag closing. Some examples from different industrial sections are agriculture (potatoes, rice, sugar), food (premixes, powders, candy, frozen products), chemical industry (pesticides, plastic, fertilizers, soap, explosives, detergents), building (cement, decorative stones, tile glue), minerals (clay, coal, salt, sand, cat litter), pharmaceuticals and medicals (cotton wool, syringes) and others (charcoal, ice cubes, screws, bolts). For each customer we will help to find solutions to the specific bag closing requirements from an extensive range of quality equipment, using the latest proven technology.

Influence of potato growth and storage conditions on the quality of fresh table potato and potato products in the central part of Russia

V.N. Zeiruk[1], K.A. Pshechenkov[1], S.N. Elansky[2,3], O.N. Davydenkova[1] and S.V. Maltsev[1]
[1]All-Russian Potato Research Institute, Moscow region, 140051 Russia
[2]Lomonosov Moscow State University, Biological Faculty, Moscow, 119992 Russia
[3]'AgroKlimatMash' research-and-production company, Moscow, 119331, Russia

Abstract

Twelve potato cultivars with different maturing times (early (Zhukovskiy ranniy, Skoroplodniy, Udacha), mid-early (Belosnezhka, Iliynskiy, Nevskiy, Effect, Bronnickiy, Golubizna), and mid-late (Belousovskiy, Malinovka, Osen)), recommended for central Russia, have been selected by the Potato Research Institute and North-Western Agriculture Research Institute and tested for their yield, quality and suitability for processing (production of potato chips and dried potato mash). The cultivars tested differ in the reaction for reconditioning and blanching. The treatment with Maxim, Kolfugo super, Vist (fumigant), and Sprout Stop preparations provided an improvement of the keeping quality, future yield, and decrease of the reducing sugars content. For central Russia, characterized by sod-podzol soils, the following potato cultivars were the most suitable for processing: Belosnezhka, Bronnickiy, Golubizna, Effect (for light clay soil), and Belosnezhka, Golubizna, Nevskiy, Effect (for moderate loam soil). Tubers, intended for the chip production, should be stored at 6-8 °C and treated with Sprout-stop in January-February. Tubers, intended for the dried potato mash production, should be grown on the moderate loam soil and stored at 2-4 °C.

Keywords: storage, processing, reducing sugars content, processing

Introduction

In recent years a tendency towards an increase of the significance of potato processing into different products is observed in Russia. It is known that the yield, quality, and suitability for processing of potato cultivars depend on the growth conditions, soil type, and storage temperature and duration. In the present paper we are trying to estimate the prospects of cultivating of twelve recommended for central Russia potato cultivars, having different maturing time, for their use in the processing industry.

Materials and methods

The experiments were carried out in 2000-2006 in the Moscow region and involved two soil types: moderate loam (loam) and light clay (clay) soils. The loam has the higher humus (2,4-2,5%) and phosphorus (182-231 mg/kg) content in comparison with clay. The actual acid content (Ph(KCl)) was 4.45-5.25. A standard cultivation technology, recommended for the central region, was used. For the loam and the clay we used the active and passive soil

cultivation, respectively. Potato tubers were stored in refrigerators at 2-4 °C and 6-8 °C (RH 85-95%), and also in potato storage facilities with active ventilation. The laboratory and field tests were performed according Dospekhov (1983) and Dospekhov (1985), respectively. The comparison of different methods (based on the test-paper application) for the determination of a reducing sugars content in tubers was done. Bertrane's method was used as a control one. The best results were obtained for the use of a glukourichrom test-paper, so we decided to use this kind of the test-paper in our experiments.
The influence of different biological and chemical preparations (Maxim, Kolfugo super, Vist (fumigant), and Sprout Stop) on the potato storage and product quality was studied. The taste of table potato and potato products was appreciated using a 9-point scale.
Twelve potato cultivars, recommended for central Russia and having different maturing time (early (Zhukovskiy ranniy, Skoroplodniy, Udacha), mid-early (Belosnezhka, Iliynskiy, Nevskiy, Effect, Bronnickiy, Golubizna), and middle late (Belousovskiy, Malinovka, Osen)) were selected by the Potato Research Institute and North-Western Agriculture Research Institute and tested during our study.

Results and discussion

Production of potato cultivars

Environmental conditions provide the most important influence on the production, structure, and quality of the crop. Compared to clay, the yield of all cultivars on the loam was higher (+0.9-4.6 t/ha). In dry years the yield on the loam was 2-3 times higher, than on the clay. These facts correspond to the results, obtained by Lorch (1948), who wrote that on the clay soil plants require more water, than on the loam soil, and Kiryukhin (1970), who wrote that in the case of a dry July-August the tuber formation is depressed.

All cultivars, except Belousovskiy, Iliynskiy, and Malinovka, formed many stems on the clay. The maximum number of stems was registered for cv. Belosnezhka: 7.0 stem/plant on the clay and 6,2 stem/plant on the loam. The maximum height of plants in the budding phase was registered on the loam. Cvs. Belousovskiy, Malinovka, and Skoroplodniy grew well on both soil types.

The yield structure is very important for tuber processing. The fraction of tubers, diameter of which was 40-80 mm, was higher at the loam soil. For cvs. Belosnezhka, Bronnickiy, Golubizna, Iliynskiy, Malinovka, Osen, and Udacha most of tubers had a diameter 40-60 mm (optimal for the chip production) if they were grown on clay soil. All tubers, which diameter exceeds 40 mm, can be used in the dried potato mash production. The maximum production of such tubers was observed on the loam.

Storage temperature

The storage temperature greatly influences the degree of mass losses, reducing sugars content, quality of chips, and the amount of waste after the abrasive peeling of tubers. This factor has no effect on the quality of the dried potato mash. For all cultivars the storage at 6-8 °C increased the total mass losses 3-4 times, compared to the storage at 2-4 °C. The storage at 6-8 °C provided a good quality of chips without any tuber reconditioning or blanching. The breathing of tubers is the main reason of mass losses at all temperatures. The differences in the mass losses between different cultivars were higher at 2-4 °C. The total mass decrease at

2-4 °C did not depend on the soil type, but at 6-8 °C it was lower, if the cultivars (Belosnezhka, Golubizna, Ilyinskiy, Malinovka, Skoroplodniy, Udacha) were grown on the loam.

Biochemical parameters

In our study we determined two biochemical parameters: the solids content and the reducing sugar content. The solids content did not depend on the storage temperature or the soil type.

The reducing sugar content, measured right after harvesting, met the standard requirements (0.15-0.31%) for all cultivars, except for cv. Belousovskiy. During the storage the reducing sugar content increased; the increase rate was maximal at 2-4 °C, and minimal at 6-8 °C (Figure1). The increase rate did not depend on the soil type.

Suitability for chip production

The amount of waste depends on the level of a tuber peeling. In the case of the full peeling (100%), the amount of waste was 20-30% (depending on the cultivar); if the level of peeling was 60-85% (such peeling is still suitable for the chip production), then the amount of waste was 10-15%. Different storage temperatures also affected on this parameter: it varied from 4.4–17% (2–4 °C) to 8.3–27.3% (6–8 °C).

After-harvest processing of all cultivars grown on both loam and clay soils allowed the production of chips of good quality (>6 points). The best cultivars were Belosnezhka (7.3-7.7 points), Bronnickiy (8-8.7), Golubizna (6.3-7.3), Nevskiy (7.3-8,0), Effect (8-8.3). During the storage at 2–4 °C, the reduced sugar content increased from 0.19-0.32% to 0.27-1.46% in different cultivars. This fact affected the quality of chips, decreasing it by 1-2 points (chips became brown).

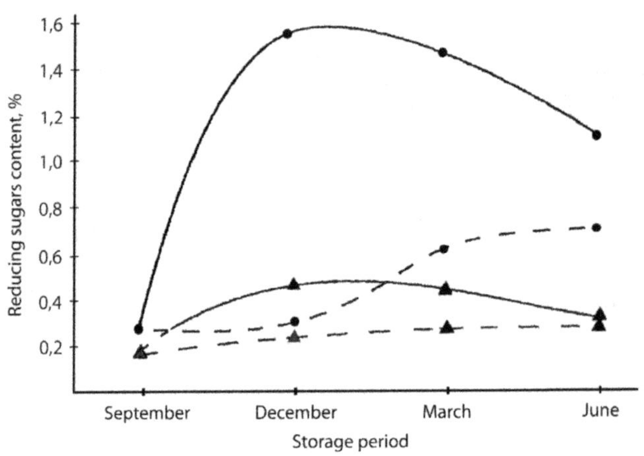

Figure 1. Changes in the content of reducing sugar during the storage of tubers (cvs. Skoroplodniy (●) and Bronnickiy (▲), grown on the moderate loam). ——— storage temperature 3- 4 °C; ------ *storage temperature 6- 8 °C.*

The best chips were produced from cvs. Bronnickiy, Golubizna, and Effect, stored at 6-8 °C and grown on both soil types, and cv. Nevskiy, grown on the loam. This storage temperature initiated early germination of tubers and decreased their turgor. In this connection, the special study of the treatment of tubers with the Sprout–Stop (chlorprofam) germination inhibitor were carried out.

Dried potato mash production

All examined potato cultivars were processed into dried potato mash of good or excellent quality (>6 points). A small decrease in the product quality was observed in the case of the spring processing of tubers, stored at 2-4 °C. Tubers of cv. The taste of tubers of cv. Udacha (cultivated on the clay) decreased during a storage period from 7.4 (autumn processing) to 5.6 points in spring. The spring processing of cvs. Belosnezhka, Belousovskiy, Golubizna, Malinovka, Osen and Effect, grown on the loam, provided the better taste (7.4-8.4 points) in comparison with those grown on the clay (5.6-7.2 points).

Resistance to tuber pulp darkening (after cooking darkening)

Cvs. Belousovskiy, Bronnickiy, Malinovka, Nevskiy, and Golubizna were the most resistant to the tuber pulp darkening independently of the cultivation or storage conditions. A small decrease in the resistance was registered at the end of the storage period for the tubers, cultivated on the loam. Fresh peeled tubers of all tested cultivars were not suitable for the storage for more then 24 hours because of their darkening.

The resistance of boiled tubers of all tested cultivars (except Belosnezhka and Effect) to the darkening was quite good independent of any cultivation or storage condition. Peeled boiled tubers did not decrease their quality after 24-hour storage. Tubers, grown on the loam, showed better results. Boiled tubers of cvs. Belosnezhka, Bronnickiy, Golubizna, Zhukovskiy ranniy, Ilyinskiy, Malinovka, Osen, and Effect demonstrated moderate or good taste.

Chip quality improvement

Reconditioning is one of widely used methods to decrease the content of reducing sugars. In our study this method included the storage of tubers at 20-22 °C for 20-22 days. The application of such procedure provided an increase in the quality of chips, manufactured from cvs. Belosnezhka, Bronnickiy, Golubizna, Nevskiy, and Effect, cultivated on both soils, Malinovka and Osen, cultivated on the loam, and Udacha, cultivated on clay.

Blanching is another method to decrease the content of reducing sugars, including the blanching of tuber's slices (1,2 mm thick) with hot water (80-95 °C) for 2 minutes with the following washing with water. The duration of the subsequent frying should not exceed 4 minutes. Blanching increased the quality of chips, manufactured from tubers of the most of tested cultivars, which have been stored at 2–4 °C for a long time. The blanching was not necessary for the most of cultivars in the case of the after-harvest tuber processing. In this case it can decrease the quality of chips, because they can become gray. The blanching was ineffective in the case of processing of potato, grown in 2002 (dry and hot year).

Therefore, the blanching machine should be included in the processing line as an optional equipment to be used as necessary.

Chemical and biological treatment of tubers

Influence of protecting and stimulating preparations on tubers during the storage period was also studied. Cvs. Nevskiy and Udacha were selected for such study. The autumn treatment of tubers increased their keeping quality. The best results were shown for Maxim and Kolfugo-super preparations (0.2 l/ton). They decreased the mass losses in 2-2.3 times. Total yield of the next season increased by 3.7-5.2 t/ha. The fumigation of potato with Vist (tiabendasole) in the dose 10g/t gave an additional yield +3.7-5.1 t/ha. The treatment with Sprout-stop (dust) in the dose 1.0 kg/t decreased mass losses up to 20% (in comparison with the non-treated control), increased the next season yield by 2.4-5.1 t/ha, decreased the content of reducing sugars, and increased the quality of chips on 2-3 points for all tested cultivars, excepting Zhukovskiy ranniy, Scoroplodniy, and Ilyinskiy.

Conclusions

For the Central Russian (sod-podzol) soils we can recommend to grow the following potato cultivars for the processing: Belosnezhka, Bronnickiy, Golubizna, Effect - for light clay soils, and Belosnezhka, Golubizna, Nevskiy, Effect - for moderate loam soils.

Tubers, intended for chip production, should be stored at 6-8 °C and treated with Sprout-stop in January-February. Tubers, intended for dried potato mash production, should be grown on the moderate loam soil and stored at 2-4 °C.

References

Dospekhov B.A. (1985) Methods of performing of field experiments. Moscow. Agropromizdat. 1985. 351 p. (In Russian).

Dospekhov B.A. (1983) Methods of potato evaluation for the suitability for industrial processing. Moscow. Agropromizdat. 1983. 57 p. (In Russian).

Kiryukhin V.P.(1970) Physiology of potato. In: N.S. Batsanov (Ed.) Moscow. Kolos. 1970. P. 27-41. (In Russian).

Lorch A.G. (1948) Dynamics of potato yield accumulation. Moscow. Selkhozgiz. 1948. 192 p. (In Russian).

Curriculum vitae – Vladimir N. Zeiruk

Vladimir N. Zeiruk studied at the Jitomir Agricultural Institute from which he received a MSc in 1972 and a PhD in biology in 1982. Since 1976 he works as a researcher in the Department of Plant Protection of All-Russian Potato Research Institute. In 2001-2005 he was a leader of the Seed-Growing Department, since 2006 he is a head of the Department of Plant Protection and Immunity.

Dr. Zeiruk is the expert in the field of seed growing, plant protection, immunity, and potato storage. His scientific interests include investigations in plant protection, ecological safety, technologies of storage and pest management.

Development of machine-assisted potato production technology in Russia

S.S. Tubolev[1], N.N. Kolchin[1], K.A. Pshechenkov[2], V.I. Starovoitov[2], N.I. Vereshchagin[3] and S.I. Shelomentsev[1]*
[1]CJSC 'KOLNAG',42, Oksky street, Kolomna, 140402, Moscow region, Russia
[2]All- Russian Potato Research Institute, 23, Lorkh Street, Kraskovo, 140051, Moscow region, Russia
[3]Central Testing Machinery Station, 1 Tsentralnaja Street, Solnechnogorsk, 141500, Moscow region, Russia.

Abstract

There is a notable tendency to convert to machine-assisted techniques in potato production. Investment in potato companies increases requiring increased supply of up-to-date complete potato production machine sets. High-yield potato varieties and an integral chain of home-and-foreign-machine-assisted potato production and storing techniques (including wide-row technologies) help all kinds of growers solve basic challenges of efficient potato production in Russia. It is practical to produce the bulk of potatoes in climatic zones most suitable for the purpose, like the Central zone, Povolzhie (the Volga zone), Western Siberia, etc. In the central zone of Russia machine-assisted 90 cm and 140 cm production techniques have proven to have the advantage in terms of reproduction, sparing use of seeds, starch content, labour costs as well as economic efficiency. There is an overview of circumstances by region in terms of quality potato production potential by means of available logistics resources.

Keywords: machine-assisted technique, complete machine sets, inter-rows, storing

Introduction

Potato consumption is increasing in Russia as it is worldwide, and the demand has to be met both quantity-wise and quality-wise, for fresh ware as well as processed food markets. This makes improving consumer-desired potato characteristics an important issue along with preserving quality during long-time storage, production ecology, reducing yield loss in the 'field-consumer' chain, delivering a wider offer of potato-derived products with minimum consumer spare time requirement.

There is considerable variation in agro-ecological conditions in Russia. Therefore potato production and storage technology should be zone-defined, taking into account specific circumstances of very different regions of the country and making possible to use the right varieties (home-bred, in the first place) for specific outlets with specific vegetation timing, a choice of row distances, planting schemes and a suitable machinery and equipment complex.

Specific features of potato production in Russia

The yearly potato production in Russia totals ca. 35 million tons and cropping an area of about 3.2 to 3.5 million ha. Unlike world practices of relatively uniform potato areas, the potato crop in Russia is grown in a wide variety of agro-ecological conditions covering lowlands and highlands all the way through from Kaliningrad to Kamchatka crossing more than eleven time zones. Potato planted areas also vary greatly latitude-wise – from the Extreme North with long daylight hours to the Northern Caucuses with a dry hot climate. This peculiarity accounts for variations in potato production, harvesting and storage techniques given specific climate and soil characteristics. A combination of the above-mentioned features dictate the design and assortment of machinery and equipment for various functions and capacity.

Another peculiarity of current potato production in Russia is a large share of the crop grown by small holders and low-scale farmers on small fields. In existing circumstances this potato production sector is likely to last long. These growers need a complex of compact machines adapted to their current specific needs and suitable for future growth as well.

At the same time there is a notable tendency of mainstream potato production to shift towards machine technology in response to big business taking more interest in agriculture, less number of employees therein, a sharp increase in labour costs and a wider improved offer of machinery and equipment. There have been set up quite a number of large-scale professional potato growers with 600–1000 hectares and more. There is increased investment in developing such farms and an increased need for appropriate potato production machines.

Agro-ecological zonation of potato production Russia can be expressed by four categories each having a choice of soil types:

- Category I – Central region, Volga-Vyatka region, the Urals, West and East Siberia. The available soil types include sod-podzol, grey forest, partially black, texture-wise they are clays, loams, light loams and loamy sands. Hereto belong the majority of potato production sites in Russia, accounting for about 50% of the total area.
- Category II – Central black soil and Povolzhie regions, the Northern Caucasus. About 80% of soils in this category are black, texture-wise clays and loams. In several regions potato vegetation can be exposed to drought. Southern regions have best chance of production early potatoes. The category accounts for about 30% of the total potato area.
- Category III – the North-West. The soils are mainly podzol and sod-podzol of different texture and poor nutrient content, for the most part stony. The category accounts for 15% of the total potato hectares.
- Group IV – the Far East. The soils are light, sod-podzol, medium and heavy loams, mainly overwet, disposed to swelling and compacting. The category accounts for about 5% of the total potato area.

Studies on machine technologies

The complete system of machinery and equipment for machine technology of potato production and storage is a combination of 2 sets: field machinery and equipment and stationary machinery and equipment. One of the basic parameters of the field machinery set is the row width of potato spots. The mainstream potato crop in our country is grown with 70 cm row distance.

There has been a long-term comparative study and trial on machine potato production with row widths of 70, 75, 90 and 140 cm (1-line or 2-line by 110 + 30 cm layout) in the central zone of Russia. Table 1 gives an indication of the results of the study.

Large-scale comparative research has shown that row distances should be dictated by agro-ecological zonation of Russia. Moreover, several planting schemes can be applied within one climate zone. In a number of areas conventional 70 cm row distances will remain in use, for instance, by small holders cropping limited growth habitus varieties on small plots of poor nutrient land in dry climate.

Study and trial of 75 cm row width technology show that compared to 70 cm row width the former helps achieve higher-volume ridges, use more powerful tractors and stimulate growth and evolvement of potato plants. For many conditions there is a good chance to use wide row distances of 90 cm and production potatoes on 140+ cm row width beds in one or two lines spaced 40 cm apart. Bed technique was originally used in the Far East with substantial rainfall late in summer nearly every year. These conditions lead to mainstream potatoes grown at row widths of 70 cm being exposed to suffocation. Harvesting is done with considerable loss. 90 cm row width gives fair results. According to DalNIISH (Far-east institute for research in agriculture) production potato at a row width of 90 cm and on beds notably improve quality of tubers and increase yield by 22–35% in Category IV farms. However, planting the potato crop into pre-formed ridges at a row width of 70 cm should not excluded on light flood lands.

Wide row distances have had maximum effect when cropping intensive potato varieties on highly cultivated soils. Beds have been effective both on overwet and dry soils. They increase field machinery capacity by 25–28%, which is especially important for sites with a short production season and little favourable weather conditions during harvesting.

According to VNIIKH (All-Russian institute for research in potato production) compared to 70 cm row width average yields cm in Category I farms increases by 7% at 75 cm row distances and

Table 1. Performance of field machinery sets for different row width potato planting technology.

Location	Row width, cm	Total hectareage	Seed demand, t/ha	Yield, t/ha	Soil type
Moscow region	75	385	2.6-4.5	23.0-28.0	Light loams
Kaluga region	70	500	3.1-3.9	14.9-32.2	Light loams
	140	300	2.5-3.2	11.9-33.8	Light loams
Moscow region	70	500	4.3	22.9	Peat
	140	500	1.5	27.7	
	90	300	3.6	42.2	
	75	800	3.1-3.3	27.9-37.8	Sandy loams
	90	500	2.7	24.2	
	140	50	1.7-1.8	21.1-31.4	

by 15,8% at 90 cm row distances. This is while total production costs are minimum on beds at a row distance of 140 cm and total 38 kRUR/ha, at 90 cm distances they total 48 kRUR/ha, and 68 kRUR/ha at 75 cm row distances.

Basic zone variations of field machine technology

Operations common for all technologies by zone include: autumn disking or breaking stubble remains and applying organic manure as well as phosphor and potash fertilizers. Instead of organic manure green manure such as rape, brassica, winter rye can be applied on poor soils. Protection measures are also the same. The other machine technology operations of potato production in a wide variety of agro-ecological conditions in Russia can be reduced to the following basic variations, depending in the first place on soil texture type.

- Light sandy loams. In autumn these are chisel ploughed with boardless ploughs or stubble breakers. In spring shallow oval-shape ridges are formed along with local application of fertilizers. In case organic manure is used it is applied in a conventional way. Straight afterwards, seed tubers are planted into the ridges. In 7–10 days' time (depending on weather conditions and weed emergence) the ridges are tilled along with hoeing the furrows and forming ridges with passive tools two times with a gap of 6–7 days and if necessary for a third time after emergence. When the haulms are 20–25 cm high the first earthing up is done along with chisel furrows ploughing and if necessary the final earthing up is done before haulm closure. Herbicides are not found in this technology.
- Loams. In the zones with stable snow cover, long frost and soil humus content of up to 3% turn underwinter ploughing is applied for a depth of 25–27 cm, and in spring they apply a speed harrow for a depth of 14 cm. On large spots in order to optimally time planting of potatoes field marking is done along with local application of fertilizers. After emergence full-volume ridges are formed and in 2–3 days' time or just before emergence the ridges are treated with herbicides. The number of protection operations depends on weather conditions and variety. The haulms are destructed 10–12 days before harvesting.

In the southern regions with unstable snow cover, fluctuating winter temperatures and humus soil content of over 3% after the underwinter ploughing in autumn the ridges are formed with passive tools along with local application of fertilizers. In spring seed tubers are planted into ridges without seed bed preparation, and the ridges may be hilled up before planting only if they are heavily silted-up after winter. Further technology is the same as in the previous production scheme. In some cases the same techniques can be applied here as on light soils.

Specific features of harvested potato yield transportation

Basic requirement for vehicles used during harvesting is minimum damage to the transported potato tubers and maximum capacity for less waiting times at unloading and loading. Tipping heavy trailers with lifted rear board significantly reduce damage to tubers.

Tipping trucks, trucks with (tipping) trailers, tractors with tipping trailers and box-carriers (when harvesting into boxes) are all used for machine harvesting. Some farms use tractor trailers with a moving floor based on the design of fertilizer spreader RUM-6. Recently bigger volume vehicles are widely used for more efficiency.

In case there is little soil content in a good yield bulk, potatoes can be loaded from under the harvester straight into boxes put on special carriers for direct transportation to the box store.

When harvesting potatoes by diggers the tubers are picked up into bags, nets or boxes and taken from the field with tractor trailers, trucks and box-carriers. Experience shows that use of diggers for harvesting results in 2.5-4 times more damage to potatoes as compared to machine harvesting. Yield loss can go up to 30% in unfavourable autumn conditions.

Specific features of potato storage

At the moment large farms in different regions of Russia upgrade the technical basis for potato storage. There are both domestic and foreign projects of state of the art stores, combining post-harvesting product handling, storage and pre-sales finishing. Available storages are renovated, domestic and foreign active ventilation systems with automatic storage mode control are installed. New bulk and box stores are built (the latter most comfortably use rigid one-piece 'euroboxes').

Harvested potato storage techniques vary according to tubers quality and type of store. Storage starts with post-harvest handling followed by putting the potato into the store. In Russian circumstances tubers are very often not fully matured by the time of harvesting, their skin is too soft and they are often disposed to suffocation. In unfavourable conditions cold and wet tubers have increased turgor and are easily damaged by harvester tools as well as during transportation. Autumn sorting alone has been shown to increase mechanical damage to tubers by 20–25% and by 5-6 times susceptibility to diseases during storage.

Therefore direct 'field-to-store' practice should be preferred to put potatoes into the store. Potato bulk from under the harvester should ideally contain less than 15- 20% soil and less than 5% diseased tubers. If the content of trash is greater than these the incoming bulk of potatoes are given basic handling – they are put into the store along with soil and undersize extraction by means of the specialized line of appropriate machinery. With low quality yields (badly affected by *Phytophthora*, suffocation and bacterial rot) a buffering stage should be applied keeping the produce for 2-3 weeks in a shed prior to loading it into the store. After the so-called buffered storage the tubers are inspected, graded if necessary to then be put for long-term storage. While loading the bulk of potatoes in the store they should be dried by ventilation if necessary and therapy storage mode should be applied.

Direct instoring practice is more feasible in the regions where potatoes have enough time to fully mature by the time of harvesting in dry and warm weather and are not affected by *Phytophthora* or other diseases. Grading and quality inspection is done at the end of the storage by way of pre-sales finishing of the tubers.

Development of early potatoes production techniques

In the southern regions of potato-production Russia (Category II) there is a tendency to develop production of early potato. Production early potatoes for sale in big cities is an important issue allowing to reduce import of too expensive early potato from abroad. However current early potatoes production practices require too much labour. Therefore it is feasible to use foreign know-how and develop machine techniques for the early potato production ensuring efficient sprouting and mechanical strength of sprouts making it possible to use the semi-automatic potato planter with manual delivery of tubers into planting belt cups. Available cup planters break up to 10–15% and more causing delayed emergence and yield decreased by 20–25%.

Application of machine techniques

Despite the challenging situation in the Russian agricultural sector potato production practices preserve development potential. This is observed in large farms classified by the Ministry of Agriculture of Russia as top potato producers (see Table 2).

These farms deal with production, storing and sales of potato including potato-derived processed food. Efficient quality potato production for different outlets is made possible by using state of the art domestic and foreign machines with the yield storage on the site. There has been gained positive experience of high potato yield production (including loamy soils) based on up-to-date machine practices.

Table 2 suggests that the majority of farms (70.0%) grow potatoes most efficiently on sites up to 300 hectares. This is while the majority of the farms (83.4%) belong to Category I and II zone-wise, enjoying conditions most favourable for potato production. With local variations techniques used are based on the production schemes described above.

Potato yield largely depends on seed tubers quality. Long-term research by VNIIKH shows that seed tubers preparation technique should include soil and trash extraction, grading, spring quality inspection, instore treatment with protectants and stimulators before planting and

Table 2. Basic performance of large Russian farms by potato hectareage.

No	General performance by farm category		Category of farms by size (ha) and performance				
	Description	**Value**	**1 -100**	**101 -200**	**201 -300**	**301 -400**	**401 -600**
1	Number of farms, units	96	11	33	37	10	5
2	Area under crop, ha/%	21,227 (total)	927/4.4	4979/23.4	8946/42.2	3408/16.1	2967/13.9
3	Average area under crop per farm, ha	221.1 (for all category)	84.3	157.5	253.2	352.9	617.6
4	Average yield, t/ha	20.8	24.7	25.2	17.7	18.3	18.1
5	Average economic efficiency, %	69.6	72.4	75.8	71.1	53.7	50.0

if necessary pre-planting warm-up. The issue of quality seed tubers supply can be solved by setting up well-functioning all-Russian system of seed tubers production.

Many large farms use foreign-bred varieties. In the Russian circumstances these varieties require at least 6-10 treatments against Phytophthora and other diseases, which fact increases pesticide stress and increase pollution. It is further aggravated by the necessity to spread more fertilizers. Given environment protection requirements as well as high costs of fertilizers it is more efficient (with application rate two times reduced and yields maintained) to apply the latter locally by means of the planter. In some areas potato plants during tuberisation and tuber bulking suffer from lack of water and relatively high air and soil temperatures. Therefore production techniques especially those used in Category II farms should provide for irrigation, including trickle (subsoil) irrigation.

Efficient use of machine techniques requires timely performance of all operations against optimal agricultural terms. This is where the issue of capacity (working width) compatibility of machines used for production and harvesting potato comes into fore. For small potato growers it is feasible to apply techniques based on two-row planters and one-row harvesters. For large growers four-row planter and two-row harvester is preferable.

Use of elevator-type harvesters unloading tubers into trailers driving along side is not likely to spread widely in Russia because of the non-availability of special heavy potato transporting vehicles (20–30 tons) with a moving floor.

It is worth mentioning that there is much development abroad in designing high-capacity self-propelled potato harvesters (two-row bunker-type with a volume of up to 12 tons and four-row elevator harvesters). Their use in Russia should be economically efficient. Conventional potato production set of field machines was pulled by category 1.4 wheel tractors. The more efficient and power-consuming machines (cultivators, planters and high capacity bunker harvesters, heavy trailers) are pulled by category 2 tractors, the latter forming a new power force for up-to-date field potato production machines. However in this country category 1.4 tractors remain basic due to a large share of heterogeneous spots and possibility to use passive tools on light soils.

Speaking about economic efficiency one of the Category II farms obtained a yield of 2822 tons of ware potatoes at 120 hectares (row width of 75 cm, light loams). The field costs are 3.49 RUR/kg. With a sales price of 5.00 RUR/kg the efficiency is 43.44%, and the payback period for optimal set of field machines is 1.61 years. Similar figures have been obtained for minimum required set of field machines: potato cost price of 3.32 RUR/kg, economic efficiency of 50.54%, machines payback period 1.17 years.

Storage costs depend on the capacity and type of the store, storage period and quality, tuber treatment and taking the product out of the store, etc. For instance, in a bulk store estimated depreciation rate on machines applied is relatively low and totals 0.46 RUR/kg for 2000-tons store, 0.23 RUR/kg for 4000-tons store and 0.18 RUR/kg in a 5000-tons store. The payback period for machinery and equipment is 1.2–2.5 years.

KOLNAG offer of machines complex for potato production

Machine techniques provide basis for a solid and efficient supply of quality ware potatoes both for the home and foreign market, and raw material for production of an assortment of potato-derived food. Estimated effect of widespread machine techniques for potato production and storage with the yearly domestic demand in potatoes for different outlets totalling 42–45 million tons (expert evaluation) show their high efficiency (without taking into account release of nearly 1.5 million hectares of arable land due to potato yields increased up to 25–35 t/ha). Share of large potato farms may increase up to 40–60% against current rate of 7%, yield loss is likely to reduce to 6–12% (i.e. up to natural loss level), share of processed potatoes should go up to 15–20%.

Realization of machine techniques for potato production and storage in the current variety of agro-ecological conditions is made possible through the producer carefully selecting the best set of appropriate types of machinery and equipment complete with replacement tools and adjustable working settings factory preset for particular operation conditions. Many machines allow automatic control of settings during operation.

JSC Kolnag offers an up-to-date machinery and equipment set for machine techniques of potato production and storage. It comprises a field set for potato production and harvesting at a selection of row widths and a set of stationary and mobile storage machinery and equipment. Composition of Kolnag set of machines (number of machines and devices, their types and available replacement tools) produced in cooperation with AVR bvba (Belgium) and Miedema bv (Holland) is laid out in Table 3.

Field set ensures soil tillage and seed bed preparation, followed by planting, treatment and harvesting quality potatoes on 100–120 ha sites with minimum labour and power cost including most promising 90-cm row distances.
Machines making part of the field set including rotary cultivators ensure mechanization of all potato production stages and seed bed preparation for effective harvester operation.

Table 3. Composition of machinery and tools complex offered by JSC Kolnag for potato production and storage machine techniques.

Description of sets within the complex	Composition of complex		
	Number of basic models of machinery and equipment	Number of types of machinery and devices	Number of available replacement tools
Field set (optimal/minimum)	6/4	6	over 20
Stationary and mobile store machinery and equipment set	9	34	over 110

The choice of available replacement tools allows to use these machines for onion and root vegetables production as well.

Stationary and mobile machines and equipment set for overall mechanization of potato store operations varying in type and capacity as seen in Table 3 is also offered in a number of modifications complete with an assortment of tools. Composition of the complex allows its best use for realisation of flexible machine techniques in various environmental conditions of Russia.

Conclusion

Russia is one of the top potato producers worldwide. There has been gained domestic experience of efficient potato production, with good use made of foreign achievements. High-yielding adapted potato varieties and up-to-date machine techniques, including wide row distances all join to solve issues of effective potato production for different outlets, fully meeting consumer quality requirements in all kinds of farms irrespective of size. A solid basis has been formed for machine techniques application on large farms – state of the art potato production machines complexes offered by Kolnag. It is necessary to extend research and development of machinery and equipment sets for potato production, harvesting and storage for different outlets as adapted to various soil and environment conditions given demand for such machines on the part of small growers. In view of ecology requirements it is feasible to more widely use local application of fertilizers and minimize number of plant protection treatment. Aiming at more effective use of machines in potato production with minimum labour it is feasible to develop potato production in areas most suitable for the purpose according to their agro-ecological conditions.

References

Anisimov B.V. Potato farming. The intensive way of farming. Картофель и овощи, No. 1, 1996, c. 3 (pp.).

Anisimov B.V. Which potato varieties to choose? Картофель и овощи, No. 5, 1996, c. 4-6 (pp.).

Kolchin N.N. and others. Technology and machinery for basic crops farming. Часть I. Картофель. – М.: ИНФРА – М, 1997. – c. 104 (pp.).

Kolchin N.N. Machinery for potato and vegetables harvesting and processing. Научно – аналитический обзор (По материалам Международной выставки 'SIMA – 2005'), А. А. Ежевский и др., гл. 1. 8, с. 121 – 134 (pp.), Москва, ФГНУ 'Росинформагротех', 2005.

Kolchin N.N., Fomin S. L. Potato storing: condition and development trends. Картофель и овощи, No.1, 2006, c. 28 – 31 (pp.).

Potato farming intensive technology/ Gained by Pshechenkov K. A. – M.: Росагропромиздат, 1989, c. 304 (pp.).

Potatoes of Russia/ Edit. by Korshunov A.V. Научное издание (Монография). Starovojtov V.I. and others, т. 2 – М.: ВНИИКХ, ООО Редакция 'Достижения науки и техники АПК', 2003, с. 321 (pp.).

Pshechenkov K.A. and Davydenkova O.N. Technology of potato seeds preparation and planting. Тракторы и сельхозмашины, No. 8, 2002, c. 33 – 36 (pp.).

Pshechenkov K.A. and Davydenkova O.N. Potato storing: storages renovation. Картофель и овощи, No. 6, 2005, c. 5 – 7 (pp.).

Roslov N.N. Potato and vegetables: storing, cleaning and processing. МСХ РФ, ФГУП 'Гипронисельпром', г. Орел, 2002 г., с. 229 (pp.).

Simakov E.A. and others. Russian potato varieties (catalogue). ВНИИКХ, Россельхозакадемия, 2005, с. 126 (pp.).

Starovojtov V.I. Wide-row technologies and machinery for potato farming. Техника и оборудование для села, No. 11, 2004, с. 16 – 20 (pp.).

Spaar D. and others. The potato/ Edit. By D. Spaar. – Торжок; ООО 'Вариант', 2004, с. 466 (pp.).

Struik P.C. and Wiersema S.G. Seed potato technology. Wageningen Pers, Wageningen, the Netherlands. 1999, pp. 383.

Tubolev S.S. Home potato farming requires modern machine-assisted technologies and machinery. Картофель и овощи, No. 6, 2006, с. 2 – 3 (pp.).

Veresharin N.I. and others. Modern machine-assisted technologies in potato farming. Техника и оборудование для села, No.8, 2004, с. 16 – 19; No.9, 2004, с. 12 – 14 (pp.) (продолжение).

Curriculum vitae – Sergey Tubolev

Sergey S. Tubolev, born on 3-d November, 1949 in Tulskaja oblast, Russia. He graduated from Tulsky Polytechnical University in 1973. After graduation he worked for the defence industry. In 1995 he was appointed the General Manager of JSC Kolnag. As a Head of the manufacturing company he is in close cooperation with his colleagues in the Netherlands, Germany, Belgium and France. He is engaged in creating of new machinery and equipment for agriculture. He is the editor, author and co-author of numerous scientific articles and presentations on conferences. He received 10 certificates of authorship. His focus is on further developing and manufacturing of potato and vegetable farming machinery and feeding equipment.

Company profile – CJSC Kolnag

Joint Stock Company Kolnag, established in 1995, produces reliable up-to-date agricultural machinery in Russia under licenses of well-known western manufacturers.

Over eleven years JSC Kolnag has run all the way from final assembly of three machines to complete production of fourteen machines and their modifications with 80% isolation level in two main divisions:

- potato and vegetable farming machinery;
- equipment for preparing and distributing animal feed.

All Kolnag-made machines have been successfully tested and certified. They are included into 'The State Register of Machines for Growing, Storing and Processing agricultural products'. Kolnag's research and development activity is based on interaction of the main stages of machine-assisted potato growing technologies: production, storage and marketing. Kolnag participates with Russian scientific institutes in various networks and projects that deal with the research on contemporary machinery technologies and new potato and vegetable machinery.

Kolnag conducts guarantee and post-guarantee service over the whole operation term of machines. Kolnag also supplies parts and spares ExWorks Kolomna, Moscow oblast. Well-qualified service specialists are available for instruction on operation particulars of the delivered machines and check-out at the field site.

JSC Kolnag is an ambitious grower, which thinks globally. That is why it enters new strategic alliances both with Russian and foreign companies. In recent years Sales Offices were established in many parts of Russia. While an export sales team covers areas of Western Europe. JSC Kolnag is an active member of Agricultural Machinery Producers Union of Russian Federation.

Seed potatoes

Seed potato in Russia: production, market and quality system development

B.V. Anisimov, A.I. Uskov*, Y.A. Varitsev and S.M. Yurlova*
All-Russian Potato Research Institute, 23, Lorch Street, Kraskovo, 140051, Moscow region, Russia

Abstract

The annual demand of potato planting material in Russia is evaluated at 9 million tonnes for the total planted area of 2,9 million hectares including agricultural enterprises, peasant farms and small private family farms. But the volume of certified seed potato is not sufficient now. As a result many potato producers use non-certified seed potato badly affected by virus, bacteria and fungus infection. That is why one of the most vital factors of yield increase and efficiency of potato production is forming the seed growing systems at Federal and regional levels. It is necessary also to bear in mind the vast diversity of climatic conditions as well as not to forget the main factors which most influence the quality of seed potato material in places of its production on the territory of the Russian Federation. The new project was developed for quality improvement of seed potato in Russia at present. Very important part of this project is the Bank of Healthy Potato Varieties (BHPV). This Bank is located in the Northern area (Archangelsk region) and maintained under the cleanest phytosanitary conditions. The fund of original free of viruses seed material based on BHPV was created in Russian regions where the level of infection load is minimal. This system allows to use the advantages of favourable conditions with the complex application of modern methods and seed potato technologies and high performance laboratory tests of virus and other pathogens. Systems of virus detection, norms and methods of laboratory tests as well as requirements to the quality and tolerance level for different seed classes (generations) in general are unified and harmonized with European ones, which are adapted by the UNECE for the European Union.

Keywords: production, market, certification, standard, ELISA

Introduction

The low potato yield in Russia is due to a complex of economic factors as well as biological peculiarities of that crop. In technological process of potato production the breeding and seed growing are taking the leading role since being a vegetative reproduced culture potato rapidly accumulates infection during the reproduction process. One of the main factors determining the low level of potato yields in all Russian regions is low quality of seeds and its infection by pathogens.

The situation worsens due to the shortage of production of planting material grown under favourable conditions free of viruses, also due to the lack of technological regulation of production of original seed as well as elite classes and reproductive seeds. As a result many producers use very old potato generations badly affected by fungus, bacteria and viruses. That is why one of the most vital factors of yield increase and efficiency of potato production is the formation of a seed growing system at Federal and Regional levels. It is also necessary to bear in mind the vast diversity of climatic conditions of our country as well as not to forget the main factors which influence most the quality of seed potato material in places of its production on the territory of the Russian Federation.

Seed potato classification and production system

International potato seed trade increasingly develops. Therefore it is necessary to unify and harmonize the Russian seed potato classification system with the European one, which is adopted by UNECE for the European Union, and with other similar national systems, which are used in most Western European countries.

According to the 'Breeder Achievements Law' (1993) and 'Federal Seed Law' (1997) in Russia seed potato is divided in the following groups (categories): virus-free stocks and micro-propagation material; original seed material; elite seed classes; reproductive seeds. Seed potato classification and production scheme of original, elite and reproductive classes in Russia is shown in Figure 1.

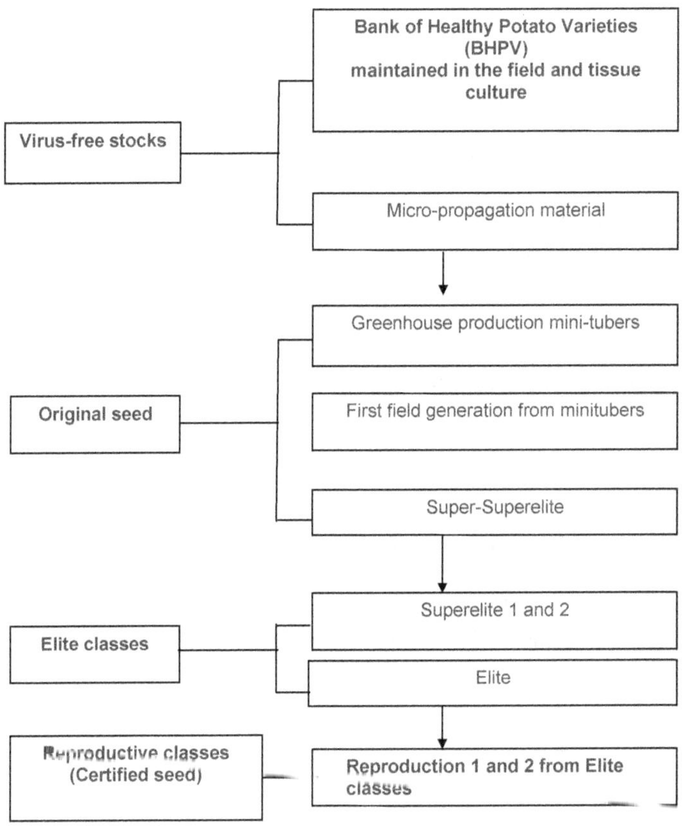

Figure 1. Seed potato classification in Russia.

Virus-free stocks

Virus-free material is produced by tissue culture and micro-propagation for the starting point of Russian seed potato production. Healthy mother potato plants are selected in the field or derived on the basis of meristem culture. The Bank of Healthy Potato Varieties (BHPV) is maintained in the field under cleanest phytosanitary conditions for viruses free material production.

Original seed

Original seed material includes three classes / generations: mini-tubers, 1st field generation from mini-tubers and Super-superelite class. Mini-tubers are produced in greenhouse or hydroponic facilities. Next two generations from mini-tubers are produced in the field. In praxis, actually only two of these classes of original seed are produced in the field because short production chains guarantee better quality of Super-superelite class.

This generation corresponds to high standard requirements regarding cultivar purity and minimally allowed levels of viral infections. In this respect the super-super-elite generation can be recognized as 'S' class (2nd field generation) of the European classification systems.

Original seed material is produced by originator or his representative person. The originator is an organization or person, which created, bred, revealed and/or provides maintenance of a variety, listed in the official List (National Catalogue) of breeding achievements, which are allowed to be used.

Elite classes

Elite classes is seed material derived from original material and corresponds to state standard requirements and other normative documents of seed production. The number of elite seed generation is defined by the variety originator. Because of potato reproduction features, the numbers of elite seed generation can't be more than 2-3 field generations after super-super-elite. Therefore super-elite-I, super-elite-II and elite seed class are included in this group. These seed classes conform to 'SE' and 'E' classes of the classification system on the basis of field generation in most European countries.

Reproductive seed

Reproductive seeds are the next field generations from elite seed classes. For certification the inclusion of this category of 1st and 2nd reproductions after elite seed class is most reasonable. There is no point in the use of more than the 1st and 2nd field generations after elite, because of progressive accumulation of viral and other diseases in the vegetative multiplied potato. As a result the qualitative characteristics are degenerated and productivity drops with each next generation. Therefore the seed material of 3rd and following generations after elite can not be certified.

In conclusion, the division of Russian seed potato material to classes/generations is corresponding to analogical classification systems of most European countries. Differences are mostly in terminology of classes and names of categories.

Cultivar resources and market

From 2001 the legislation allows the certification of seeds of varieties specified in the official List of varieties admitted for cultivation and use in the Russian Federation. The quantity of varieties in the list is increasing constantly. There are 200 potato cultivars present in the Russian Official List in 2006 which consists of 121 Russian cultivars (about 60%), 15 Belarusian, 4 Ukrainian, 25 Dutch, 25 German, 8 Britain and 1 Finnish cultivar.

Prevalence of the early and medium late maturing cultivars meets the requirements of the agro-climatic conditions of most country regions. The vegetation period of potato plants is limited by the late cultivars that do not have enough time to reach maturity. Immature tubers can be seriously damaged at harvest and heavy losses may happen at the storage of such tubers.

The analysis shows that the leading potato cultivars in Russia are Nevsky and Udacha. In the commercial seed production these cultivars take the primary position as to the original, elite and reproduction seed material production and they comprise more than 60% of certified seed of Russian potato varieties. The share of planting material of Dutch and German varieties and that of other countries is more than 40% (Tables 1, 2).

All foreign cultivars included in the State register are of prime interest mainly because they are resistant to the golden potato nematode and to some viruses and their complex, and they are suitable for potato processing (for dry potato puree, French fries, crisps, etc). Since a few years ago Red Scarlet is very popular variety in Russia. This new early maturing variety with red skin and resistance to diseases would be very successful and very promising for the Russian potato market.

The annual demand of potato planting material in Russia is estimated at 9 million tonnes for all categories of potato producers, including agricultural enterprises, peasant farms and small private family farms. But the volumes of production of certified seed potato is not sufficient now. The volumes of certification of potato planting material are increasing annually and here things look promising.

Table1. Volumes of certification of potato planting material in Russia, thousand tons (according to the data of the State Seed Inspection of Russian Federation, 2006).

	Certified seed	Original seed	Elite classes	Reproductive seed
Total	874.2	18.8	60.2	795.2
Russian cultivars	472.2	13.4	38.6	420.2
Foreign cultivars	402.0	5.4	21.6	375.0

Table 2. Varieties, leading in seed production volumes in 2006.

Russian varieties	Quantity, t
Nevsky	263,519
Udacha	105,766
Elizaveta	20,203
Zhukovsky ranny	18,326
Volzshanin	7,729
Foreign varieties	**Quantity, t**
Romano	77,735
Rosara	54,760
Red Skarlet	47,377
Saturna	22,498
Karatop	22,255

Thus varieties selected by Russian breeders are the main of source in potato production of the Russian Federation. The foreign varieties on the Russian market of potatoes constantly increase. And it is a positive factor showing the integration of the national seed potato market in the international seed potato market.

Seed potato prices in Russia are 1.5 times higher than table potato prices especially when traded via wholesale buyers (intermediaries). In the spring time the prices go up to 20-35 RU per kg because the cultivar's popularity influences the price more than the seed potato category and class. National seed potato prices are very often higher than foreign prices due to a wide range of intermediaries who first buy them. For example the foreign company seed potato price for the class A (which is the 1st generation after elite) comprises 350-380 EUR per tonne (i.e. 13-15 RU per kg). This is a reasonable price for the Russian potato producers through direct deals on the market because selling seed potato through intermediaries always doubles prices.

Nowadays it is most important to create an infrastructure to achieve a rational seed potato market.

Quality control

The quality control, including field inspection, laboratory tests and final tuber inspection before definitive acceptance and packing of seed potato, is required for all seed potato classes.

The main aspects of quality control are:

- Varietal purity and diseases control in the field: it includes 3-time field assessment during growing period (Elite classes – 2 assessments and reproductive classes – 1 assessment);

- All original seed material produced from micro-plants and mini-tubers are tested by ELISA (post harvest control);
- Lot inspections are realized on the basis of visual observations means of tuber analyses.

The observations of super-super-elite, super-elite and elite classes are made by the official accredited specialist of State seed inspection. All seed potato lots before definitive certification are visually inspected by cutting the tubers for diseases and mechanical and physiological injuries.

The requirements of the quality of plantings and tubers that are used in Russia are shown in Tables 3 and 4.

Laboratory tests

Potato viruses are tested in laboratory by ELISA and other laboratory methods. The scheme of laboratory control for different generations (seed classes) is shown in Table 5.

This scheme of laboratory control is effective to obtain real information about seed potato quality and to increase it.

To initiate the seed potato programs each basic plant of BHPV and *in vitro* material is tested at presence of complex potato viruses, potato spindle tuber viroid, *Clavibacter michiganensis* subsp. *sepedonicus, Pectobacterium atrosepticum* by modern high-sensitive methods: ELISA, immunofluorescence microscopy (IFM), PCR, NASH, biological test etc. Before cloning all basic plants are tested repeatedly at presence of complex potato viruses and bacteria by ELISA and IFM.

Table 3. Tolerance levels of seed crops in Russia (field inspection).

	Requirements for seed classes, max %		
	OS	**ES**	**RS**
Field inspection*			
Variety purity	100	100	97
Severe virus (PVY-mosaic and Leaf-roll)	0	0.5	1.5
Mild mosaic (PVX and PVM-mosaic)	1-5	3.0	9.0
Blackleg	0	0	0.5
Laboratory test by ELISA**			
PVY and PLRV	0.5	-	-
PVX, PVS, PVM	4.5	-	-

**Phytophthora infestans* does not exist. The field must be free from Golden Nematode and Wart diseases
**Only for 1st field generation

Table 4. Tolerance levels of seed potato tubers in Russia (Lot inspection').*

	Requirements for seed classes, max %		
	OS	**ES**	**RS**
Variety mix	0	0	0.5
Blackleg	0	0	0.5
Ringrot	0	0	0.5
Phytophthora	0.5	0.5	2.0
Dry rot	0	0.5	1.0
Stem nematode	0	0	0.5
Scab (covered more than 1/4 of the tuber)	0.5	1.5	3.0
Rhizoctonia (covered more than 1/8 of the tuber)	0.5	1.0	2.5

*Seed potato tubers must be free from MOP-TOP and Rattle-viruses and from quarantine organisms. A Lot should not contain more than 3% by weight of tubers smaller the minimum size or more than 3% larger the maximum size indicated.

Table 5. Norms and methods of virus detection for different generations (seed classes)

Generation (class)	**Testing norms**	**Methods**
In vitro material	100% of plants	Elisa-test, NASH, PCR etc
Mini-tubers in greenhouse	100% of plants	Elisa-test
1-st field generation	200 plants per plot	Elisa-test
Super-super-elite	200 tubers (postharvest test)	Elisa-test
Super-elite	Visual symptoms assessment	
Elite and Reproductive classes	Visual symptoms assessment	

To test all plants growing in the greenhouse by ELISA average probe is used. For example, to test 1000 greenhouse plants it is necessary to select 200 probes of 5 plants. The plants in probes with positive reaction on PVY, PLRV and bacteria are destroyed.

For viral and bacteria laboratory control of the 1st field generation seed tubers by ELISA, 200 leaf samples per 0. 1 ha of origin plot are selected. In this case only one positive reaction on PVY and PLRV in sum is supported. There is using of visual symptoms assessment for breaking in the plots. For viral and bacterial laboratory control of super-super-elite material by ELISA, 200 tuber samples per 0.5 ha of origin plot or per 10 tonnes of origin material are selected. In this case real tolerance levels are: 18 positive reactions (9%) on PVX, PVS, PVM and 2 positive reactions (1%) on PVY and PLRV in total.

For viral and bacterial control of super-elite, elite and reproductive classes visual symptoms assessment is used. Quarantine pests are not accepted and each seed potato lot is tested in the laboratory for potato cyst nematodes.

Current trends in the Russian seed potato production

For quality improvement of seed potato in Russia the new project was started in 2004. A very important part of this project is the Bank of Healthy Potato Varieties (BHPV). This Bank is located in the Northern area (Archangelsk region) and maintained under the cleanest phytosanitary conditions.

The fund of original free of viruses seed material based on BHPV was created in the Russian regions where the level of infection pressure is minimal. The bank is used for Elite and certified seed classes production by all types of potato producers, including Agricultural enterprises, peasant farms and small private farms.

It was suggested that the seed production system be based on BHPV. This system allows to use the advantages of favourable conditions with the complex application of modern methods and seed potato technologies and high performance laboratory tests of virus and other pathogens.

For the most optimal prophylactic effect to be obtained in seed potato cultivation the following requirements are met:

- Guarantee of a complete absence of quarantine pathogens (potato wart disease - *Synchitrium endobioticum*, golden nematode of potato – *Globodera rostochiensis*, brown bacterial rot - *Ralstonia solanacearum*.
- Absence of soil viruses (TRV and PMTV) and their vectors (*Trichodoris* spp., *Spongospora subterranea*.
- Minimal risk of wide spread of subterranea phytopathogens in the field conditions (absence of PRLV, PVY, PVM virus pestholes and their vectors for a 2 km around).
- Minimal likelihood of bacteria spread (blackleg - *Pectobacterium atrosepticum*, ring rot – *Clavibacter michiganensis*).

The possibility of viral infection of potato plants and virus transmission into tubers are determined by place and conditions of growing, level of infection pressure (amount of insect vectors and infection sources), resistance of cultivars and other factors.

Taking into consideration a wide diversity of agro-climatic conditions, infection loading level and number of insect vectors, we can mark the three main groups of regions:

1. The northern, North-western and North-eastern regions have most favourable conditions for cultivation of quality seed potato. For example, the Leningradskiy region is situated on the Baltic Sea coast. It has cool summer weather and relatively low background of insect vectors. These conditions conduct to minimum the expansion of most heavy virus infections. The vegetative period in this region is very short - from end of May till middle of September (100-110 days), however this region has a very long light day (till 23 h). It creates

very good conditions for fast potato growth and maturity. These agro-climatic conditions make this region favourable to cultivate high quality seed potato.
2. The middle part of Russia (Central, Central black-earth and Volgovyatka regions, Ural, Siberia, Far East) has a wide diversity in the soil content and fertility, rainfalls, sum of active temperatures, frost-free period etc. However these regions can be characterized like regions with moderate infection pressure and can be considered favourable areas to organize own quality seed potato production.
3. The South and South-East regions have less favourable conditions to cultivate potato and organize own quality seed potato production. These regions have hot and dry climate during vegetative period and high infection pressure level (excepting of highlands in the North Caucasus area). Therefore the rates of accumulation of viral infections with each following field generation are much higher than in the Central regions. A high rate of spread of infection decreases potato production. Quality of seed potato becomes much worse after 2-3 vegetative periods (or only after 1 vegetative period at the susceptible varieties).

To ensure a transfer to the adaptation of a uniform seed potato certification scheme in Russia, the following activities are planned:
- Introduction of a compulsory set of requirements for virus control of original and elite seed potato categories, using laboratory test methods on the basis immune-enzyme analysis/ ELISA-test.
- The measures are necessary to develop a net of regional testing laboratories with modern equipment, technical means and diagnostic kits for a full range of laboratory tests for viruses, bacteria and other phytopathogens which are heavily decreasing the quality of seed potatoes.
- To organize the centres of phytosanitary monitoring in the regions on the base of research organizations with involvement of plant protection service specialists and regional State seed inspectors. Their activities should include the phytosanitary monitoring of movement dynamics of virus transferring insects, definition of their varieties and notification of seed production farms of the dates of mass aphis migration with the purpose of definition of the optimum dates for leafy tops removal on various seed potato plantations of various classes and categories.
- To arrange on the basis of the All-Russia Potato Research Institute and Certification Centre 'Test Potato Service' the regular short-term trainings, practical trainings and consulting services for the specialists with the purpose to improve the practical skills in recognizing variety properties and disease symptoms on potato plants and tubers, as well as trainings of specialists in post-harvest control of seed potatoes by means of ELISA.

Conclusion

The development of seed potato production facilitates the conditions to improve the quality of potato seed material, to increase potato production and to integrate the Russia into the international potato market.

Unification and harmonization with international requirements of Russian system of seed potato production, quality control and certification of seed material as well as transfer of modern potato technology is very important for the development of potato production in Russia.

References

Anisimov, B.V. (2005). Production of potato and seed potato in Russia. In: Potato in progress. Wageningen Academic Publishers, the Netherlands, 352-358.

Simakov, E.A. and B.V. Anisimov (2006). Quality improvement and market development of ware and seed potato in Russia. In: Potato developments in a changing Europe. Wageningen Academic Publishers, the Netherlands, 161-174.

Schuhman, P. and B.V. Anisimov (2006). Die aktuelle Large der Kartoffelwirtschaft in Russland. Kartoffelbay, 5/2006 (57. lg): 240-245.

Curriculum vitae – Boris V. Anisimov

Boris V. Anisimov studied at the Agricultural University in Stavropol where he received a Master of Science degree in 1959. In 1966, after a short period of practical work, he was invited to the All-Russian Potato Research Institute, where in 1970 he received a PhD degree in Biology. In 1970-1980 Boris Anisimov worked in the Potato Research Institute as a Head of the Department of Seed Potato Research. In 1981-1986 he worked as a Director of Scientific Pathological Laboratory (SPL USSR) of the Ministry of Agriculture, Ambo, Ethiopia. In 1987-1992 he worked as a Head of Breeding and Seed Potato Centre. In 1993-2003 - Head of Potato Section Program Development at the Ministry of Agriculture of the Russian Federation. Since 2003 until present-Deputy Director of All-Russian Research Potato Institute.

Boris V. Anisimov was an initiator and coordinator of seed potato research production and seed certification program development in Russia, the breeding and seed potato program development at the Potato Research Institute, was a team-leader responsible for the breeding and seed potato program (1970-2003).

At the present the is an initiator of the development of a new concept for seed production in Russia for the period 2005-2010. Boris Anisimov is an active member of Russian Academy of Natural Science since 2001. He took part in national and international potato congresses as a member of the Organizing Committee including the Congress in Finland 1999, the Netherlands Congress 2005 and Potato European Congress in Germany 2006/

Curriculum vitae – Alexander I. Uskov

Alexander I. Uskov studied at the Timiriazev Agricultural University in Moscow where he received a Master of Science degree in 1979 and a PhD in Biology in 1984. In 1979-1994 Alexander I. Uskov worked in the All-Russian Potato Research Institute as a researcher. Since 1995 until present– Head of the Department of Biotechnology. A.I. Uskov is a specialist in biotechnology, protection from plant diseases and seed potato production.

His current research interests include problems of creating free-pathogen basic plants of new potato varieties, increasing methods of the rapid multiplication of virus-free original material and basic seed potatoes, laboratory quality control and certification of seed potato material.

Profile – All-Russian Potato Research Institute (VNHKH)

The All-Russian Potato Research Institute (VNIIKH) was founded in 1930 in order to improve yield and quality of potatoes in Russia. The main objectives of All-Russian Potato Research-Institute are coordination of the national potato research and development programs; improvement of genetic Variability and breeding of new varieties; production and delivering original seed potato of high biological quality; development and promotion of modern potato technologies (growing, harvesting, storage); national and international cooperation development. The Institute includes the following research departments: Genetics, Breeding, Biotechnology, Seed potato, Agrotechnics and Fertilizers, Plant protection, Technology and Mechanization, Storage and processing. Marketing and management, Scientific and Technical Information.

For 75 years the All-Russian Potato Research Institute has been responsible for improving the quality of potatoes in the Russian Federation. The Institute's core business in the coordination of the national potato research programme, the improvement of existing potato varieties and breeding of new varieties. Other key tasks are the production of high quality starch potatoes and the development and promotion of modern technologies in the field of breeding, harvesting and storing of potatoes.

The main results of All-Russian Potato Research Institute are the following: The breeders of the Institute created more than 90 potato varieties, amongst which there are new high yielding disease-resistant and very popular varieties, such as Jukovskiy Ranniy, Udacha, Lukianovskiy and others. The Institute annually produces about 80,000 micro-plants *in vitro,* over 1,000,000 mini-tubers on which basis the Institute's experimental farms produce about 10,000 tonnes of original seed material (pre-basic seed material of the varieties recommended for growing in different regions). Up to 2,000 diagnostic kits for testing potatoes for viruses and bacterial pathogens are produces annually for the seed certification programme development. Fifty young specialists and farmers and 120 regional trainers annually attend the practical training course organized by Institute.

Production of minitubers of the varieties 'Spunta' 'Nicola' and 'Cara' from healthy micro propagated material in Cyprus

S. Gregoriou, G. Minas, T. Kapari, N. Seraphides and L. Papayiannis
Agricultural Research Institute, 1516 Nicosia, Cyprus

Abstract

Good quality propagating material is a basic input in agriculture and horticulture and can substantially improve crop quality and productivity. Plant tissue propagation methods have been developed for several crops. This allows rapid multiplication and production of disease-free plants. The main propagation method for potato crops throughout the world is the vegetative seed tuber. Plantlets of the varieties Spunta, Nicola and Cara were produced *in vitro* on a modified MS medium. They were hardened for 20 days at 20/10 °C (day/night) temperatures, 80% RH and illumination at 20000 lux for 16 hours. Planting was done in pots in a mixture of two parts of potting soil and one part of vermiculite. Fertilizers were given through a fertigation system. All microplants were tested for the PLRV, PVY, PVA, PVS, PVX and PVM potato viruses. Yellow sticky traps and plant indicators were used to monitor the aphid population in the net house. Two liftings at 70 and 100 days after transplanting were performed. The majority of the plants had only one stem. The mean number of tubers per pot was 13.0, 10.3 and 7.7 for the varieties Cara, Nicola and Spunta, respectively. The average weight of tubers was 126, 90 and 112 g per plant for each of the above varieties, respectively. The date of lifting had no effect on the number of tubers but affected significantly the tuber weight and their dry matter.

Keywords: seed, varieties, plantlets, virus, aphids

Introduction

The usual means of propagation for potato crops throughout the world is the vegetative seed tuber, which may be planted whole, as in most of Europe, or after cutting into pieces, as in North America and Cyprus. Seed tubers can be infected with numerous virus diseases and can carry many fungus diseases which may affect ware crop growth and the health of the progeny tubers. For effective potato production the availability of an annual supply of healthy tubers is an essential requirement (Allen *et al.*, 1992; Ranalli *et al.*, 1990; Struik and Wiersema, 1999).

The use of micro propagation completely changed the technique of potato seed production (van der Zaag, 1990; Jones, 1988; Ranalli, 1997). Micro propagation has the ability to increase the production of propagation material in protected environments over short time. This is useful, especially in countries where there are no vector-free production areas for producing high quality potato seed tubers.

Micro propagation multiplication steps are sped up using *in vitro* plantlets, micro tubers or minitubers (Ranalli, 1997). *In vitro* plantlets are produced from either single node cuttings or from multiple stem nodes. Taking single node cuttings from *in vitro* plantlets and growing them in tuberizing medium routinely produces microtubers (4-10 mm). Minitubers (5-20 mm) are produced *in vivo*. Minitubers can be obtained from *in vitro* plantlets or microtubers and plants raised from both propagates can be subjected to a destructive harvest or to repeated harvest (Lommen and Struik, 1992).

In Cyprus about one fifth of the seed planted in autumn and spring crops is produced locally. It appears that under Cyprus conditions the best approach to produce part of the requirements of seed potatoes was a single multiplication of imported stocks (certified or foundation seed). Locally produced seed performs equally satisfactorily as imported seed. Ware crops grown from local seed were as healthy as crops grown from imported seed (Vakis, 1980).

Seed potatoes can be infected by many different viruses that can reduce yield and tuber quality. Virus diseases can often be diagnosed by mosaic patterns on leaves, stunting of the plant, leaf malformations, and tuber malformations. Symptoms are not always expressed due to interactions between the virus and the potato plant, growing conditions, such as fertility and the weather, or the age of the plant when infected. The serology detection technique ELISA (Clark and Adams, 1977) is mainly used to diagnose the existence of the following virus diseases in the mother tubers and in the miniplants: Potato leafroll virus (PLRV), Potato virus Y (PVY), Potato virus S (PVS), Potato virus A (PVA), Potato virus X (PVX) Potato virus M (PVM).

Green peach aphid *Myzus persicae*, potato aphid *Macrosiphum euphorbiae* and buckthorn aphid *Aphis nasturtii* may transmit PLRV and PVY. Aphid management with a combination of cultural and chemical practices relies to a large extent on our understanding of the biology of the insect. There are almost 4,000 aphid species in the world (Greer, 2000). Since the females of many aphid species are able to produce live young without mating aphid populations can rise in a short period of time.

Cara, Nicola and Spunta are the three main varieties in Cyprus and are cultivated in more than 90% of the area used for the potato production. Spunta is a very early variety used mainly for ware production. Nicola is a salad variety with medium earliness and Cara is a very late variety exported in Europe to be used as 'baker' potato. The aim of the experiment was to find ways for rapid multiplication of potato seed under the Cyprus climatic conditions.

Materials and methods

Pre-sprouted potato tubers of the three varieties, which were under thermotherapy for forty days, at 36 °C and 95% humidity, were used as mother propagating material. Plantlets were produced *in vitro* on a modified MS medium, supplemented with 3% sucrose, 4.5 mg/l BAP, 0.009 mg/l IBA, 55.7 mg/l ascorbic acid and was solidified with 2.5 g/l phytagel (Minas *et al.*, 2006). They were hardened in transplanting trays of 77 small cells (4x4x7cm, w,l,d) filled with potting soil, for 20 days at 20/10 °C (day/night) temperatures, 80% RH and illumination of 20000 lux for 16 hours in a SANYO MLR 351 H growth chamber.

After hardening and before planting (Figure 1), the plantlets had a height of 13.8±0.4, 14.1±0.5 and 16.3± 0.4 cm for the varieties Cara, Nicola and Spunta, respectively. The leaf area was at 56±2cm^2 for all varieties.

Planting was done in a mixture of two parts of potting soil and one part of vermiculite in pots of 18 cm diameter on the 28th of December in a net house (Figure 1). For the first two weeks, during the day the misting system was turned on for two minutes in every hour, according the intensity of the light. There were two liftings 70 and 100 days after transplanting. Each plot consisted of 16 pots. The four pots in the middle were used for the data recorded. Drippers of 2 l/h capacity on 16 mm diameter polyethylene line were used for irrigation. Fertilizers were applied through a fertigation system with a Dosadron D300 fertilizer injector at the rates of 70 ppm N, 20 ppm P and 100 ppm K, including micronutrients. During harvest, plants were separated in the canopy, roots and tubers. Fresh and dry weights were recorded.

Pre-sprouted potato tubers of the three varieties, which were under thermotherapy, were used as mother propagating material and the plants produced by *in vitro* culture were planted in the screen house and were indexed by ELISA and macroscopic examination for the PLRV, PVY, PVA, PVS, PVX and PVM viruses.

A double-door entrance hall and fine-mesh screens on all openings of the screen house were used to exclude aphids blown in from outside. Four yellow sticky traps and twelve indicator plants were used to monitor insect pest populations. Yellow sticky traps were uniformly distributed in the screen house and placed approximately 10 cm at the top of the plants to detect winged aphids (Figure 1). Indicator plants were chosen randomly throughout the screen house and were inspected for aphids on a weekly basis. Upon detection of aphids the plants were sprayed with insecticides.

The use of plantlets produced from single-nodes cuttings was chosen instead of microtubers as microtuber production takes 1-2 months longer than production of plantlets. Besides, microtubers must be harvested, stored, and their dormancy needs to be met before they can

Figure 1. Plant ready for planting after hardening and in full growth in the screen house. Yellow sticky traps were placed approximately 10 cm above the top of the plants.

be planted (Donelly *et al.*, 2003). On the other hand, microtubers are easier to transport and handle than plantlets, are less delicate requiring less after care when planted in a green house and they can also be recycled back into tissue culture once they become non-dormant (Khuri and Moorby, 1996). In our case, the use of plantlets could have a serious advantage, as they grow very fast and there can be more than one planting in the screen house.

Results

The plants had only one stem. The mean number of tubers per pot was 13.0, 10.3 and 7.7 for the varieties Cara, Nicola and Spunta, respectively (Table 1). The number of tubers was not affected by the date of lifting. Cara and Spunta gave the same number of tubers per pot in both liftings but there was an increase from 9.6 to 11 tubers per pot for Nicola. The mean tuber weight per pot for the three varieties was significantly affected by both the variety and the date of lifting. The tuber weight was 126 g, 90 g and 112 g for Cara, Nicola and Spunta, respectively. Between the first and the second lifting there was an increase of 76, 53 and 45% for each one of the above varieties, which is reflecting the maturity of the varieties. The dry matter of tubers increased for all varieties between the first and the second planting. The number of stolons per plant decreased between 70 and 100 days after planting. The reduction was 1.6, 1.0 and 1.0 stolon for Cara, Nicola and Spunta. This probably due to the resorption of some of the stolons and tubers. The number of stolons was related to the number of tubers.

The size of tubers produced by each variety is very important for the next generation of tubers (Table 2). The variety Cara 70 days after transplanting gave 50% of the tubers in the smaller size up to 5g. In the second lifting 50% of the tubers were over 10g (Table 2). In both liftings, 70 and 100 days after planting, fifty percent of the tubers produced by the variety Nicola were of the smaller size (<5g), whereas 25% of the tubers of the same variety in the first lifting and 37% in the second lifting were >10g.

Plants of the variety Spunta produced higher proportion of tubers >10g. The proportion of tubers in this fraction was 50 and 69% for the lifting at 70 and 100 days, respectively. The effect of the size of tubers on the progenies in the second generation has to be examined in

Table 1. Field characteristics of potato varieties Cara, Nicola and Spunta at two different lifting dates, 70 and 100 days after planting.

	Variety			Lifting	
Records	**Cara**	**Nicola**	**Spunta**	**70 days**	**100 days**
Number of tubers	13.0a	10.3b	7.7c	10.2a	10.5a
Weight of tubers	126.0a	90.0c	112.0b	85.0b	133.0a
Dry matter	18.2a	16.3c	17.3b	16.5b	18.1a
Number of stolons	7.6a	6.3b	5.9b	7.2a	6.0b
Tubers/stolon	1.8a	1.7a	1.3b	1.4b	1.8a

Table 2. Tuber size distribution of potato varieties Cara, Nicola and Spunta at two different lifting dates, 70 and 100 days after planting.

	Tuber size (g)			
	1-5	5-10	10-20	20-40
Cara - 70 days	6.31	3.31	3.06	0.32
Cara - 100 days	3.53	3.00	4.63	1.87
Nicola - 70 days	4.50	2.75	1.88	0.50
Nicola - 100 days	5.75	1.25	2.19	1.87
Spunta - 70 days	2.06	1.94	2.94	1.00
Spunta - 100days	1.75	0.56	2.13	3.00

the field. The small tubers (<10g) may be replanted for a second time in a net house, in pots or in the soil.

Seventy days after planting, the fresh weight of the canopy was one third of the total fresh weight, for all varieties but in the second lifting there was a reduction in the percentage of the canopy fresh weight by 17, 16 and 10% for the varieties Cara, Nicola and Spunta, respectively (Figure 2). The fresh weigh of the root was only a small fraction of the fresh mass of the plant accounted for 4.3, 1.5% and 1.2% for Cara, Nicola and Spunta, respectively. The reduction in the fresh weight of roots according to the total weight was significant only for the variety Cara and it was up to 2%. Tubers composed the major part of the fresh weight. Nicola and Spunta had more than two thirds of their weight in the first lifting and they reached up to 80% in the second one 100 days after planting. There was an increase in the percentage of the tuber weight of 19, 16 and 10% for the varieties Cara Nicola and Spunta, respectively.

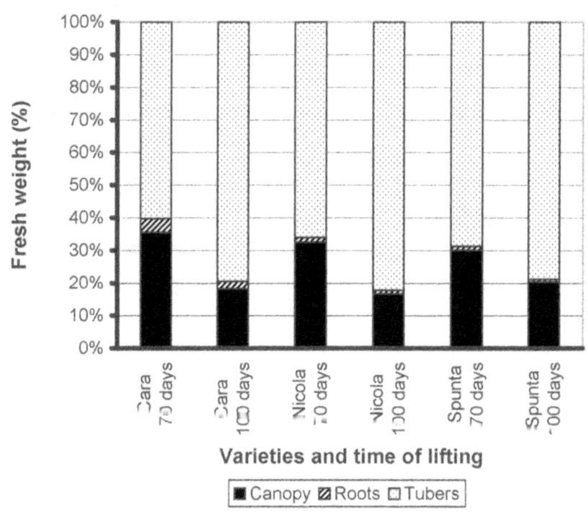

Figure 2. Percentage fresh weight of the canopy, roots and tubers of the varieties Cara, Nicola and Spunta, 70 and 100 days after planting.

Figure 3. Plant of the variety Cara 100 days after planting.

Viruses and aphids

One mother tuber was infected by PVY and was discarded from the beginning. One plant showed mild necrosis symptoms on the leaves and stunting and when indexed by ELISA was found to be infected by PVY. The infected plant and six neighbouring plants were discarded. All other plants were found free from all examined viruses.

Winged aphids of the species *Aphis gossypii* were first detected on traps during March, while young aphids were located on plants at the end of the cropping season, in May. Aphid population was very low throughout the growing season, while appearance of *Myzus persicae*, the main vector of potato viruses did not appear on the plants.

Discussion

Tissue cultured plantlets from tissue culture arc very delicate and require special growing conditions, especially before planted in the field. They are usually planted in beds for the production of minitubers to be used as the starting material for the second generation. Generally, they are transferred into growth champers at high plant density or into an aphid-proof screen house and are subjected to destructive or repeated harvest (Lommen and Struik 1992; Gregoriou and Moorby 2004). After hardening of the plantlets it is possible to plant them at normal climatic conditions without any cooling or heating. Planting in large pots (30 pots/m^2) offers the opportunity to grow better plants and produce large size tubers which render vigorous progeny after planting in the field for production of the next generation of potato seed.

There were differences between the varieties regarding the number and weight of tubers. Spunta as an early variety resulted in early bulking and a majority of tubers in the range of 10 to 40g. Nicola, (mid-late) and Cara (late) yielded tubers smaller than 10g. As they produced a greater number of tubers than Spunta further experiments will be needed to investigate the difference in yield per tuber size (5, 10, 20 and 40g) in the field the coming year. It could be valuable if the tubers of the lower classes be planted once again in the screen house. Ranalli *et al.*, 1990 also reported such differences between varieties Spunta and Nicola in trials with plantlets in the screen house and the in the field.

As the pre-basic material needs to be multiplied for a few generations in open field, it is very important to find varieties and cultural techniques in the screen house that produce an adequate number of tubers having the appropriate size. These may be planted in the field to produce profitable yields and a faster generation of potato seed tubers for the growers.

In Southern Europe and especially in the Mediterranean area and countries like Cyprus the growing conditions are favourable for the growing of potato almost nine months in a year. There are three main growing seasons, the summer crop (September-November), the winter crop (October–February) and the spring crop (December to January- March to June). The bulk of potatoes are grown as a spring crop mainly for export. The potatoes produced in autumn are mainly for home consumption. The growing season of the potato crop is not limited by frost in the spring or cold in the winter but rather by very high temperatures and water stress during the summer months.

In programming more potato seed production cycles the major problem faced is breaking tuber dormancy. The method used bears an impact on both plant growth and the tuber production. Various chemicals have been used to terminate seed tuber dormancy, as for example, Rindite (Nasiruddin and Blake, 1997), ethylene chlorohydrin (Struik and Wiersema, 1999) and gibberellic acid by haulm application (Ittersum and Scholte, 1993) or immersion of tubers before planting. Shortening the duration of dormancy and advancing growth vigour are often unreliable and until now they have not been used commercially. The use of temperature to shorten dormancy period and to advance growth vigour is cultivar specific and time demanding which is a disadvantage for countries like Cyprus where several potato growing seasons per year are possible.

Monitoring virus and aphid populations is critical for the production of healthy propagation material and it should command the choice of growing site and the time span of the plantation in the field.

The danger from water stress is eliminated in Cyprus as the potatoes are grown under irrigation and/or fertigation systems. This gives growers the opportunity to maintain ideal growing conditions for the plants and to intervene when necessary.

The project is continued with experiments on the break of dormancy by haulm application of gibberellic acid and immediate planting of the progeny tubers. Further evaluation is carried out in the field regarding the performance of different size seed tubers and of the locally produced versus imported seed.

Acknowledgements

Research supported by Cyprus Research Promotion – Special grand – Rapid multiplication of potato seed in Cyprus. The Agricultural University of Athens is participating in the project.

References

Allen, E.J., P.J. O'Brien and D. Firman 1992. Seed tuber production and management. In: P.M. Harris (Ed.), The Potato Crop. The scientific basis for improvement. Second edition, Chapman and Hall, London, pp. 247-292.

Clark, M.F. and A.N. Adams 1977. Characteristics of microplate method of enzyme-linked immunosorbent assay for detection of plant viruses. J. Gen. Virology 34: 475-483.

Donelly, D.J., W.K. Koleman and S. Coleman 2003. Potato microtuber production and performance: A review. Amer. Potato J., 80: 103-115.

Greer., L, 2000. Integrated Pest Management for Greenhouse Crops. ATTRA, Florida p. 12.

Gregoriou, S. and J. Moorby 2005. Effect of growth regulators paclobutrazol and B-9 on the growth and yield of seed tubers combined with repeated harvests. 16th Triennial Conference of the EAPR, Bilbao, Spain, pp 945-949.

Ittersum, M.K. and K. Scholte 1993. Shortening dormancy of seed potatoes by haulm application of gibberellic acid and storage temperature regimes. Amer. Potato J., 70: 7-19.

Jones, E.D. 1988. A current assessment of *in vitro* culture and other rapid multiplication methods in North America and Europe..

Khuri, S. and J. Moorby 1996. Nodal segments or microtubers as explants for *in vitro* microtuber production of potato. Plant Cell Tiss Org Cult 45: 215-222.

Lommen, W.J.M and P.C. Struik 1992. Production of potato minitubers by repeated harvesting. Plant productivity and initiation, growth and resorption of tubers. Netherlands Journal of Agricultural Science 40: 342-349.

Minas G.J., S. Gregoriou, T. Kapari-Isaia and L.C. Papayiannis 2006. Seed potatoes production in Cyprus starting from in-vitro culture of apical meristems. Acta Horticulturae (in press).

Nasiruddin, K.M. and Blake, S., 1997. Effect of rindite on storage behaviour, dormancy break and sprout growth of potato microtubers (cv Desiree). *Amer. Potato J.*, 74: 325-330.

Ranalli, P. 1997. Innovative propagation methods in seed tuber multiplication programmes. Potato Research 40: 439-453.

Ranalli, P., E. Forti, G. Mandolino and B. Casarimi 1990. Improving production and health of seed potato stocks in Italy. Potato Research 33: 377-387.

Struik, P.C. and S. G. Wiersema 1999. Seed potato technology. Wageningen Press, Wageningen.

Van der Zaag, D.E. 1990. The implications of micropagation for the future of seed potato production systems in Europe. Abstracts, 11th Triennial Conference of the European Association for Potato Research, Edinburgh, U. K., pp. 28-45.

Vakis, N.J. 1980. Production of seed potatoes in Cyprus. Technical Bulletin 34. Agricultural Research Institute, Nicosia, 12p.

Curriculum vitae – Sophocles Gregoriou

Sophocles Gregoriou is married with two children and was born on 09 November 1951 in Nicosia, Cyprus. From 1957-1969 he received his elementary and high school education after which he served in the army from 1969 to 1971.

Then he was employed as Agricultural Research Technician at the Agricultural Research Institute and worked till 1984 when he got a full scholarship from the Federal Republic of Germany to study Vegetable in the faculty of Horticulture at the University of Hanover. He got his diploma in 1989 and worked for five year in the Vegetable department of the Institute dealing with the cultivation of out- and indoor crops. From then on he was responsible for the evaluation of potato varieties and new cultural practices for the improvement of potato industry. He cooperated with many breading organizations and trading companies of Europe. His work is extended in the production of 'bunnies' potatoes, the set skin of the tubers and organic production of potatoes. He is the official researcher for carrying out the experiments on the Value of Cultivation and Use for the registration of the varieties in the National Catalogue.

Since 2005 he is the Coordinator of a project on 'Rapid multiplication of potato seed' in which the University of Athens is participating. He is the author and co-author of more than forty scientific papers and Technical Bulletins.

Company profile – Agricultural Research Institute

The Agricultural Research Institute was established in 1962 as a cooperative project between the Government of Cyprus and the United Nations Development Fund, with the Food and Agriculture Organization of the United Nations

The Agricultural Research Institute undertakes applied and basic research within the wider domain of plant and animal production. Its mission is to provide high quality scientific research using methods that are financially, environmentally and socially sustainable. Its contribution to the solution of actual problems and to the introduction of new technological methods and approaches in agricultural production is greatly valued, both locally and abroad.

The Institute is organized into two divisions. The production division includes the disciplines of crop breeding and genetics, fruit tree and vegetables production, viticulture, floriculture, animal production, biotechnology and tissue culture. The support division provides services and scientific support to the production division. Among its disciplines are plant protection, insect toxicology and pesticide residues, soil fertility, plant nutrition, irrigation, environmental impact, experimental statistics and agricultural economics.

Experimental work in potatoes covers the areas: evaluation of varieties and clones, biological production, plant nutrition and fertigation, quality evaluation and storage, production of ethanol and potato seed production. The Institute is the official Technical Examination Office for the registration of agricultural and vegetable varieties in the National Catalogue. After joining the European Union the Institute was assigned the task of maintaining the Accountancy Data network in Cyprus. The Institute was selected by the European Commission as Centre of Excellence in Agriculture and Environment.

Means of quality improvement of seed potatoes in South Russia

V.N. Samodurov and V.N. Maistrenko
Krasnodar Research Institute of Vegetable Crops and Potatoes, 350921 Krasnodar, Belozerny

Abstract

The main questions of quality improvement of potato seed material during elite and reproductive seed-production in South Russia are studied in the article. Species, number and dynamics of plant aphids flight on potatoes are studied. It was determined that preharvest removal of vegetable tops at the end of flowering allows to significantly limit an opportunity of late plant infection and to avoid infection transition into tubers of a new yield. There is a significant increase of virus infection at late removal of vegetable tops. It was determined, that virus infection of seed potatoes directly depends on treatment against virus infection carriers, as well as on terms of vegetable tops removal.

Keywords: vegetable top removal, virus, aphid, seed production

Introduction

The increase of potatoes production under modern conditions can be achieved only when the main requirement of domestic market – quality improvement of a seed material is taken into consideration.

For solving this problem under modern economic conditions serious improvement of elite and reproductive seed-production is necessary.

In the process of potato seed-production under conditions of Krasnodar territory, injurious effect of such phytopathogenic viruses, as VSLK, UVK (various strains), MVK and some others, causing heavy and average forms of quickly progressing virus infection of the majority of cultivated improved varieties is especially great. Greater injurious effect of phytopathogenic viruses under conditions of Southern potato growing is explained first of all by the fact that they quickly spread, for example by *Myzus persicae*, and are almost always transferred vegetative to propagating progeny through tubers.

Therefore one of the main tasks in elite and reproductive seed potato cultivation is application of a complex obligatory agricultural means such as:

- Insecticide treatment against plant aphids – virus infection carriers;
- Early removal of vegetable tops by chemical or mechanical means when maximum seed marketability of tubers is achieved, with consideration of data concerning dynamics of virus carriers' expansion (flying generation of plant aphids) under certain climatic conditions.

Non-observance of these tasks is a serious reason for the fast deterioration of varietal and seed quality in the process of elite and reproductive potato seed-production and as a result of it efficacy with each following reproduction decreases.

Because of hot and dry weather during vegetation period and, as a rule, high level of infection in Southern Russia rates of increase of virus infections with each following reproduction here are significantly higher than in Central Russia. High infestation rate reduces efficacy and worsens seed qualities of potatoes. As a result it is necessary to pay greater attention to the crop to get qualitative seed material of standard seed fraction with minimum virus infection.

In this article some questions of quality improvement of potato seed material in the process of elite and reproductive seed-production are studied.

Methods

The experiment was carried out in 2004 - 2006 in a foothill zone of Krasnodar territory by Krasnodar research institute of vegetable crops and potatoes on the base of 'Rastenievod'. In the result of experiment plant aphids flight – virus carriers, optimum terms of preharvest removal of vegetable tops, which allows significantly to limit an opportunity of late plant infection and to avoid infection transition into tubers of a new yield and efficacy of insecticide treatment against virus carriers had been determined.

The number of plant aphids was determined by means of Moericke traps. Insecticide treatment was carried out two times (on the third decade of May and at the beginning of the second decade of June).

Preharvest removal of vegetable tops was carried out by means of mowing in three stages: at the end of flowering, two weeks after flowering, at the beginning of dying off of bottom leaves. Removal of vegetable tops at the end of flowering was carried out in the third decade of June that coincided with a high level of plant aphids flight. Removal of vegetable tops in two weeks after flowering was carried out on the second decade of July, dying off of the bottom leaves was in 20 - 25 days after high point of plant aphid flight.

Viral control over studied experiments was carried out by means of immune-ferment analysis on the basis of testing tuber tests. Not less than 100 tuber tests were tested of each variant. (recommendation on application of immune-ferment analysis for potato virus diagnostics, 1990).

The results of phytosanitary monitoring of plant aphids – virus carriers on potatoes

The research carried out in 2004 showed, that plant aphids appeared on trial sites on the third decade of May. The degree of plant aphids expansion was not high and amounted to 60 individuals per trap in the third decade of May and the first/second decade of June.

During vegetation period of 2004 plant aphids flight began on the third decade of May (22 individuals per trap). The mass plant aphids flight was observed on the first decade of June - 38 individuals per trap. The maximum quantity of virus carriers was observed on the third decade of June - 117 individuals per Merike trap. But on the first decade of July intensity of plant aphids flight decreased almost twice and made 65 individuals per trap. Intensity of plant aphids flight, virus carriers on potatoes, during 2004 vegetation period amounted to 362 individuals (Table 1).

The next two years of research showed, that the period of plant aphids begins on the second/third decade of May and continues till the first decade of August. The number of plant aphids in 2005 was the same as in 2004 and amounted to 361 individuals per Merike trap. Hot and dry weather of 2006 promoted increase in plant aphids number (554 individuals per trap during vegetation period) almost a half of which was peach plant aphid (Figure 1).

There are different approaches in determining 'critical point' of injurious effect on the basis of observations of dynamics and intensity of plant aphids flight in modern practice of potato seed production in different countries. Usually indexes of determining number and intensity of peach plant aphid, as it is one of the most active virus carriers, are used for this purpose. For example, in Holland 'the critical point' is 50 individuals of peach plant aphid per trap during the whole observation period, in Germany 10-20 individuals of peach plant aphid on 100 leaves or 50 individuals per trap, after what it is recommended to remove vegetable tops not later than in 7-14 days. 50 individuals of peach plant aphid per Merike trap is taken as 'critical point', this index show relative injurious effect of certain plant aphid species. Thus relative injurious effect of peach plant aphid is accepted as 1 point, brimstone butterfly, buckthorn as 0,4 points,

Table 1. Species and the number of plant aphids, virus carriers on potatoes.

Year	Total plant aphids number per trap, individuals	Species of a plant aphid, %			
		Peach	Brimstone butterfly, buckthorn	Black bean	Other species
2004	362	45	28	19	8
2005	361	46	22	15	17
2006	554	46	19	17	18

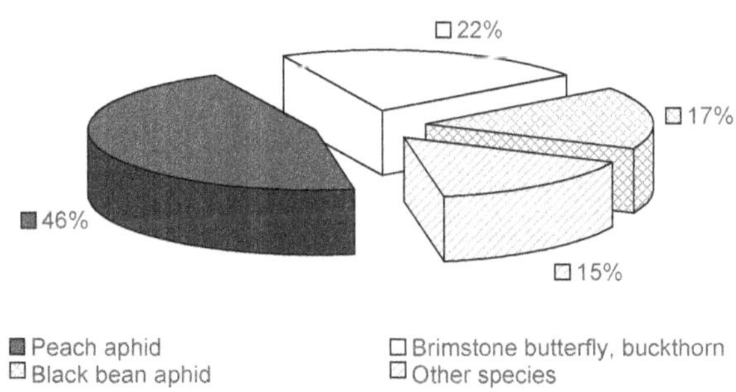

Figure 1. Percentage of plant aphids number during potato vegetation period, per Merike trap.

black bean aphid - 0,1 points, other plant aphids species - 0 points. All points are summarized. Getting 50 points means 'critical point'. (Banadysev, 2003).

In our research total relative injurious effect reached 'critical point' during different periods of different years (Figure 2; Table 2).

In 2004 - 2006 'critical point' at the level of 50 points was observed on the second decade of June, namely on the 15th of June, 2004 and 2005, and on the 11th of June, 2006 correspondingly.

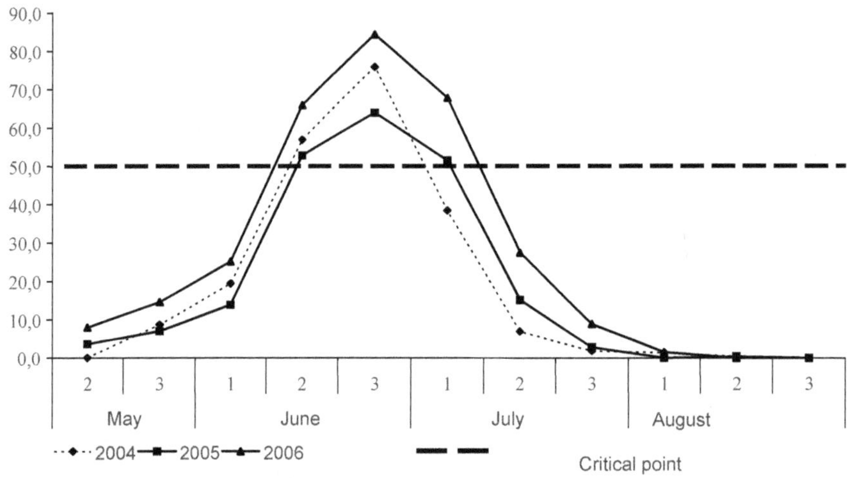

Figure 2. Injurious effect of plant aphids during different years, points.

Table 2. Species, distribution and injurious effect of plant aphids – virus carriers on potatoes.

Plant aphid species	**Injurious effect of plant aphids, points**							
	May	**June**			**July**			**August**
	2	**1**	**2**	**3**	**1**	**2**	**3**	**1**
Peach plant aphid	8.0	16.0	48.7	62.0	41.7	12.3	3.3	1.0
Brimstone butterfly, buckthorn	1.3	2.4	7.7	10.8	10.1	3.5	1.1	0.4
Black bean aphid	0.8	1.1	2.1	1.6	0.8	0.7	0	0
In total	10.1	19.5	58.5	74.4	52.6	16.5	4.4	1.4

Influence of insecticide treatment and terms of vegetable tops removal on potatoes infestation with viral diseases

Insecticide treatment against plant aphids – virus carriers began on the first decade of June. Konfidor insecticide on the basis of reactant with a dose of 0,25 litres/hectares was applied. Protection actions were taken three times. The first treatment was carried out at the beginning of the first decade of June, the next ones in 10 - 12 days, two of them coincided with treatment against Colorado beetles.

Vegetable tops were removed in three terms. The first term, at the end of flowering, 26th of June, which coincided with high time of plant aphids flight when maximum number of virus carriers was caught in Merike trap and in nine days after 'critical point of injuriousness'. Beginning of the second decade of July – was the second term of removal when plant aphids number significantly decreased and the third decade of July – was the third term of vegetable tops removal (Table 3).

The results of post-harvest control regarding the latent infestation of tubers by virus diseases showed that in 2004 after insecticide treatment virus infestation rate decreased in comparison with the results of non-treatment. Thus, maximum infestation rate (without insecticide treatment) was 90%, which is almost three times higher than with insecticide treatment with the same term of vegetable tops removal, namely 64%.

With each following term of vegetable tops removal virus infestation increased irrespective of insecticide treatment (Figure 3).

Table 3. Potatoes infestation according to the terms of vegetable tops removal and insecticide treatment against virus carriers (2004).

Terms of vegetable tops removal	**Virus infestation, %**					
	In total	**Including**				
		XBK	**SBK**	**MBK**	**YBK**	**VSLK**
With insecticide treatment						
June, 26	10	0	0	6	4	0
July, 11	27	0	2	14	11	0
July, 22	64	0	5	36	23	0
Without insecticide treatment						
June, 26	32	2	5	12	10	0
July, 11	52	4	8	17	21	2
July, 22	90	4	8	47	31	3

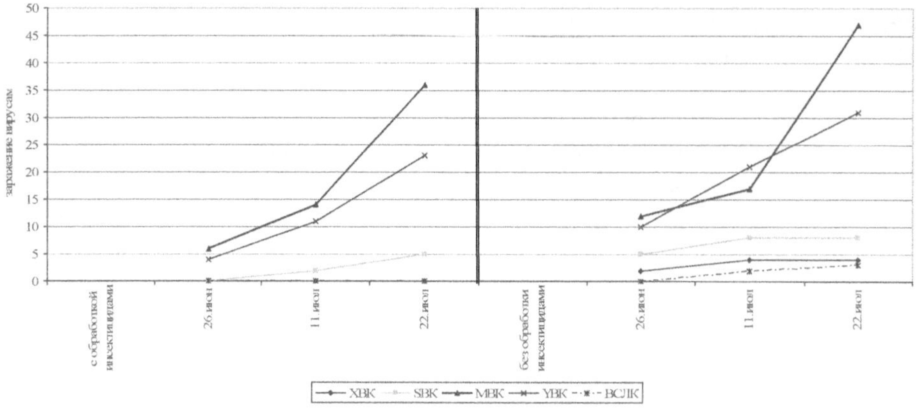

Figure 3. Increase of virus infection depending on terms of vegetable tops removal and treatment against virus carriers.

At removal of vegetable tops at the end of flowering virus infection varied within 10%, at the last term of vegetable tops removal it increased six times, and at removal of vegetable tops in two weeks after flowering it increased 2,5 times and made total 27% on all viruses.

Mainly increased infection by M virus and made 36% with insecticide treatment, which is by 10% lower than without the treatment.

Conclusions

Species, number and dynamics of plant aphids flight on potatoes are studied. It is determined, that under Krasnodar territory conditions the 'critical level' of plant aphids flight takes place at the end of the second/beginning of the third decade of June, and insecticide treatment reduces 2 times tubers infestation.

It is also determined, that preharvest removal of vegetable tops at the end of flowering allows to significantly limit an opportunity of plant late infection and to avoid infection transition into tubers of a new yield. There is a significant increase in virus infection at later removal of vegetable tops. Especially increases infection by M virus, which in case of vegetable tops removal at the beginning of dying off of bottom circle leaves, amounted to 36% and was 30% higher than in case when vegetable tops were removed at the end of flowering.

Analyzing the obtained data it is possible to draw a conclusion that virus infection of seed potatoes directly depends on treatment against virus carriers as well as on removal of tops.

References

Anisimov, B.V., 2003. Phytopathogenic viruses and their control in potato seed-production. Potatoes and vegetables, 25-27.

Banadysev, S.A., Yashenko N.P. Plant aphids are virus carriers of seed potato diseases. Proc. of the Conference dedicated to 90th Birthday of scientist Adamov I. I. Minsk, 45-47.

Banadysev, S.A., Yashenko N.P., Malinovsky S.V., etc, 2003. Instructions for insects registration. Minsk, 34.

Vershinin, B.M., Lyalko R.V., 2001. It is necessary to improve seed-production system. Potato and vegetables, 28-29.

Zykin, A.G., 1970. Plant aphids are potatoes virus carriers. Leningrad, 72

Instructions of research on potatoes. 1967-263.

Instructions on immune-ferment analysis of virus diagnostics, 1990.

Selection and seed-production of high-quality seed potato in the Northwest of Russia

A.I. Osipov, T.A. Shelabina, Z.Z. Evdokimova and N.I. Tsarkov
State Research Enterprise, Leningrad Research Institute for Applied Agricultural Science (LRIAAS), Russian Academy of Agricultural Science, Russia

Abstract

The Northwest of Russia keeps a leading place in areas under crops and total area of potatoes in the Russian Federation. However, recently there appeared a negative tendency in this branch of agriculture. This tendency results in a reduction of a share of areas under crops by large agricultural producers and growing of these areas in a small private sector.

This situation is a result of a state policy absence on stimulation of the production development of this culture and insufficient attention to the development of scientific research and innovative processes as a whole. On some basic parameters of potato production in the Northwest of Russia has an opportunity to be highly competitive and become a basis for stable development of the region.

For many years the Northwest takes leading place in areas under crops and croppage of potato in the Russian Federation. According to historical realities, soil - climatic conditions and presence of highly authoritative school of selectors of potatoes, this region, and in particular Leningrad region, is justified to consider potato production as a strategically important culture capable of making a competition inside the country and abroad and to provide employment of rural population and sustainable development of the region.

Keywords: genetic resources, research, seed production, breeding

Recent developments

Recently there appeared a negative tendency in the given branch developing. It results in a reduction of a share of areas under crops at large agricultural producers and increase of these areas in a small private sector. In 2001 areas under crops constituted 56.2 thousand ha, in 2005 they decreased to 48.3 thousand ha. During the same period areas under crops of large agricultural enterprises have decreased from 11.2 to 6.6 thousand hectares. In all categories of economies average yield of potatoes in 2005 was 12.7 t/ha. While at the agricultural enterprises – 15.5 t/ha. Such division of labour makes it impossible to provide production competitive, first of all, to provide standard quality potatoes meeting modern requirements to consumer properties, and specialized requirements for further different ways of processing. Small-scale system of production is poorly susceptible to scientific and technical progress; it is not able to provide compliance of technological discipline. Further conservation of the given situation will inevitably lead to an even more significant reduction in competitiveness of production, a shutdown of a significant part of rural population socially useful work that will result in subsequent depopulation of villages, to dropping of soil fertility, switching and distribution of dangerous fungal diseases, golden nematode of potato disease, bacterial diseases, and diseases of virus and viroid nature. Despite of the increasing demand for food potato, for example, in the system of hyper- and superstores in Saint Petersburg, the given segment of the market is being filled more and more with imported potato.

It is necessary to admit, that the originated situation is the consequence of a state policy absence in the matter of stimulation production development of this culture, and insufficient attention to research and development of scientific and innovative process as a whole.

Absence of modern innovative forms of production organization and investment, out-of-date material base, imperfect system of production, storage and processing lead to quick loses of positions in this historically developed specialization of the Northwest region.

Scientists - selectionists are not sure in their future as they do not see correct system of adequate compensation for their work and prefer to register their own small companies on seed material production that are ineffective, but in a way allow their owners to guarantee their future.

Such decentralization makes the development prospects of widely known Leningrad scientific school of potato selectors not evident. The scientifically-based selection process aimed at decades of years in the future can be carried out only at really expressed state support combined with fair principles of authors of varieties compensation and flexible system of scientific refinancing of research depending on achieved results. Unfortunately, at the moment such stimulus and preconditions are absent. It is necessary to mention, that such a crisis is typical not only for the Northwest, but also for the country as a whole, and this concern has been underlined at the session of the Rosselhozakademy Presidium discussions on the given question. An important problem is a necessity of choosing new guiding lines in selection.

Needed developments

The state loss of production control of high-quality seeds will inevitably lead to replacement of varieties of domestic selection by foreign analogues. However the situation is not hopeless. Taking into consideration the most important indexes of potato production in the Northwest of Russia this production has an opportunity to be highly competitive and be a basis of stable development of the region.
The advantages allowing to reach this goal are:

- Highly authoritative school of global level selectionists.
- Inexhaustible genetic resources of culture concentrated in Russian Plants Institute.
- Concentration in the region of the most powerful scientific research institutes of Russia, capable to provide creation of modern systems of agriculture focused on quality of production and competitive prices.
- Significant agricultural land resources, allowing to provide rational accommodation of cultures in a crop rotation, guaranteeing quality of seed production and preservation of soil fertility.

- Settled traditions of cultivation of the given culture, developed at an able-bodied population.
- Favourable investment climate in the Northwest.

Making the best of the mentioned advantages is possible only in case of making a system combining the state support of scientific research; control over production of high-quality seeds with stimulation of private initiative of food potato production, focused on scientifically grounded forecast of demand and providing refinancing of scientific research will be created.

In our opinion these purposes could be promoted by creation of a large, research-and-production association (agricultural holding), uniting selection research works, activities on seeds improvement, primary seed-growing and production of high-quality food potato of existing and perspective varieties. Thus the received income should provide not only the investments payback, but also make it possible to spend a part of money resources for agricultural science development.

Despite of negative tendencies in the development of potato growing in the region the areas related to increase of a share of the areas in a private sector, and also deterioration of technical equipment in public sector, demand for a potato grows. Agricultural commodity producers, both in Leningrad region, and in the Northwest region of Russia desperately need high-yielding varieties of early term maturing potato tubes suitable for mechanized cultivation and cropping, complex stability to the most nocuous pathogenic agents.

In the state register of Russia in 2005 200 varieties of potato were allowed to use in production, 133 varieties of them or 66.5% were of domestic selection. The analysis of high-quality structure testifies to insufficient supply the Northwest region of Russia with early and middle term maturing varieties of a potato.

The share of early ripening varieties of potato in the general assortment should make about 30%. However the area under crops with these varieties in Leningrad region is much less than required.

Moving of the basic potato production to a private sector results in significant change of requirements to the varieties. In connection to the fact that almost all private plots are infected with golden nematode of potato, the basic value of newly grown varieties of potato becomes resistance to this disease. Points of the region infected with golden nematode of potato in Leningrad region are found in Boksitogorsk, Volosovo, Volkhov, Vyborg, Gatchina, Kirishi, Lodeinoe Pole, Luga and Tikhvin areas. Totally more than 125 points in 45 economies are registered in the Leningrad region with the area about 2.5 thousand ha.

Pests and diseases

Management of golden nematode of potato disease by application of agro technical actions, using chemicals and cultivation of varieties resistant to golden nematode of potato disease, is the major condition of reception high and steady crops on infected sites. The basic actions in the system of control with golden nematode of potato disease are varieties resistant to golden nematode of potato disease. Besides golden nematode of potato disease in Northwest significant damage to potato crops is caused by the Irish potato famine fungus. The increased humidity and moderate temperatures of air during the vegetative period in Leningrad region carry it to a zone with high probability of occurrence Irish potato famine fungus where flashes of this disease are probable each three years out of four.

When the late blight disease of potato brakes out it causes losses a potato crop up to 40% and higher. In this case the basic way of control against the given infecting agent is raising and cultivation varieties of potato resistant to late blight disease of potato.

Losses of crop as well cause virus bacterial diseases, control with which should be carried out by means of steady varieties use. On plantings of potato in the Northwest region for some last years a new vermin could be noticed that is a completely new to these places – Colorado beetle. As a result of it one of the requirements to the new varieties of potato in the near future has to be resistant quality to this vermin.

Potato breeding

Selection of new potato varieties in the institute is conducted in a direction of creation high-yielding, early maturing varieties possessing stable resistance to cancer, hook-worm, late blight, potato scab, bare patch, and to viruses.

The success of selection work in these directions depends on presence of initial parental forms, their genetic level of scrutiny, methods of hybridization, an estimation and selection perspective grade samples. Previous long-term work carried out under the direction of outstanding selector Slipover E. A. has resulted in creation of such widely known varieties of a potato as Neva, Elizabeth, and the Petersburg, which high adaptable potential provides them distribution in various ecological zones.

Within more than 50 years genetically various hybrid material was created in which wild-growing varieties were used. That provides stability to various diseases and gives many economically useful attributes.

Wide scope of a potato gene pool used in the laboratory of potato selection, its profound study to the basic selected attributes, allows creating varieties early maturing potato with stability on years, with high quality tubers, resistant to cancer, potato hook-worm disease and other fungal pathogens. Work in these directions will proceed and be improved.

Expansion of research in selection to increase the contents of dry substances, starch and fibre in tubers are planned. This direction is an integral part of varieties selection, suitable to industrial processing for food from potato. A perspective direction in the given work is selection for stability to stressful factors, and also creation of sustainable genotypes.

In 2005 the institute has received patents on early maturing varieties of potatoes: In commemoration of Osipova, Belgorodskiy, and Kholmogorskiy. Prepared for State tests golden nematode of potato disease resistant varieties early- and middle- term of maturing with increased contents of starch: Colour, Farmer, Madam, and Dobrinja. They possess a complex of economically valuable attributes and high adaptable potential.

There are 30,000 genotypes of potato of interspecific origin in the laboratory. The genotypes have been selected according to the basic economic parameters and representing big interest for practical selection. Further targeted research of complex interspecific hybrids demanding additional material and financial investments are needed.

Seed production

As the potato is subjected to many diseases of a virus, fungal and bacterial origin, the majority of which are transferred through tubers from reproduction to reproduction, it is necessary to put into practice regular varieties improving which is carried out in a system of seed-growing. Process of seed-growing is subdivided into three stages: original, elite and reproductive.

In our institute during long period of time original seed potato of two classes is made: initially health-improved and super-super elite.

In seed-growing the first stage of creation initially health-improved material is the most significant and cost expensive. At this stage laborious work with use of diversified methods, receptions and selections on preservation of economically-significant attributes of investigated varieties is conducted.

It is extremely important to use modern diagnostic means for the control of virus and bacterial cleanliness of material produced. In fact, quality of the created initially health-improved potato in many aspects will determine success of all seed-growing cycle including growing of elite material.

Today the institute is in the process of improvement of 9 varieties of potato allowed to plant are included and 6 new perspective hybrid numbers. The quantity of own initial material makes depending on a year from 5.0 up to 7.5 thousand pieces of clones, and the occupied area - from 2.5 up to 3.5 ha.

Besides research works the Institute delivers original potato of the class super-super elite to the basic elite agricultural producers of the Leningrad region. Volumes of the delivered material delivered in 2001-2005 are shown in the Table 1.

In connection to the changed socioeconomic conditions of managing, the requirements showed by consumers to the quality of delivered seed potato began to become tougher. Since 2001 Leningrad region the new regulations on potato seed-growing and new normative requirements to quality and sowing parameters are entered in view of the international standards. At the same time, the problem of seed potato quality these years continues to become aggravated because of distribution of heavy forms of virus diseases, BBKK, bacteriosis, and also new forms and races of fungus diseases.

Table 1. Volumes of production and deliveries of original seed potato.

Year	Area, ha	Volumes of production and realization, tons			
		In total	Seeds, t		Other realization
			Reproduction	Realization in elite households	
2001	10.3	138.2	22.3	95.0	20.9
2002	10.8	130.0	26.6	82.0	21.4
2003	12.6	192.0	40.3	120.0	31.7
2004	14.1	217.0	42.0	124.4	54.6
2005	9.8	168.0	20.3	105.5	42.5

Alongside with the modern diagnostic means revealing latent contamination virus and viroid type diseases and diseases of bacterial character, at a stage of creation of a fore-basic seed potato it is necessary to use in control against activators of fungi diseases modern means of plants protection such as protectants, fungicides, insecticides, herbicides, desiccants, the chemicals stimulating growth and development of plants.

It is simply impossible to create high-quality seed material and to meet the requirements of the Russian Federation State Standard without using in seed-growing process the last mentioned means of plants protection.

Now in many regions and economies of the Russian Federation, together with the above mentioned modern methods of conducting potato elite seed-growing are widely applied. They are based on the application of biotechnological methods of improvement of varieties from viruses and other infections, clonal duplication formative tissue plants, and also use of various technologies of mass production of virus free tuber as an initial material in elite seed-growing.

For the purpose of selection green houses are used, namely for: accelerated duplication of the improved seed material of potato; raising virus-free plants by biotechnological methods; production of micro clones and cultivation of plants in tubes; indexations of clones; the display analysis of the latent virus infection; diagnostics of plants on a bacterial infection.

We are sure that, the most effective instrument of necessary amounts of seeds production, control over their quality and acceleration of duplication of perspective forms are building of a factory on mini tubers production with productivity of 1, 5 million tubers a year. The high-quality material produced here can be multiplied in a network of skilled - industrial economies and other enterprises certificated for these purposes.

Now *LRIAAS* with a network of research and production facilities produces about 3500 t of potato seeds. Production efficiency of the enterprise declines and grants from the regional budget do not restore necessary production costs, storages and realizations. The organization of a holding will allow solving the problem of efficiency increase of the specified enterprises.

In the long term the enterprises which can be included in the holding, and also those cooperating with it on a contractual basis, will be capable to provide needs of the Russian market with high-quality seeds to 50%, or for ½ of the sown area of the country that is important in development of Leningrad region economy.

A complex of organizational and technical actions with debt capital and budgetary investments will provide development of modern material base of the created holding.

The planned volumes of production and quality of potato seed will enable not only to provide region with seed material, reducing import of foreign selection varieties which are not adapted to climatic conditions of Russia but also to export seed material to the neighbouring countries.

Curriculum vitae – A.I. Osipov

Anatoloy Ivanocich Osipov, born April 18 1951, graduated from the Leningrad Agricultural Institute in 1973. He studied agronomy, agro-chemistry and soil science. In 1978, he finished a post-graduate course of the Leningrad Agricultural Institute and got his Candidate of science and in 1991 he became a Doctor of Agricultural Science. From 1993 he was deputy Director of the All-Russia Research, Design and Project-technological Institute of Chemical Soil Melioration.

In 1995 he became director of the All-Russian Research, Design and Project-technological Institute of Chemical Soil Melioration. As a result of integration of the All-Russian Research, Design and Project-technological Institute of Chemical Soil Melioration and the Agrochemical Research Institute in 1998, dr. Osipov becomes a deputy Director of the Intergated Institute. From 2003 on he is Directorof the State Research Enterprise, Leningrad Research Institute for Applied Agricultural Science (LRIAAS), Russian Academy of Agricultural Science.

A.I. Osipov is one of the leading scienticts in the Russuan Federation in the sphere of agricultural chemistry and chemical soil melioration. He published 170 works dealing with ecological safe methods of chemicals application in order to ensure reduction of chemicals losses, ecological harmless processing, norms of manuring lime rests application as fertilizers, efficient application of echologically safe and energy saving technologies etc.

Profile – Leningrad Research Institute for Applied Agricultural Science

The State Research Enterprise, Leningrad Research Institute for Applied Agricultural Science belongs to the system of scientific enterprises allied to the Russian Academy of Agricultural Science. It is situated in the village Belogorca in the Gatchina district, Leningrad region,

Russian Federation. The history of the Institute began in 1925 when it moved to Belogorca. It was an experimental agricultural station with responsibility for the research institutions administration for five regions of the North-West and the neighbouring regions of the Russian Federation. In February 1956 the Supreme Soviet made a decision to establish the 'North-West Research Agricultural Institute' based on this experimental agricultural station.

In 1998 the 'North-West Scientific and Research Agricultural Institute' was acknowledged as scientific institution responsible for the North-West region of the Russian Federation. In 2003 it was renamed to 'State Research Enterprise, Leningrad Research Institute for Applied Agricultural Science of the Russian Academy of Agricultural Science'. In total the Institute has about 170 workers. 3 doctors of science and 25 candidates of science work there. The institute consists of 15 laboratories.

The Institute has post graduate courses with the following specialties: agricultural chemistry, selection and seed-farming, vegetable-growing, plant-growing, economics, economical activity management in the Institute. The Institute's area of expertise is agriculture specialized in plant growing, farming, selection and seed-farming, vegetable-growing, grass-farming, agricultural chemistry and economics. More than 100 cultivars were submitted for the State Public Register of cultivars (selective achievements), which contains the cultivars allowed for cultivation. Of those cultivars, 43 cultivars are registered as allowed. They are divided as follows: 25 grass, 9 potato, 2 spring wheat, 1 winter rye, 1 spring oats and 2 spring rape cultivars.

The Institute continues its research in the field of new productive agricultural cultivars using traditional and non-traditional methods of selection, initial seed-farming, working out a regional system of seed-farming and invigoration of potato on the basis of new biologically active agents, fungicides and anti-virus chemicals, working out regional norms of fertilizer usage and so on.

Potato seed production in the middle Volga region

F.F. Zamalieva, Z. Stasevski, G.F. Safiulina, R.R. Nazmieva, V.B. Karimova, Z.Z. Salikhova, I.V. Pikalova, E.A. Gimaeva, S.G. Vologin, E.A. Prishchepenko, L.A. Garanina, E.F. Davletshina and G.D. Kadyrova
Tatar Institute of Agriculture, Agriculture Biotechnology Department, Orenburgskiji street 48, 420140 Kazan, Tatarstan, Russia, kartofelsemennoi@yandex.ru

Abstract

Disease-free seed potatoes production started in Tatarstan in the middle of eighties of the last century. For a long time large scale of disease-free seed potatoes introduction was restrained by fast virus re-infection of potatoes in the field. Main sources of virus re-infection of the disease-free seed potatoes in Tatarstan conditions are virus infected potato plants inside seed potatoes plots. According to our data weeds are not really dangerous sources of PVY infection for disease-free seed potatoes. High numbers of aphids cause large scale virus transmission from infected plants to healthy in potato fields. Large scale of PVY re-infection of disease-free seed potatoes in middle Volga region conditions can be explained by predominance of transmitters of this virus. The continental climate of Tatarstan has appeared unfavourable for reproduction of a peach aphids (*Myzus persicae* Sulz.) the most dangerous transmitters of PLRV, PVY, PVM. Data on monitoring aphids and viruses was used for developing management system of disease-free seed potatoes virus re-infection. Disease-free seed potatoes met the international quality requirements were produced by aid of system of biological and chemical control of potato virus re-infection. Tatar Institute of Agriculture has produced and delivered seed potatoes (class SE) to a many consumers of Tatarstan, of other regions of Russia and the CIS countries which virus infection rate as ELISA test show does not exceed 4% since 2005. Technology of fast seed potatoes multiplication was developed. Tatar Institute of Agriculture has produced 2500 tons of SE class disease-free seed potatoes annually for 3 last years. Disease-free seed potatoes provide high level productivity of potato: without irrigation up to 35-45 tons/hectare, on irrigated fields up to 70 tons/hectare.

Keywords: virus diseases, aphids, seed production

Introduction

New approaches of seed potato growing have been used in Tatarstan for more than 20 years. For a long time complete virus-free seed potato introduction was restricted by fast virus re-infection of potatoes under field conditions and absence of approachable seed-potatoes multiplication technology. Virus re-infection amounting up to 50-60% during 2-3 years was detected at this time (Figure 1). Potato virus Y (PVY) was the most widespread between other potato viruses in fields of Tatarstan.

The aim of this study was to detect the main causes of virus-free potatoes degeneration. Potato diseases controlling measures were developed and introduced. An original seed potatos multiplication technology was introduced.

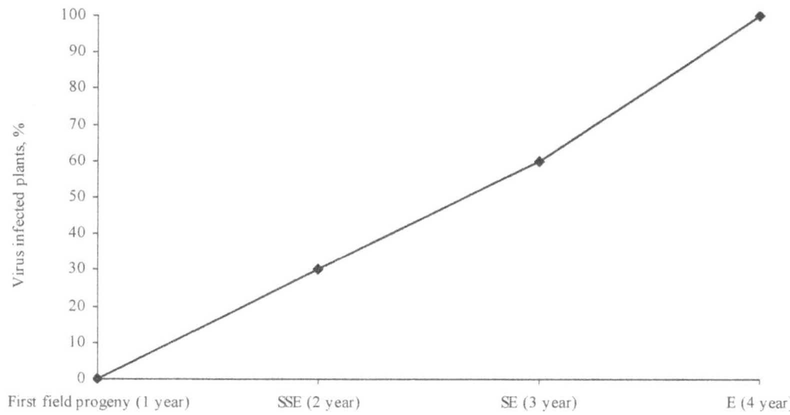

Figure 1. Virus re-infection rise of virus-free potatoes. First field progeny – first field progeny from greenhouse mini-tubers, SSE – super-super-elite, SE – super-elite, E – elite.

Fast virus re-infection had been caused 'inside' from virus infected *in vitro* potato plants and 'outside' transmitted by aphids for many years before virus detection and controlling measures was improved. 'Inside' infection was eliminated by introducing molecular virus detection methods, such as enzyme-linked immunosorbent assay (ELISA) and polymerase chain reaction (PCR). Potato plants testing for PVY, potato virus M (PVM), potato virus X (PVX), potato leaf roll virus (PLRV) and potato spindle tuber viroid (PSTV) have been introduced in Tatar Institute of Agriculture (TIA) since the late of 90-s. 70 *in vitro* potato plants lines from TIA gene bank were examined and 43 (60%) lines containing one or more virus were eliminated.

'Outside' virus infection is transmitted by aphids from infected potatoes, weeds and trees. According to our data weeds and trees are not really dangerous sources of virus infection in Tatarstan condition. Similar results was demonstrated in Vlasov (Vlasov, 1974) study.

Farmers in Europe had long noticed that crops grown in maritime or upland areas usually degenerated less rapidly than those grown in lowland continental areas. West France, North Germany, the Netherlands, Scotland and Ireland farmers supply new seed tubers to areas where home-saved seed degenerated and became unproductive within one or two seasons. National seed potato production programmes, based on clonal selection and seed multiplication in selected regions, started in Europe over 60 years ago (Basky, 2002; Robert *et al.*, 2000).

The main Russian potato research institutions are developing the concept of seed potatoes production in Russia. Under the concept the virus-free potato gene bank might be located in Solovetskije islands. Original potato seeds (mini tuber, super-super-elite (SSE) and super-elite (SE)) multiplication are planned in the northern part of Russia as well as the Archangelsk region. Healthy seed potato production is based on knowledge four components of virus epidemiology: the causal viruses, their insect vectors and plant hosts and the environment in which these interact. Long-term survey and detailed gained data study gave possibility to produce healthy potatoes in the Middle Volga region.

Material and methods

For a long-term survey, an area in Middle Volga region of Russia was selected at Sokury (Tatarstan, 30 km to the south from Kazan). Yearly surveys were made at Sokury between 1993 and 2005.

Aphids were caught in the water-filled yellow trays (YWT) (Moericke, 1955). Trays were placed along the edge and in the middle of the seed potato field. Trays were used from the time of plant emergence until haulm destruction (usually between June and August). Aphids from YWT were collected twice weekly and were preserved in 70% ethanol until being identified and counted. All species were identified with the key of Ivaniuk et al. (2005).

The total number of aphids caught in YWT during the yearly survey, the individual and total number of virus vectors and the cumulative vector intensity was calculated. The cumulative vector intensity is an index that describes vector abundance and their propensity to transmit potato virus Y (Irwin and Ruesnik, 1986). The propensity is the highest for the most efficient vector and the lowest of the least efficient vector.

In this study, the following vector propensity values were used, based on the results of: *Myzus persicae* Sulzer 1.0, *Macrosiphum solanifolii* Ashm 0. 10, *Acyrthosiphum pisum* Harris 0.05, *Aulacorthum solani* Kaltenbach 0.4, *Aphis nasturtii* Kaltenbach 0.42, *Aphis fabae* (Scop.) 0.1, *Aphis frangulae* Kaltenbach 0.42 (Siegvald, 1984; Ivaniuk *et al.*, 2005). Cumulative vector intensity for the season was calculated by accumulating species-specific vector intensity values at each trapping date.

Virus infection of potato plants was determined during visual inspection and collected leave testing with ELISA. Virus infection of progeny tubers was assessed during post-harvest inspections. Tubers were selected and sprouted in laboratory. The sprouts were tested with ELISA (one tuber – one sample). Bioreba ELISA test kits for PVY, PVM, PVX and PLRV detection was used.

Results

During the long-term experiment between 1993 and 2005 years the total annual number of aphids caught in YWT varied from 82 to 5000 (Figure 2).

The total number of *Myzus persicae* Sulcer the most dangerous transmitter of persistent PLRV and non-persistent PVY, PVM was not significant and varied from 0.5-5. The total number of *Aphis nasturtii* Kalt, *Aphis frangulae* Kalt vectors of PLRV, PVY and PVM was up to 684 and 784 (16% from total number of virus vectors), respectively. PVY transmitting *Aphis fabae* (Scop.) and *Acyrthosiphum pisum* Harris the total number was the highest up to 2407 (up to 57%) and 2070 (up to 44%). Thus the biggest amount of vectors is able to transfer PVY, then PVM and PLRV was detected in Tatarstan conditions. PVY is the most widespread virus in Tatarstan condition due to vide rage of PVY vectors.

The influence of environmental factors on aphid population dynamic, total number of aphids and cumulative vector intensity was studied and taken into account then aphids-born virus control

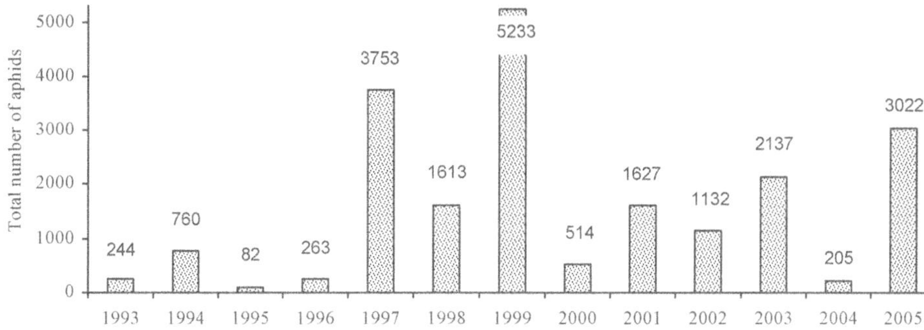

Figure 2. Total annual number of aphids in Sokury (Tatarstan, Russia) 1993-2005.

measures was designed. Temperature regime of May and early June determine directly the date of aphid emigration flight. There was significant positive relationship between accumulated air temperatures and the date of aphid emigration flight. The earliest aphid emigration flight from primary host was observed by the first decade of June. No relationship was observed when cold temperature with night frosts came back during spring and/or early summer time. Sharp temperature falls have long-lasting negative effect for the start of aphid distribution.

Thus, two significant environmental factors play a key role in aphid population development was revealed for Tatarstan conditions. Cold temperature was deleterious for insects during multiplication on primary host usually took place in May. Rains and strong wind restricted aphid emigration flight from primary host usually took place in June. Temperature falls following warm is frequent event in Tatarstan in May. The aphid population size directly depends on the number of overwintered aphids, which survived spring cold and was able to reproduce progeny. Therefore the cold temperature in May significantly decreased the total number of aphids.

The limiting role on the aphid population development plays the environmental conditions during winged forms emigration flight too. Long-term rains and strong wind prevented winged insect flight from primary host to new host. The progeny of overwintered aphids starved during long-term stay on primary plant. They have been washed away by rain and/or gone by wind when the June was unfavourable. The first generations losses in June play a crucial role in total number of aphids and drastically reduce insect population.

During the long-term survey between 1993 and 2005, the total number of aphids varied with no cycle and mostly depended on June-July weather conditions. The most important was availability of favourable temperature (18-23 °C) and humidity. Droughts have reduced the number of vectors usually. Host-plant presence and its juicy had the influence on aphid population dynamics.

In favourable environment conditions we detected two aphid flight peaks during the third decade of June to first decade of July and during second decade of July to third decade of July, respectively. There was a positive relationship between favourable environment conditions and intensity of aphid flight peaks.

The critical threshold of total aphid number means cumulative vector intensity caught in YWT from the beginning of cropping season. The critical threshold equal 50 individuals of *Myzus persicae* Sulzer. The forecasting for the risk of virus spread and determination of haulm destruction dates were empirical and based on cumulative vector intensity in YWT. Haulm destruction date is recommended within 12 days after the critical threshold was exceeded.

The total number of aphids and cumulative vector intensity was low and the critical threshold was not exceeded in 1993, 1995, 1996, 2000, 2002, 2004 years (Table 1).

The total number of aphids and total number of virus vectors caught in YWT the critical threshold was reached varied between 438-1280 and 338-782 (61-96%), respectively (Table 1).

The risk of virus spread was low in 6 years (1993, 1995, 1996, 2000, 2002 and 2004) from 13 years-long observation. The critical threshold was reached late (at first decade of August) in 1994 and the risk of virus spread in seed potatoes fields was low too. The total number of aphids and cumulative vector intensity was high and the critical threshold was exceeded early in 1997, 1998, 1999, 2001, 2003, 2005 years (Table 1). The mentioned situation is unfavourable for seed potatoes growers. There was not reached economically significant tuber yield at the early haulm destruction date.

Table 1. Total number of aphids, total number of virus vectors, critical threshold dates in Sokury (Tatarstan, Russia) 1993-2005.

Year	Total number of aphids	Critical threshold date	Total number of aphids at critical threshold date	Total number of virus vectors at critical threshold date	
				Number	%
1993	244	Total number of virus vectors below critical threshold			
1994	760	Early first decade of August	748	467	62
1995	82	Total number of virus vectors below critical threshold			
1996	263	Total number of virus vectors below critical threshold			
1997	3753	Early third decade of June	550	338	70
1998	1613	Second decade of July	438	351	80
1999	5233	First decade of July	493	360	73
2000	514	Total number of virus vectors below critical threshold			
2001	1627	Early second decade of July	809.5	525	65
2002	1132	Total number of virus vectors below critical threshold			
2003	2137	Early third decade of July	1280	782	61
2004	205	Total number of virus vectors below critical threshold			
2005	3022	Early third decade of July	462.5	443	96

The influence of potato planting date on virus re-infection was studied in 2001. Potato planted 25 May was enough mature at the critical threshold and virus infection was not detected in plant material. The process termed 'mature plant resistance' means that the older the plants at the time of inoculation, the less virus is translocated downwards to the tubers. At the planting date 5 June seed potatoes were infected most of all (cv. Belojarskij rannij (5%), cv. Nevskij (13.6%)). If potato was planted late (14 June), plants were very small at the time of vector aphid dissemination flights. There was not continuous green in the field. Virus infection was not detected in plant material as well as in early planting. However tubers yield were very low (Table 2). Thus early seed potatoes planting (till 25 May) reduces the risk of virus re-infection and allows to harvest the highest tuber yield.

The influence of the haulm destruction date on virus re-infection was studied in 2000-2001 (Table 3, 4). The total aphid number was low (514) but the initial level of crop infection was high (25%) in 2000. The latest haulm destruction date (26-08) drastically increased seed potatoes infection (98.5%) (Table 3).

The total aphid number was high (1627) and the initial level of crop infection was low in 2001. A three week-long delay of the haulm destruction date resulted in a significant seed potato infection raise of cv. Nevskij (1.7%) and cv. Belojarskij rannij (9.9%) (Table 4).

Table 2. The influence of potato planting date on virus re-infection (Infection, %).

Planting date	cv. Belojarskij rannij			cv. Nevskij		
	Virus infected plants, %	Yield, t/ha	Yield decrease, %	Virus infected plants, %	Yield, t/ha	Yield decrease, %
25-05	0	18.3	0	0	14.7	0
05-06	5.0	9.8	46.4	13.6	6.6	55.1
14-06	0	5.7	68.8	0	2.9	80.2

Table 3. The influence of the haulm destruction date on virus re-infection and tuber yield (cv. Nevskij) 2000.

Haulm destruction date	Virus infected plants, %	Yield, t/ha	Yield decrease, %
26-07	25	25.5	38.5
1-08	33	30.5	26.5
7-08	80	35.0	15.6
14-08	98.5	40.0	3.6
26-08	96.0	41.5	0

Table 4. The influence of the haulm destruction date on virus re-infection and tuber yield (cv. Nevskij, cv. Belojarskij rannij) 2001.

Harvest date	cv. Belojarskij rannij			cv. Nevskij		
	Virus infected plants, %	Yield, t/ha	Yield decrease, %	Virus infected plants, %	Yield, t/ha	Yield decrease, %
19-07	3.0	12.0	49.3	0	10.0	56.7
30-07	4.5	19.3	18.5	1.5	18.0	22.1
13-08	9.9	23.67	0	1.7	23.1	0

Direct relationships were found between seed potatoes re-infection and the dates of the total number of aphids in YWT, the total number of virus vectors, cultivars resistance and the initial level of crop infection. The initial level of potato infection plays key role in aphid-borne virus spread in Middle Volga region.

It is essential in the high grade seed potatoes production in high level of virus vectors area to avoid infected tuber planting. The initial level of crop infection plays a crucial role in aphid-borne virus spread. While high virus vector intensity provides fast re-infection. The seed potatoes have to be planted early as much as possible. Early yield and maturity of seed potatoes, it means the early haulm destruction date, can be reached by low planting density (45-55 thousand tubers per hectare).

The rules controlling spread of non-persistent viruses (PVM, PVY, PVX,) require to grow seed potatoes at least 1-2 km away from other potato fields. The control of persistent PLRV spread can be carried out with the aim of regular seed potatoes treatment by Bi-58 and Konfidor.

TIA gene-bank of virus-free *in vitro* potato plants contains 38 cultivars, presented by 69 cultivar lines. TIA seed potatoes production facilities are as follows: 60 thousand *in vitro* potato plants multiplied during autumn-winter time, 1000 square metre greenhouse for mini-tubers obtaining, 2 hectare of first field progeny from greenhouse mini-tubers, 15 hectare of super-super-elite (SSE), 100 hectare of super-elite (SE). Every seed potatoes reproduction is tested for virus (PVY, PVM, PVX, PLRV and PSTV) infection spread. *In vitro* plants are tested with ELISA and PCR. Greenhouse mini-tubers are tested with ELISA. Fields reproductions are visually inspected at least three times. Leaves collected at the potato early flowering stage and sprouts from harvested tubers are tested with ELISA.

The seed potatoes multiplication number for most widespread cv. Nevskij was determined. For *in vitro* plants planted in greenhouse multiplication number were equalled 3. For greenhouse grown mini-tubers – 7, for others reproduction (SSE, SE, E) – 8.5.

The introduction of new high grade seed potatoes and regular seed tubers exchange in Tatarstan farms resulted in an average yield increase from 10.6 ton per hectare in 2001 to 18.5

ton per hectare in 2006. TIA produces 2500 ton seed potatoes class SE annually. Following reproduction are grown in Tatarstan farms (50%), in other Russian Federation regions (40%) and CIS countries (Kazakhstan, Armenia) (10%) (Table 5).

In the last 10 years, increasing knowledge of the epidemiology of potato viruses has intensified their control. The new control strategy is based on continued surveillance to identify virus infection and vectors changes. Introduced potato devastating virus control measures has influenced significant virus infection reduction under 4% (ELISA test results of see potatoes SE class) during last 5 years (Figure 3).

Table 5. Seed potatoes production at TIA, following reproductions multiplication area.

Year	TIA production of SE, ton	Following reproductions multiplication area		
		Elite, hectare	First reproduction, hectare	Second reproduction, hectare
2001	500	-	-	-
2002	500	170	-	-
2003	2000	170	1200	-
2004	2500	700	1200	8400
2005	2500	830	4900	8400
2006	2500	830	5800	34000

Figure 3. TIA produced seed potatoes (SE) virus re-infection intensity decrease.

Conclusion

The data presented in this article have shown the possibility of high grade (virus-free) seed potato production in the Middle Volga region. There virus vectors intensity might be able to influence virus spread in some years.

Acknowledgements

A special acknowledgement goes to the director of the Tatar Institute of Agriculture Dr. R. Gareev for supporting the project. The authors thank all the project participants for assistance in laboratory, greenhouse and fieldwork.

References

Basky Z. (2002). The relationship between aphid dynamics and two prominent potato viruses (PVY and PLRV) in seed potatoes in Hungary. Crop Protection 21: 823-827.

Ivaniuk V.G., Banadysev S.., Zuromskij G.K. (2005). Zascita kartofelia ot boleznej, vreditelej i sorniakov. Belprint, Minsk, Belorussia, 427-475.

Irwin, M.E., Ruesnik, W.G. (1986). Vector intensity: a product of prospensity and activity. In: McLean, G.D., Garrett, R.G., Ruesnik, W.G. (Eds.), Plant Virus Epidemics Monitoring, Modelling and Predicting Outbreaks. Academic Press (Harourt Brace Jovanovich, Publishers) Sydney, Orlando, San Diego, New York, London, Toronto, Montreal, Tokyo, 13-33, 550pp.

Moericke V. (1955). Über die Lebensgewohnheiten der geflügelten Balltläuse (Aphidae) unter besonderer Berücksichtigugng des verhaltens deim Laden. Z. Angew. Entomol. 37: 29-91.

Semenovodstvo kartofelja. Kontrol kacestva. Sertifikacija. (2002) In: A. V. Korsunov, B. V. Anisimov (Eds.) Metodiceskoje posobije. Moskva, Russia.

Robert Y., Woodford J.A.T., Duray-Bourdin D. G. (2000). Some epidemiological approaches to the control of aphid-borne virus diseases in seed potato crops in northern Europe. Virus Research 71: 33-47.

Vlasov J.I. (1974). Prirodnyje ocagi i arealy virusov. Trudy VIZR. 41: 3-6.

Curriculum vitae – F.F. Zamalijeva

Fanija Faizrachmanovna Zamalijeva studied at Kazan State University (Tatarstan, Russia) department of Biology and Soil Research. She has been working at Tatar Institute of Agriculture (TIA) since 1987. She received a candidate of biological science degree from All Russia Plant Protection Institute (Sankt-Petersburg) in 1995 named 'Soil-born viruses and virus-free seed potatoes production improving in Tatarstan'. She is a leader of Agriculture Biotechnology Department (ABD) at TIA at present time. She led a research program close related with seed potatoes production in Tatarstan conditions. As a result of investigations an original seed potatoes production technology in Middle Volga region was created and introduced. Under ABD scientific supervision has been produced more than 6000 tons of high grade seed potatoes for last five years. The introduction of this high grade seed potatoes and regular seed tubers exchange at Tatarstan farms resulted in an average yield increase from 10.6 ton per hectare in 2001 to 18.5 ton per hectare in 2006. Zamalijeva is the author of over 50 scientific papers and editor of several editions of 'Recommendation for growers: Virus-free seed potato production'.

Profile – Tatar Institute of Agriculture

New adaptive and highly disease resistant cultivars, high grade seed, advanced agricultural technologies and advanced agricultural animal feeding technologies are the basis of the twenty-first century nature-friendly agricultural production. TIA has been carrying out scientific investigation in Middle Volga Region since 1920. We combine scientific research with large-scale introduction of new cultivars and new technologies. TIA delivers high grade seed of agricultural crops to a many customers in Tatarstan, in other regions of Russia and in the CIS countries.

Science based seed potato production development in the Archangelsk region

O.D. Kononov[1], L.A. Popova[1] and L.N. Golovina[2]
[1]Archangelsk research institute of agriculture, Popova, 17, 163000 Archangelsk, Russia
[2]Kholmogor experimental station of cattle breeding and farming, village V.Matigory, Arkangelsk region, Khalmogor district, Russia

Abstract

Due to climatic conditions of the Archangelsk region it is possible to harvest a large crop of potatoes (up to 30-40 tons per hectare) and to limit the spread of harmful pathogens. That is why some districts can be called 'pure zones', producing improved original and elite seed potato in such volumes that they satisfy the needs not only of the Archangelsk region but of other regions of the Russian Federation as well. Carrying out ecological tests of perspective hybrids and samples of potatoes makes it possible to create new varieties adapted to the climatic conditions of the European North.

Keywords: original and elite seed-breeding, ecological tests, meristemic plants, clones, diagnostics

Introduction

Effective functioning of a science based seed production system that provides more rapid development of new varieties in agriculture and generation of high quality seeds is one of the key factors determining successful performance of the tasks set by the agriculture complex of the Archangelsk region aimed at potato-breeding and raise of potato production competitiveness.

The territory of the Archangelsk region is 589.9 thousands square km, the extent from the north to the south is 1050 km, and from the west to the east is 1275 km. Although the territory of the region is rather big there are few agricultural lands – 728. 9 thousands of hectares that makes 1.8% of all land in the region. Ploughed fields occupy the territory of 277.8 thousands hectares that makes 43.8% of agricultural lands. Areas under crops are 26.0 thousands of hectares. The bulk of the cropping is in a private sector (22.6 thousands of hectares); the amount of the plantation at agricultural enterprises and farms does not exceed 9% of all the territory. Climatic and fytosanitary conditions of the Arkhangelsk region are favourable for growing a high quality seed potato. Cold winters, cool summer, long daylight hours (up to 23 hours a day) create favourable conditions for rapid potato growth and development.

The following factors determine favourable fytosanitary conditions for seed-growing of potato on the northern territories:

- secure absence of quarantine pathogenic fungi;

- minimal risk of plant virus infection in field conditions due to the absence or minimum of virus carriers, the absence of virus infection centres within a radius not less than 2 km and the absence of soil viruses and their carriers;
- minimum probability of harmful bacterial spread.

Seed production concepts

The developed concept for original, elite and reproductive potato seed-growing development in Russia presupposes the creation of sound varieties bank in isolated conditions of northern territories, the creation of original high quality seed material of commercial varieties and its further increase in the process of elite seed-growing at basic farms. In favourable fytosanitary, soil and climatic conditions it is planned to create a bank of sound potato varieties that will be the basis of an original technology for an improvement and a rapid seed increase (Anisimov, 2004).

To harvest stable high yielding crops of potato it is necessary to create new varieties and introduce them into production. High yielding crops of these varieties should be combined with an ability to withstand a complex of pests and diseases as well as unfavourable environment. The modern selection presupposes the combination of these factors with other characteristics in one breed: with high content of dry substance and starch, good consumer properties. The final stage of the selection process that precedes the transfer of new perspective varieties and hybrids into state breed testing is their testing and assessment according to main economic characteristics in different ecological and geographical zones.

Experimental stations of the Archangelsk agriculture research institutes researching potato breeding play an important part in solving the problems of new potato variety development that meet the requirements of an agriculture industrial complex and a processing industry as well as the production of original and elite material. There are qualified specialists, equipped laboratories for potato seed growing. The staff of the institute and of the experimental stations carries out ecological researches of perspective selected potato samples, sorts out the best samples adapted to northern conditions and researches technologies of cultivating district and perspective potato varieties for seeds and consuming. They also carry out the seed-growing process, study the influence of soil differences on the crop capacity and the quality of different potato varieties, create demonstration fields. The researches carried out make it possible to work out technological regulations for a seed production suitable for the conditions of the Archangelsk region.

Ecological testing of perspective hybrid potato samples selected by the Leningrad agriculture research institute is carried out at the Kholmogorsky experimental station of cattle breeding and farming. Annually at the nursery for ecological tests more than 20 breed samples are estimated according to the duration of a plant development stage, their resistance to diseases, the ability to give crops in 60-70 days after planting, the crop capacity and yield structure during the main harvesting, the content of dry substance and starch in tubers, the keeping capacity.

As the result of the ecological tests the following varieties of intensive type have been distinguished: 'Kholmogorsky' - from the group of 'early', 'To the memory of Osipova' - from the group 'middle early'. They were introduced into the state register of selection achievements allowed to be used in the production since 2005.

Perspective samples have been tested for the last years and the following hybrids were distinguished: from the group of 'early ripe' – 9547/2 (Lada) and from the group of 'middle early' – 95130/12 (Sudarynya) – non-resistant to nematodosis. Hybrid samples have been tested on the resistance to diseases in the conditions of natural infectious background and showed high and very high level of resistance to late blight of potato during dry years of experiments, middle resistance (5 points) during years with high humidity level. The productivity of the hybrid 95130/12 (Lada) from the group 'early ripe' was 4.6 tons per hectare higher than the productivity of the standard breed 'Pushkinets' in 60 days after planting; from the group 'middle early' the productivity of the hybrid 95130/12 N is 1.2 tons per hectare higher than that of the standard breed - Elizabeth, in 70 days after planting the productivity of these hybrids was 8 and 4.3 tons per hectare higher than that of the standard varieties.

During the main harvesting the productivity of the distinguished hybrid samples was much higher than that of standards: in the group of 'early ripe' it is 10.4 tons per hectare higher; in the group of 'middle ripe' it is 12.5 tons per hectare higher. The content of dry substance and starch in the hybrid 9547/2 (Lada) was 0.9 and 0.1% respectively higher than that of the standard breed 'Pushkinets' from the group of 'early ripe'. In the group of 'middle ripe' the content of dry substance in the distinguished non-resistant to nematadosis hybrid 95130/12 N (Sudarynya) is 2.8% higher than that of the standard 'Elizabeth' and the content of starch is 1.9% higher. The distinguished during the tests hybrid samples will be referred to the state breed testing.

By now the improved schemes of potato seed growing have been developed and are successfully used. They include effective ways and methods for the improvement and rapid reproduction of original and elite potato, methods of diagnostics of virus infection, complex systems of plants and tubers protection from fungi, bacterium and virus diseases.

The task set to the staff of experimental stations of the institute is to provide agricultural enterprises of al kinds with high quality seeds. To solve this problem it is necessary to develop original and elite seed-growing. Kholmogory experimental station of cattle breeding and farming and Kotlas experimental seed-growing station produce elite district and perspective seeds for the Archangelsk region. In practice the system of seed-growing is built on the basis of a particular scheme, the duration of which depends on the ways of getting basic material, its productivity rate, climatic conditions of the regions and some other factors.

At the Kholmogory experimental station of cattle breeding and farming the scheme of getting elite seeds is being carried out on the basis of the five-year scheme: hothouse mini-tubers, the first field reproduction, the nursery of super- super elite seeds, the nursery of super elite seeds, the nursery of elite seeds.

Annually test-tube plants are bought at the Leningrad agriculture research institute in the biotechnology department. To get sound potato basic material the method of multiple cutting of meristematic plants (grown in a sterile test-tube culture in amount of up to 30 thousand) is used. In the same summer it is possible to get from test-tube plants up to 170 thousand of mini-tubers grown in film-gauze hothouses on the territory of 0.25 hectares. From these mini-tubers during the second year the first field reproduction is grown on the territory of 2 hectares, during the third year the super-super elite is grown on the territory of 5.5 hectares, during the fourth year the super elite is grown on the territory of 12.0 hectares, during the fifth year the elite is grown on the territory of 31 hectare.

At Kotlas experimental seed-growing station seeds are grown on the basis of six-year scheme of elite potato production from improved basic material by the method of clone selection: the nursery of basic material selection (4000 clones), the nursery of first-year clone testing (1.5 ha), the nursery of second-year clone testing (2.5 ha), the nursery of super-super elite (8.0 ha), the nursery of super elite (13 ha), the nursery of elite (21 ha).

Primary potato seed-growing in the region is carried out according to types included into the state register of selection achievements: the early ripe – Kholmogorsky, Snegir; middle early – Rozhdestvensky, Sante, Nevsky, Elisabeth, To the memory of Osipova; middle ripe – Peterburgsky, Bronnitsky.

To change and to renew varieties in the due time it is necessary to produce annually up to 250 tons of elite potato seeds. Experimental stations offer this volume of elite potato seeds to agriculture producers. At experimental stations more than 20 varieties of potato are being studied. From these red tuber varieties are produced and they are the most popular among the population of the region: from the early varieties Alena, Kholmogorsky, Zhukovsky are the most popular, from the middle early these are Raduga, Ilyinsky, Lila; from the middle ripe – Troll, Malinovka.

Quality control

The control system of the state seed inspection is an essential part of the seed-growing system. The introduction of the quality and certification control system of potato seeds is very important and necessary not only at the level of elite types but also in the process of original seed-growing beginning with the production of improved mini tubers. At every stage of reduplication serologic analysis is used for diagnostics of X, Y, S and M viruses using the method of drop agglutination, clearing and cull.

Annually at agriculture enterprises of the region there appear serious problems with seed material quality diagnostics because of the absence of efficient virus control and comprehensive fytosanitary monitoring. In the Archangelsk region an accredited experimental laboratory for virus diseases and certification of seed material is required. It will provide secure protection from falsification and usage of low quality seed material for reduplication.

The basis of getting sound seeds is the choice of a predecessor of the potato crop, putting into the soil of organic and mineral fertilizers, growing plants according to the technology, timely mechanical and chemical processing. To grow elite it is necessary to use absolutely sound tubers as the basic material in combination with necessary agricultural methods that limit spread of infectious diseases in field conditions, provide high quality elite seeds. All this will forward the production of high quality potato at agricultural enterprises of the region.

UNECE standards for certification, marketing and commercial quality control of seed potatoes and early and ware potatoes

P.G. Bianchi[1], W. Schrage[2] and S. Malanitchev[3]
[1]Head, General Affairs, Ente Nazionale Sementi Elette, Via Ugo Bassi 8, 20159 Milan, Italy, Chairman of the UNECE Specialized Section on Standardization of Seed Potatoes
[2]Manager MN Seed Potato Certification, Minnesota Department of Agriculture, 312 4th Ave. NE, East Grand Forks, MN 56721, United States, Vice-Chairman of the UNECE Specialized Section on Standardization of Seed Potatoes
[3]Economic Affairs Officer, UNECE, Palais des Nations, CH-1211 Geneva 10, Switzerland, Secretary of the UNECE Specialized Section on Standardization of Seed Potatoes

Abstract

The commercial quality standards of the United Nations Economic Commission for Europe (UNECE) Working Party on Agricultural Quality Standards help to facilitate international trade, encourage high-quality production, improve profitability and protect consumer interests. UNECE standards are used by governments, producers, traders, importers and exporters, and other international organizations, and cover a wide range of agricultural products. The UNECE Standard for Seed Potatoes is a global reference intended to facilitate fair international trade by creating a harmonized certification system, promoting its use and defining harmonized quality requirements for seed potatoes. It is the only international reference covering all aspects related to seed potato certification: varietal identity and purity, genealogy and traceability, diseases and pests, external quality, sizing and labelling. The Standard considers issues falling under the World Trade Organization (WTO) Agreements on Technical Barriers to Trade and on Sanitary and Phytosanitary Measures. The UNECE Standard for Early and Ware Potatoes defines minimum requirements of commercial quality of potatoes traded internationally.

Keywords: trade, quality standards, certification

Development of agricultural quality standards is part of UNECE activities

UNECE, based in Geneva and set up in 1947 by the United Nations Economic and Social Council, is one of the five regional commissions of the United Nations.[1] As a multilateral platform, UNECE brings together 56 countries located in the European Union (EU), non-EU Western and Eastern Europe, South-Eastern Europe, the Commonwealth of Independent States (CIS) and North America to promote pan-European economic integration and cooperation as well as prosperity within and outside the region through:

- Policy dialogue.

[1]The other four are: the Economic and Social Commission for Western Asia (ESCWA, Beirut), the Economic and Social Commission for Asia and the Pacific (ESCAP, Bangkok), the Economic Commission for Africa (ECA, Addis Ababa) and the Economic Commission for Latin America and the Caribbean (ECLAC, Santiago).

- Negotiation of international legal instruments.
- Development of regulations, standards and norms.
- Exchange and application of best practices as well as economic and technical expertise.
- Technical cooperation for countries with economies in transition.

The areas of expertise of the UNECE include such sectors as energy, environment, human settlements, population, statistics, timber, trade, and transport. All United Nations Member States, as well as professional and other non-governmental organizations, may participate in its work.

History and aims of the UNECE Working Party on Agricultural Quality Standards

Agricultural quality standards covering a wide range of agricultural products - fresh fruit and vegetables, dry and dried produce, seed potatoes, meat, cut flowers, and eggs and egg products - are developed by the Working Party on Agricultural Quality Standards and its Specialized Sections under the UNECE trade programme.

In 1949, the UNECE Committee on Agricultural Problems established the Working Party on Standardization of Perishable Foodstuffs to determine common standards for perishable foodstuffs and to 'study steps to be taken on the international level in order to secure the general adoption of standards and control systems'. The responsibility of the Working Party was later extended to cover non-edible horticultural produce and quality development, which is reflected by its present name, the UNECE Working Party on Agricultural Quality Standards.

Through its four Specialized Sections on Standardization of Fresh Fruit and Vegetables, Dry and Dried Produce, Seed Potatoes and Meat, the Working Party provides a forum where countries can discuss all the commercial quality issues that may arise in their domestic markets and have implications for international trade. The UNECE Standards for Seed Potatoes and for Early and Ware potatoes are developed and maintained by the Specialized Sections on Seed Potatoes and Fresh Fruit and Vegetables respectively. The Working Party also offers assistance to countries in transition by organizing workshops on harmonizing national with international commercial standards. Figure 1 shows the position of UNECE work on agricultural standards within the overall United Nations structure.

Figure 1. Agricultural quality standards within the United Nations structure.

At the time when the UNECE initiated work on agricultural standards in 1949, countries were using national quality standards, which existed for some products and not for others, to regulate trade within their borders. National regulations represented non-tariff barriers to the rapidly growing post-war international trade. UNECE standards were created by harmonizing existing national commercial quality standards to produce a common trading framework for sellers and buyers to facilitate fair international trade and prevent technical barriers to trade. They also promoted high-quality, sustainable and profitable production, and created market transparency for producers and consumers.

Governments, producers, traders, importers and exporters, consumers and international organizations may all use UNECE standards. As recommendations they are, in principle, not obligatory for practical application, but are referred to, for example, in many EU regulations making them mandatory in all EU countries at export, import and retail stages. Some other countries have used them, eventually with amendments and modifications, as a basis for their national standards. UNECE standards are widely used as references in international trade contracts. The Organisation for Economic Co-operation and Development (OECD) adopts UNECE standards as OECD standards and develops explanatory material to interpret them.[2] The Joint FAO/WHO Codex Alimentarius Commission develops its standards for fresh fruits and vegetables drawing on the related work by UNECE.

Any country can initiate work either on the creation of a new or the revision of an existing standard. Figure 2 illustrates both processes. The proposal made to the relevant Specialized Section must contain a justification for why this new work or revision is necessary. The reasons mainly reflect technological developments and changes in trade practices. If the Specialized Section agrees, work on a revision can begin; in the case of a new standard, the proposal is transmitted to the Working Party for approval.

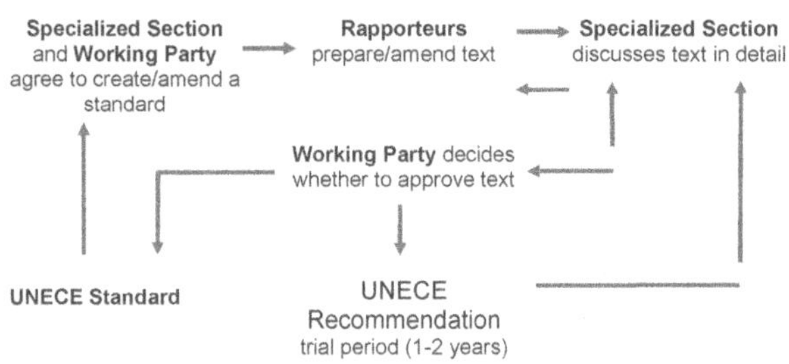

Figure 2. Process of developing or revising UNECE agricultural quality standards.

[2]More information on the work of the OECD Scheme for the Application of International Standards for Fruits and Vegetables can be found on: www. oecd. org/agr/fv.

The task of drafting a revised text of an existing standard or creating a new standard is normally assigned to a working group, led by a rapporteur and composed of several delegations. The rapporteur works on the text taking into account comments from the delegations, and presents the text to the Specialized Section for consideration. During the session, a standard may be revised on the basis of the proposed text or referred back to the rapporteur for further discussion. In the intersession period, delegations can send proposals for amendments or comments on the draft to the rapporteur and the secretariat. Once the text is accepted by the Specialized Section, it is forwarded to the Working Party for adoption either as a new or revised UNECE Standard, or for adoption as a UNECE Recommendation, with a trial period of one or two years for testing in actual application. Decisions are taken by consensus. The Specialized Sections and the Working Party meet on an annual basis.

The UNECE Standard for Seed Potatoes

History

Work on the UNECE Standard for Seed Potatoes began in 1958. At the 9th session of the Working Party, varying views were expressed concerning the nomenclature of the different categories of seed potatoes. The Group of Experts from the Federal Republic of Germany, the Netherlands and the United Kingdom of Great Britain and Northern Ireland was charged with preparing an analysis of existing national regulations and drafting recommendations for international standardization. Provisional recommendations were adopted in 1960 at the Working Party's 10th session for trial and revision according to the resulting experience. The first version of the text was adopted by the Working Party in 1963 at its 16th session. The Standard has been regularly updated since then.

Goals and scope

The aim of the Standard is to act as a global reference intended to facilitate fair international trade by:

- Creating a harmonized certification system.
- Promoting its use.
- Defining harmonized quality standards for seed potatoes.

It is the only international reference covering all the aspects related to seed potato certification, namely:

- Varietal identity and purity.
- Genealogy and traceability.
- Diseases and pests.
- External quality.
- Sizing and labelling.

The UNECE Standard for Seed Potatoes is part of the framework of complementary international agreements regulating varietal identity, phytosanitary requirements, commercial quality and the marketing of crops traded internationally. The Standard considers issues falling under the WTO Agreements on Technical Barriers to Trade (WTO-TBT) and on Sanitary and Phytosanitary

Measures (WTO-SPS).[3] These agreements recognize that countries have the right to establish scientifically justified protective measures for human, animal or plant life, and health as well as for the environment. On the other hand, they seek to ensure that technical negotiations and standards, as well as testing and certification procedures, do not create unnecessary obstacles to trade.

Regarding plant protection, the WTO-SPS Agreement identifies the International Plant Protection Convention (IPPC) as the international forum for setting phytosanitary standards and harmonizing phytosanitary measures affecting trade. The Convention is a legally binding agreement, but the standards developed and adopted by the IPPC are not legally binding. However, national measures that are based on international standards do not require supporting justification.

National and regional plant protection organizations, such as the European Plant Protection Organization (EPPO) and the North American Plant Protection Organization (NAPPO), work together to help IPPC contracting parties meet their obligations.

In 1995, NAPPO approved a regional standard for phytosanitary measures specifying requirements for the importation of potatoes into a NAPPO member country. The standard identified pest-risk management measures, including federal or state/provincial seed potato certification systems. It also established common criteria for limited generation certification schemes and diagnostic testing. The NAPPO standard does not deal with so-called quality pests, that is, pests affecting the commercial quality of agricultural products. In this sense, the UNECE Standard complements that of the NAPPO because it does address the issue of commercial quality of seed potatoes.

In 1999, the EPPO published a recommended certification scheme for seed potatoes. This scheme focused on defining conditions and requirements for the production of the initial material, principally derived through micro propagation, and the production of disease-free seed potatoes from the initial stock. This included defining the organisms, for which tests should be conducted, and the appropriate test procedures. The requirements for the certification of various categories of seed potatoes (Pre-basic, Basic and Certified) were aligned, as much as possible, with those of the UNECE Standard.

Council Directives 2000/29/EC (phytosanitary)[4] and 2002/56/EC (marketing)[5] set up certification and marketing rules for seed potatoes in the European Community. Each EC Member State is obligated to integrate the Community texts into national legislation. At the same time, for internal seed production, countries can apply stricter rules and standards. This EC Marketing Directive drew on the results of the work on the UNECE Standard in 1966.

[3]Texts of these agreements can be found at: www. wto. org.

[4]Council Directive 2000/29/EC of 8 May 2000 on protective measures against the introduction into the Community of organisms harmful to plants or plant products and against their spread within the Community (OJ L 169, 10. 7. 2000).

[5]Council Directive 2002/56/EC of 13 June 2002 on the marketing of seed potatoes (OJ L 193/60, 20. 7. 2002).

Application of the Standard

The UNECE Standard adopted by the Working Party is recommended to countries for application. Countries applying this Standard should notify the UNECE secretariat of their Designated Authority (DA) which is responsible for the application of the provisions and tolerances of the Standard. Application means the use of the UNECE Standard for export and import, that is for

Export: All seed potatoes certified and labelled for export by the DA meet at least the requirements of the Standard.

Import: Seed potatoes certified and labelled according to the UNECE Standard are accepted for import by the DA for parameters covered by the Standard, except where additional or more stringent requirements are applied by the country in respect to diseases and pests, if:

- The same requirement is applied to domestic production, and
- These requirements are justified to prevent introduction or spread of these diseases and pests which do not exist there or which seem particularly injurious to the crops in that country or in any part of its territory.

The DA shall notify the UNECE secretariat of each additional or more stringent requirement, together with technical or scientific justification for it. The responsibility for the quality of the lot remains with the owner.

The application of the Standard is without prejudice of any other legislation concerning industrial or commercial property, protection of crops, and the health of persons and animals.

Structure and contents of the Standard

The 30-page text of the UNECE Standard for Seed Potatoes covers the definition of the produce as well as provisions concerning variety, quality, sizing, tolerances for sizing, presentation and labelling. It also contains a number of annexes setting minimum conditions to be satisfied by the crop, by lots and by the succeeding crop (direct progeny) derived from seed potatoes. A comprehensive list of potato diseases and pests, with illustrative photographs, is regularly updated as a supplement to the Standard.

Varieties

Varieties are accepted into the Standard only if an official description and a reference sample are available from the DA. The variety should be distinct, uniform and stable according to the guidelines of the International Union for the Protection of New Varieties of Plants (UPOV)[6], and have a denomination allowing its identification.

Quality requirements

The Standard defines minimum requirements for seed potatoes at the export control point, after preparation and packaging. Seed potatoes shall be substantially free from injurious diseases

[6]The International Union for the Protection of New Varieties of Plants (UPOV) is an intergovernmental organization established by the International Convention for the Protection of New Varieties of Plants. The objective of the Convention is the protection of new varieties of plants by an intellectual property right.

and pests and from any defects likely to impair their quality as seed. They shall be substantially dry outside and, in general, of normal shape for the variety.

Classification
The Standard defines minimum requirements for three categories of seed potatoes which are in descending order of quality: Pre-basic, Basic and Certified. Additional choice of quality is offered by the inclusion of two classes within each of the three categories. The application of classes is optional. The definitions are set out below:
- Pre-basic seed: Seed potatoes of generations prior to Basic seed. The Standard establishes different requirements for class Pre-basic TC (tissue culture) - seed directly derived from micro propagation - and for class Pre-basic - seed not directly derived from micro propagation.
- Basic seed: Seed potatoes descended directly from Pre-basic or Basic category seed or produced under the special provisions of a national certification scheme. They are mainly intended for the production of Certified seed potatoes and are classified as either Basic I or Basic II, according to the minimum requirements for each class.
- Certified seed: Seed potatoes descended directly from Pre-basic, Basic or Certified seed. They are mainly intended for the production of potatoes other than seed potatoes and are classified as either Certified I or Certified II, according to the minimum requirements for each class.

Within these categories, countries producing seed potatoes are free to create national classes which are subject to specific national requirements.

Minimum conditions to be satisfied by the crop
The field shall not be contaminated by *Globodera rostochiensis* nor by *Globodera pallida*. The crop shall be free from:
a. *Synchytrium endobioticum* (Schilb) Perc.
b. *Clavibacter michiganensis* spp. *sepedonicus* (Spieck. and Kotth.) Skapt. and Burkh.
c. *Ralstonia solanacearum.*
d. Potato spindle tuber viroid.
e. Tomato Stolbur.

The Standard specifies maximum acceptable limits for the proportion of growing plants not true to the variety, affected by blackleg and showing symptoms of virus diseases. For example, the percentage of growing plants affected by blackleg should not exceed:
a. 0 per cent for a Pre-basic category seed crop
b. 0.5 per cent for a class Basic I crop and 1 per cent for a class Basic II crop
c. 1.5 per cent for a class Certified I crop and 2 per cent for a class Certified II crop.

Minimum quality conditions for lots of seed potatoes
The Standard defines the following tolerances for defects and diseases allowed for seed potato tubers. For surface diseases, firstly an allowable surface area (in per cent) is specified and then a standard for tubers exceeding the allowable surface area is set.
a. Presence of earth and extraneous matter is tolerated up to the maximum of 1 per cent by weight for Pre-basic category seed and 2 per cent for Basic and Certified categories.

b. Dry and wet rot, where not caused by *Synchitrium endobioticum, Clavibacter michiganensis* and *Ralstonia solanacearum*, are tolerated up to 0.2 per cent for class Pre-basic, and 1 per cent for Basic and Certified. No rots are allowed in class Pre-basic TC.
c. External defects (e.g. misshapen or damaged tubers) are tolerated up to 3 per cent in all categories.
d. Common and netted scab tolerance: common and netted scab are allowed in all categories and classes except class Pre-basic TC; no more than 5 per cent of tubers may exceed an allowable surface cover of 33.3 per cent.
e. Powdery scab tolerance: powdery scab is allowed in all categories except class Pre-basic TC; no more than 1 and 3 per cent of tubers may exceed an allowable surface cover of 10 per cent in class Pre-basic, and in Basic and Certified categories, respectively.
f. Rhizoctonia tolerance: Rhizoctonia is allowed in all categories except class Pre-basic TC; no more than 1 and 5 per cent of tubers may exceed an allowable surface cover of 1 and 10 per cent in class Pre-basic, and in Basic and Certified categories, respectively.
g. Shrivelled tubers. Excessively dehydrated and wrinkled tubers are tolerated up to 0.5 and 1 per cent in class Pre-basic, and Basic and Certified categories, respectively. Shrivelled tubers are not allowed in class Pre-basic TC.

Total tolerances for (b) – (g) are 3 per cent for class Pre-basic TC, 5 per cent for class Pre-basic, and 6 per cent for Basic and Certified categories.

The seed potatoes shall be free from *Globodera rostochiensis* (Woll) and *Globodera pallida* (Stone), *Synchytrium endobioticum* (Schilb.) Perc., *Clavibacter michiganensis* spp. *sepedonicus* (Spieck. and Kotth.) Skapt. and Burkh., *Ralstonia solanacearum* (E. F. Smith) E. F. Smith, Potato spindle tuber viroid, Tomato Stolbur and *Meloidogyne chitwoodi and fallax* and *Ditylenchus destructor.*

Minimum conditions to be satisfied by direct progeny (succeeding crop) of seed potatoes
The Standard establishes the following tolerances for direct progeny with regard to varietal purity and virus diseases:
a. Plants not true to the variety and of other varieties should not exceed 0 per cent for class Pre-basic TC, 0.01 per cent for class Pre-basic, 0.25 per cent for Basic category seed or 0.5 per cent for Certified category seed.
b. Plants showing symptoms of mild or severe virus diseases should not exceed 0 per cent for class Pre-basic TC, 0.5 per cent for class Pre-basic, 4 per cent for Basic category seed or 10 per cent of severe virus disease for Certified category seed.

Labelling
The Standard prescribes that each container of marketed seed potatoes shall bear on the outside an official label indicating the designated authority, country and/or region of production, lot number, month and year of closing, variety, category and class, size and declared net weight. The colour of the label should be white with a diagonal purple line for Pre-basic seed, white for Basic seed, and blue for Certified seed. Reference to the UNECE Standard may be included on the label.

Minimum conditions to be satisfied in the production of Pre-basic TC seed potatoes
The Standard prescribes that:

a. The parent material must be true to type for the variety and the initial plantlets must be free from pathogenic organisms.
b. The facilities and procedures used should be subject to official approval by the designated authority. The record-keeping system should document the source of the material and the volume of production.
c. The growing medium should be pest-free.
d. All reasonable husbandry practices for the prevention or spread of pests and diseases must have been effectively carried out.
e. The growing crop must have been kept free from *Synchitrium endobioticum*, potato viruses, bacterial diseases and from deviations of variety or type.
f. The satisfaction of the above conditions shall be established by official inspection and/or testing.
g. Confirmation of varietal purity or trueness-to-type will be dependent on inspection of the crop derived from this pre-basic material.

Further development of the Standard

The UNECE Standard for Seed Potatoes is constantly updated to reflect changes in seed production and marketing, the evolution of harmful organisms and development of new techniques for their detection. The most recent amendments to the Standard, agreed by the Specialized Section at its March 2007 session, concerned: (i) the adoption of assessment keys for common and netted scab, and for powdery scab; (ii) revision of the text on the purpose of comparative trials; and (iii) changes to provisions on sampling and homogeneity of the lot.

At its future sessions, the Specialized Section will consider: (i) procedures for assessment of non-compliance at destination; (ii) sampling procedure for inclusion in the Standard; (iii) Potato Tuber Necrotic Ringspot Disease; (iv) Erwinia; (v) Potato Cyst Nematode; (vi) images for the list of diseases and pests; (vii) exchange of information on crop inspection practices; (viii) control under official supervision and accreditation systems; (ix) promotion of the Standard; (x) electronic certification; (xi) exchange of information on post-harvest tests; and (xii) vigour. Researchers and industry experts are invited to contribute to the work of the Specialized Section and to advise certification authorities, for example, on how to improve certification measures to control potato pests more effectively.

Advantages of using the Standard

The advantages of using the Standard either as a regulation or as a reference is that common terminology and standards can be applied to enable users to compare the quality of seed potatoes produced in a range of locations more easily. For example, seed potatoes are certified as class Elite in many countries, but the conditions and standards applying to their certification differ widely. This means that a buyer has to examine each national regulation to be assured of the quality, and that he needs to check annually for any changes.

Applying an international reference such as UNECE Standard means that this necessity is removed. Seed potatoes from any seed potato certification system can be compared with the UNECE Standard in order to determine their acceptability in the country or area. For example, if seed potatoes harvested in third countries complied with European Community Plant Health rules, then the marketing of seed potatoes according to the UNECE Standard should facilitate the establishment of equivalence and acceptance of trade.

It has to be recognized that the UNECE Standard is a commercial quality standard and that phytosanitary requirements are not fully addressed. However, for most pests and diseases of plant health concern, such as cyst and root knot nematodes or potato wart, the Standard does specify a zero tolerance because certification schemes are part of an integrated risk mitigation system delivering healthy seed potatoes.

In the United States, local use of seed potatoes is much greater than export. There are different seed potato certification programs in the U. S., because they are the responsibility of the States, while plant quarantine is a shared responsibility between the Federal and State authorities. The U. S. seed potato certification agencies have an interest in participating in the activities of the UNECE Specialized Section, because it develops a reference standard to which each State can compare its own standards. The U. S. Seed Potato Export Standard represents the common denominator of the State certification programs. However, because the UNECE Standard is more specific, an exporter can make a more detailed comparison with the requirements of an importer, when using it as reference. Also, the different seed potato certification programmes in the U. S. have used the expertise of the Specialized Section to their advantage when facing new issues, such as Potato Mop Top virus and new strains of potato virus Y (PVY).

A forum for discussion

The Specialized Section on Seed Potatoes is a forum where different countries can find common solutions to common problems. It is an open environment of professionals. One good example is the discussion on new strains of PVY, where the Specialized Section was used to exchange information and have scientists from different countries communicate with each other in order to approach the problems in a common fashion. The result was that the Specialized Section agreed on the following position on PVY:

a. New strains and variants of PVY are occurring worldwide and seem to be more readily transmissible than earlier strains.
b. Strains and variants differ in their ability to cause leaf and/or tuber symptoms.
c. Control of PVY is still only possible through strict PVY tolerances within a seed potato certification scheme, through breeding and through adoption of best crop management practices.
d. An option to control PTRD (Potato Tuber Necrotic Ringspot Disease) on susceptible varieties may be a specific certification tolerance for superficial necrosis, until an appropriate detection method becomes available.
e. Applying a zero tolerance for widely distributed viruses, such as PVY, is not effective, practical or feasible.

UNECE Standard for early and ware potatoes

The UNECE Standard concerning the marketing and commercial quality of Early and Ware Potatoes is one of the 50 standards developed and maintained by the Specialized Section on Standardization of Fresh Fruit and Vegetables. Its 2006 version is a combination of two formerly separate Standards, one for early potatoes and one for ware potatoes, first published in 1961 and 1967 respectively.

The Standard has the following structure reflecting the main aspects of commercial quality:

I. Definition of produce
II. Provisions concerning quality
   A. Minimum requirements
III. Provisions concerning sizing
IV. Provisions concerning tolerances (Quality tolerances, Size tolerances, Tolerances of other varieties)
V. Provisions concerning presentation (Uniformity, Packaging, Presentation)
VI. Provisions concerning marking (Identification, Nature of produce, Origin of produce, Commercial specifications, Official control mark)

Like other commercial quality standards, the Standard for Early and Ware Potatoes serves as a reference for trade contracts and helps prevent confusion and misunderstanding between buyers and sellers. It allows buying a product without a visual check before ordering and simplifies control at delivery. The Standard makes price comparisons for products from different sources easier. It also greatly facilitates sales and contract operations, at reduced risk and cost.

Standard definitions create similar perception of different qualities. Traders and growers can sort their products in the same way according to standard specifications. The UNECE Standard, though without setting any price differentials, indicates the existence of a differentiated consumer demand associated with different prices for different grades of produce. This in turn prompts the cultivation of higher-quality and more profitable products.

Participation

To influence the content of the UNECE standards and to ensure that all aspects of production and trading are taken into account, all interested countries are invited to contribute to the development of these standards by taking part in the activities of the UNECE Working Party on Agricultural Quality Standards and its Specialized Sections. International producer, trade and consumer associations are also invited to participate. National expertise is welcomed through the national delegations, that is, through the competent national authorities. Information on UNECE standard-setting meetings is available at: www.unece.org/trade/agr/meetings.

References

UNECE Standard concerning the certification and commercial quality control of Seed Potatoes. 2006 edition. United Nations Economic Commission for Europe. www.unece.org/trade/agr/standard/potatoes/pot_e.htm.

UNECE Standard concerning the marketing and commercial quality control of Early and Ware Potatoes. 2006 edition. United Nations Economic Commission for Europe. www.unece.org/trade/agr/standard/fresh/fresh_e.htm.

EPPO Certification scheme for seed potatoes. European and Mediterranean Plant Protection Organization. http://archives.eppo.org/EPPOStandards/PM4_CERT/pm4-28-e.doc.

NAPPO Regional Standard for Phytosanitary Measures (RSPM). Requirements for Importation of Potatoes into a NAPPO Member Country (RSPM No. 3). North American Plant Protection Organization. www.nappo.org/Standards/REVIEW/RSPM3Rev-e.pdf.

Curriculum vitae – Willem Schrage

Willem Schrage was born in the Netherlands, where he farmed three years before continuing his education. He attended Wageningen University, where he obtained his MSc in crop protection in 1976. He joined the International Potato Centre for six years and was stationed in Turkey and Tunisia. The emphasis of his work in the Mediterranean region was the advantages local seed potatoes could have due to physiological age. He immigrated to Canada, where he was employed by the New Brunswick Department of Agriculture as seed potato extension specialist. After which he served the potato industry in that province for nine years as manager of the provincial potato marketing board, until he moved to Minnesota in the USA, where he is still the manager of that state's potato inspection program. He writes regularly on national and international seed potato topics. He is secretary of the 'Certification Section' of the Potato Association of America. He represents the US state seed potato certification agencies at the UNECE 'Specialized Section on Seed Potatoes', of which he is vice chair.

Curriculum vitae – Pier Giacomo Bianchi

Born in Italy in 1958, was graduated as Agronomist at the University of Milan in 1984. He works at the seed certification agency in Italy (ENSE) since 1985 and he is Senior Researcher and responsible for the general affairs unit since 1989. The main expertise concerns seed legislation. Since 2000 he acts as Chairman of the Specialised Section on standardisation of seed potatoes of the Economic Commission for Europe of the United Nations. He seats at the Standing Committee for Seed at the European Commission in Brussels on behalf of Italy since 1989 and he assisted the Commission Services as seed expert on several occasions. He was part of the EU delegation to assess the equivalence regime for the import of soybean seed from USA and Canada and attends several meetings at the Community and international level (OECD, UPOV, CPVO). He is Member of the Board of Appeal of CPVO (The EU Variety Office). He has experience in Twinning project for capacity building in the seed area in Malta and Poland and attended several TAIEX (technical assistance exchange office of the EU) events since 1998 (screening, peer reviews and advisory meetings, seminars for new staff and Members of Parliament) in MT, PL, SK, CZ, SL, H, EE, LT, LV, CY, BG, RO, HR, TR, FYROM, B&H, NORTHERN CY, or in Brussels. He participated in workshops within the programme of the European Union for the 'development of the seed sector in Uruguay' in 2003 and 2005. He is Author of a study 'on the rules on marketing of seed and plant breeders' rights in the EU' prepared in the framework of the Project 'Biodiversity Brasil' managed by the Istututo Agronomico per l'Oltremare of Florence in 2003.

At the national level he is responsible for the national network for listing of new varieties to the seed marketing catalogue. He is Secretary to the Council of Administration of ENSE. He is author of several presentations at national and international conferences, papers and columns. He is author of the leading article of 'dal seme', the quarterly review of ENSE.

Curriculum Vitae – Serguei V. Malanitchev

Serguei V. Malanitchev was born in Russia. He studied in the Moscow Institute of Economics and Statistics from which he received a diploma in 1977. He also received a *Diplôme de Hautes Etudes Internationales* in international economics in 1991 from the Graduate Institute of International Studies in Geneva. He joined the United Nations Economic Commission for Europe (UNECE) in 1985 where for twenty years he worked on economic statistics and analysis. This work included research activities, organization of international meetings and technical assistance to countries in transition, notably in the area of agricultural statistics. He published a number of research papers and edited several UNECE publications on statistics. Since 2006 he services the UNECE Working Party on Agricultural Quality Standards and its Specialized Sections that focus on the development of commercial agricultural quality standards, including standards for seed potatoes and early and ware potatoes.

Profile – USA State Seed Potato Certification Agencies

In the USA Seed Potato Inspection is under state statute; shipping point inspection potatoes, including seed is carried by state inspectors, under license of the US Department of Agriculture, Agricultural Marketing Service.

In Minnesota the seed potato certification is part of the Plant Protection Division of the Minnesota Department of Agriculture.

US certification agencies are represented at the Specialized Section of the UN/ECE by a representative, selected by the Certification Committee of the National Potato Council. It is one of the activities of the Seed Potato Export Program of the US Potato Board.

Profile – Ente Nazionale delle Sementi Elette

The Ente Nazionale delle Sementi Elette (ENSE), which was set-up on September 11, 1954, has been acknowledged as a legal entity by decree by the President of the Republic n° 1461 - 12/XI/1955. According to law n°1096 - 25/XI/1971 'Rules and regulations governing the activities regarding seeds', which assimilated the relevant community directives, ENSE was acknowledged as a Public Corporation and was included in the list of Research Institutes under the supervision of the Ministry of Agricultural, Food and Forest Policy (MIPAAF).

By appointment of MIPAAF, ENSE carries out the necessary tests, laboratory analyses and official certification activities on seed products according to Law n° 1096 25/XI/1971, to the other regulations ruling this matter and international conventions and agreements subscribed by the Government. This assignment was granted to the Institute by the DM 17/2/1972 as regards agricultural seeds and by DM 02/11/1976 as regards vegetable seeds.

Research and experimentation aim at studying the issues regarding certification in-depth by formulating or implementing new methods to assess seeds as regards technology and variety.

Furthermore, the Institute studies the impact of new techniques applied to seed production on the procedures used for official inspection.
Experiments include the necessary evaluation tests supporting new varieties entries to the national List. Community comparison trials are also organised in co-operation with the EU to harmonize the certification process for seeds among the member countries. The Institute participates to Work Groups in national and international Bodies involved in seed-related activities (ECE/ONU, ISTA, UPOV, UE, OECD, MIRAAF, Regional Committees) also as a technical support to the Ministry.

ENSE employ 95 permanent staff and 60 seasonal staff. The headquarter is in Milan. Departments are located in Verona, Bologna, Battipaglia, Palermo, including the Seed Testing Station and Experimental Farms.

Company profile – United Nations Economic Commission for Europe

The United Nations Economic Commission for Europe (UNECE), based in Geneva, is one of the five regional commissions of the United Nations. As a multilateral platform, UNECE brings together 56 countries located in the European Union (EU), non-EU Western and Eastern Europe, South-Eastern Europe, the Commonwealth of Independent States (CIS) and North America to promote pan-European economic integration and cooperation as well as prosperity within and outside the region through: (i) policy dialogue; (ii) negotiation of international legal instruments; (iii) development of regulations, standards and norms; (iv) exchange and application of best practices as well as economic and technical expertise; (v) technical cooperation for countries with economies in transition. The areas of expertise of the UNECE include energy, environment, human settlements, population, statistics, timber, trade, and transport. Work on agricultural quality standards is carried out under the trade programme. All United Nations Member States, as well as professional and other non-governmental organizations, may participate in its work.

Agronomy

Technology of potato growing in the Czech Republic

J. Čepl and P. Kasal
Potato Research Institute, Dobrovskeho 2366, 580 01 Havlickuv Brod, Czech Republic

Abstract

Potato growing has a long lasting tradition in the Czech Republic. During the last years the growing technology has been subjected to a substantial change. Specialized potato growers use technology of de-stoning. Compared to the conventional technology higher yields were recorded in the field trials due to better soil loosening prior to planting and reduced soil compaction during the whole vegetation period.

Keywords: growing technology, soil preparation, fertilization

Introduction

Potato growing has a rich tradition in the Czech Republic. It is particularly associated with the progress of alcohol and starch industry shortly after World War I. However, seventy years ago potato areas were more than ten times larger and reached almost 0,5 mil. ha in the Czech Republic (Table 1).

Besides processing the substantial part of potato production was intended to animal feeding at that time. After World War II potato areas remained on the level of 450 thousands ha. In the 1960s a sharp decrease was recorded i.e. below 200 thousands ha to 100 thousands ha in 1990. The last stage of area decline started after 1990 (Table 2).

Table 1. Potato area development in the Czech Republic during 1935-1990.

Year	Area (ha)
1930	441,454
1935	492,871
1945	445,071
1950	468,220
1955	437,193
1960	390,237
1965	390,237
1970	229,084
1975	165,233
1980	127,865
1985	126,915
1990	109,299

Table 2. Total areas, yields per hectare and potato production in the Czech Republic after 1990.

Year	Harvest area (ha) incl. seed potatoes	Yield (t/ha)	Production (t)
1995	77,869	17.08	1,330,119
1996	85,710	21.00	1,800,272
1997	72,625	19.30	1,401,663
1998	71,855	21.15	1,519,768
1999	71,455	19.69	1,406,832
2000	69,198	21.33	1,475,992
2001	54,137	20.88	1,130,477
2002	46,914	23.57	1,105,960
2003	43,489	19.35	841,465
2004	42,141	23.57	993,203
2005	41,207	28.05	1,155,996
2006	38,549	21.70	837,000

Technology of potato growing in the Czech Republic

After 1990 considerable changes were carried out in the whole system of potato growing and consumption in the Czech Republic. The potato industry of the Czech Republic has markedly changed and gradually reached level approaching this branch to the parameters comparable to EU-countries. At present, modern potato growing technologies, the best European varieties and potato handling ensuring quality product of adequate price for consumers are applied. The relations of potato grower and customer have been improved and potato growers have been more specialized. In addition to potato area reduction a shift of table potato growing was done towards warmer and higher-yielding localities and growth in potato products on the market. Distribution of production according to individual utilization purposes is shown in Table 3.

Table 3. Potato utilization (consumption in thousand t).

Index	1998/99	1999/00	2000/01	2001/02	2002/03	2003/04	2004/05	2005/06	2006/07
Potatoes for human consumption	780	770	780	778	778	775	770	770	765
Potatoes for starch production	75	80	151	131	175	99	144	166	116
Potatoes for planting	240	240	210	170	145	135	135	135	105
Potatoes for feeding and losses	430	355	438	189	140	27	57	110	47
Export	8	3	13	4	5	21	67	121	18
Total utilization	1,533	1,458	1,630	1,291	1,262	1,082	1,206	1,358	1,096

The potatoes are very popular in the diet of the Czech population, the total per capita consumption ranges between 75 and 80 kg/year.

Since 1995 a considerable reduction in areas of late table potatoes (by 48.5%) and a reduction in potato production (by 28.2%) has been recorded. In contrast, a 39.4% increase of yield per hectare was observed. The estimated potato area is aggregated from the area of agricultural sector and the estimation of area of households. The portion of households in total area of late table potatoes is 25%. The most potato areas are concentrated in the Bohemian-Moravian Highland.

The production of early table potatoes is particularly concentrated in two main potato production regions in the Czech Republic. The first one is the region of Elbe River and the second one is South Moravia. In part of production areas (about 2,000 ha) irrigation is applied. The harvest of irrigated potato fields begins in the first June decade (in late May, when the non-woven fleece is applied).

Starch potato areas have been recently stabilized. Mean tuber yields (incl. seed potatoes) neared to 30 t, average starch content was higher than 18% in last five years and average starch yield per ha was about 5.5 t.

Potato propagation areas are also being decreased. In the year 1995 the area of potato propagation achieved nearly 8 thousands ha (9.4% of total potato area), in the year 2003 the area sank below 5 thousands hectares to 4,692 ha, i.e. 10.6% of total potato area for the first time. In the year 2006 propagation area was reduced to 3,630 ha (9.4% of total area).

An enhancement of the production intensity is the basic condition of Czech potato growing competitiveness. Comparing documented hectare potato yields in the Czech Republic and the top EU-countries we find that potato yields of the Czech Republic has long-term level about 70% of potato yields in EU-15.

Limiting intensification factors for potato yield formation, where reserves exist, are at present:

- sufficient water supply during growing period (increasing irrigated potato areas and/or using;
- drip fertigation could be a solution);
- increase in certified seed potato percentage.

Other factors are technology of soil preparation, using potato varieties with good performance, integration and optimization of commercial fertilizer and pesticide inputs.

In the National List of Varieties, 188 potato varieties are registered. So-called supporting varieties, i.e. varieties with about 100 ha of propagation area from the total 3,630 ha, are Impala, Marabel, Ornella, Dali, Adéla, Rosara, Princess, Laura, Ditta. These varieties are grown on about 50% of areas.

The fertilizer consumption (net nutrients per hectare) sharply declined after 1989. In 1990 it was almost 200 kg/ha; however, at present the consumption is 90 kg/ha (with 70 kg for nitrogen). In relation to phosphorus and potassium, the consumption reaches low volumes (12 kg/ha and 8 kg/ha). It was due to financial state of recently transformed agricultural subjects and also due to the fact that P and K soil availability was rather high in consequence of high nutrient rates in commercial fertilizers in the 1980s.

Considering data on nutrient availability, a state-supported program exists for soil nutrient analysis using Mehlich-III extraction method in the Czech Republic, once in 6 years on each plot. Potato growers receive current data especially on P, K, Mg and Ca content and pH value. Methodologies issued by PRI for potato nutrition with commercial fertilizers with P, K and Mg content consider this soil nutrient content basic aspect for recommendation of P, K and Mg rates; utility type of potato growing, growing period duration and rate of manure are further important factors.

Nitrogen rate is determined in relation to utility type of growing, growing period duration and rate of manure or data on inorganic nitrogen content in soil prior to planting are used; and from basic rate adjustments are done by actual $N\text{-}NO_3 + N\text{-}NH_4$ content. A leaf analysis between periods of flower bud beginning and flowering could be used for check of nutrient status; when needed, supplemental nutrition could be performed, most often foliar fertilization with urea or magnesium as bitter salt.

The recommended mean rate of basic nutrients according to actual conditions is 110 kg N/ha, 70 kg P_2O_5 and 120 kg K_2O.

Agricultural enterprises specialized in potato growing use the technology of de-stoning. In the Czech Republic all agricultural enterprises specialized in potato production have moved to this technology since beginning 1990s. A brief history of development and application of de-stoning technology in the Czech Republic, principles and technical solutions is reported by Fér and Čepl (1997). The origin of the technology could be found in Scotland, from that country the technology was spread across the Europe incl. Nordic states (Andersson, 1997).

De-stoning technology only slowly expanded at the beginning. A change of the technology was initiated by customers of table and seed potatoes that began to prefer potatoes grown under de-stoning technology. A growing procedure of de-stoning concerns soil preparation (especially spring; however, also autumn cultivation could be adapted to the technology), fertilization, planting, weed management and harvest.

A de-stoning line is composed of three mechanization units for soil preparation and planting. There are in regular a two-row furrower, a clod and stone separator and a two-row planter

The first measure is furrowing with a two-body furrower. The goal is to concentrate top soil into large ridges with the base of 150-180 mm (at least double of interrow distance). Further, stone and clod separation is done, which is the most demanding process in relation to energy and time. Separators are developed as strips, stars or combined. It is not possible to prefer only one working body to other; no difference was detected in the separation efficacy during the trials.

Farmer's experiences show higher performance of star-shaped separators in less favourable conditions, i.e. heavy soils or higher soil moisture. It is necessary to take into account more frequent replacement of stars in soils containing more stones. After separation, planting using two-row planters is performed.

In addition to elimination of on average 60% of stones from the space of a ridge, and lower mechanical tuber damage at harvest and further handling, a substantial positive effect is expected on soil physical characteristics and yield and tuber quality indices. Field trials were established focused on determination of this effect in PRI in Havlíčkův Brod.

In the literature there are not many data on effect of mechanical practices on change in soil physical characteristics in potato growing. Most works are aimed at water stress or limit values of soil water content generally or with respect to irrigation (Costa *et al.*, 1997) in the field of water regime. Nazarenko (1991) studied effect of different soil cultivation (passive and active bodies) on soil volume weight. In soil cultivation with active apparatus, soil was less compacted in the whole growing period and had more favourable characteristics of air regime; it was exhibited by positive effect on tuber, dry matter and starch yield. De-stoning technology is particularly represented by values of cultivation practice effectiveness on soil stone content and performance of individual set types (Andersson, 1997). Diviš and Špína (1998) aimed at evaluation of soil physical characteristics and found that higher tuber yield was reached in applying technology of de-stoning and soil kept a character of very loose soil during the whole growing period.

Following ways of autumn and spring soil preparation for potatoes were variants of the trial. Variant indication and technology description:

- +PCT: Conventional technology of growing with ploughing (autumn – skimming, ploughing on full depth of topsoil, spring – smoothing, loosening with a cultivator, planting, prior to plant emergence – harrowing, ridging, herbicide).
- +PDS: Technology of de-stoning with ploughing (autumn – skimming, ploughing on full depth of topsoil, spring – furrowing, clod and stone separation, planting, prior to plant emergence – herbicide).
- -PCT: Conventional technology without ploughing (autumn – sowing of a frost damaged intercrop white mustard, spring – crushing of mustard residue, loosening with a cultivator, planting, prior to plant emergence – harrowing, ridging, herbicide).
- -PDS: Technology of de-stoning without ploughing (autumn – sowing of frost damaged intercrop white mustard, spring – crushing of mustard residue, furrowing, clod and stone separation, planting, prior to plant emergence – herbicide).

Following hydrophysical parameters were measured – soil volume weight (g/cm^3), soil moisture (% volume), relative soil moisture (% relative), porosity (% volume), maximum capillary water capacity (% volume). Measurement of soil resistance (MPa) with a digital penetrometer was performed after planting and before harvest in the whole topsoil profile

Analysis of variance was applied for the statistical evaluation. All mentioned results are mean values of the four-year study. The results of tuber yields are shown in Table 4.

Table 4. Effect of various soil preparations on tuber yield.

	Variant			
	+PCT	**+PDS**	**-PCT**	**-PDS**
tuber yield (t/ha)	40.86	48.59	39.44	44.92
statistical values	**for var. total**	**for var. autumn soil preparation**[1]	**for var. spring soil preparation**[2]	**for years**
F test	44.17**	7.78**	52.12**	29.16**
P 0.01	3.06	2.51	2.61	7.05
P 0.05	2.43	1.86	1.91	5.15

[1]evaluation between variants with ploughing and no-ploughing
[2]evaluation between conventional soil preparation and de-stoning

The highest and statistical significant yield on average of years compared to other variants was found in de-stoning technology with ploughing (+PDS) 48.6 t/ha, further there was followed statistical highly significant yield for all other variants of no-ploughing technology of de-stoning (-PDS) 44.9 t/ha. No significant differences were recorded for conventional technology of ploughing (+PCT) with 40.9 t/ha and no-ploughing conventional technology

(-PCT) with 39.4 t/ha. Factor of ploughing had a highly significant effect on tuber yield (F = 7.8**), variants with ploughing (+P) had tuber yield of 46.8 t/ha and without ploughing (-P) 40.2 t/ha over the years. However, factor of de-stoning (F = 52.5**) had a stronger effect with the yield of 46.8 t/ha in de-stoning variants (DS) and 40.2 t/ha in conventional technology (CT) variants.

Soil penetrometric resistance measured after planting (Figure 1) was lowest in +PDS technology in topsoil profile. Due to proper topsoil sifting with separator, soil resistance in both variants with de-stoning was in dependence on measurement depth always generally lower than in variants with conventional technology. Application of autumn ploughing in +P variants brought also lower resistance than no-ploughing system in –P variants.

This finding was also fully confirmed in soil resistance measurement before harvest (Figure 2) in the framework of DS and CT variants, where elimination of tractor trips leaving out cultivation practices in DS played a role. Between the technologies +P and –P there was not found an unambiguous trend, to the depth of 200 mm lower resistance was for –P; however, in the profile of 200-320 mm it was inverse.

Generally we can conclude with respect to characteristics of soil physical properties that observed variants affected them statistically significantly only from the viewpoint of spring soil preparation (CT or DS), factor of ploughing (+P and –P) had no significant effect. All

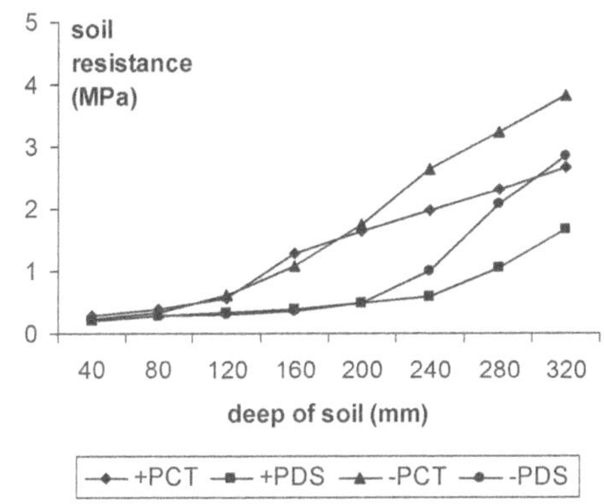

Figure 1. Penetrometric soil resistance after planting.

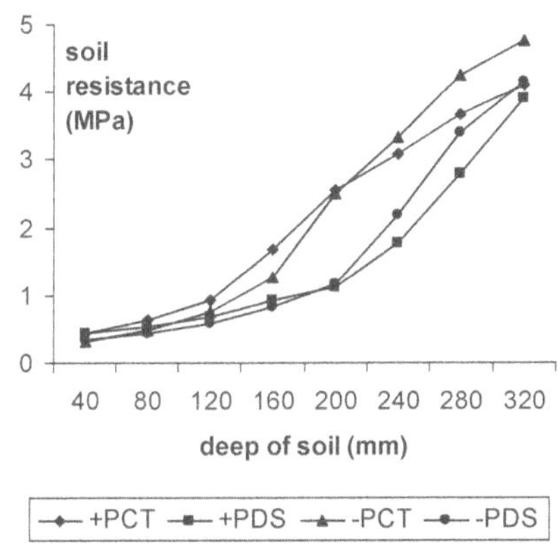

Figure 2. Penetrometric soil resistance before harvest.

parameters characterizing soil compaction were more favourable for DS technologies; it was inverse in characteristics of water regime.

In DS technology highly significant higher yields, lower penetrometric resistance after planting and before harvest and associated lower volume weight of soil were found. Values of porosity were also statistical significantly higher in DS variants. On contrary, volume and relative soil moisture and maximum water capacity were significantly lower in DS variants compared to CT. Autumn soil preparation represented by a ploughing variant (+P) and a no-ploughing variant (-P) did not show an unambiguous statistical significant difference.

In de-stoning technology reserves exit in the manner of fertilizer application. Prior to potato planting high rates of nitrogen mineral fertilizers are often applied. The rate of nitrogen sometimes exceeds 100 kg/ha. Potato plants can this nitrogen utilize not before 4-6 weeks. In addition to mineral fertilizers, farmyard manure is mostly applied in potatoes. An amount of nitrate nitrogen in soil prior to plant uptake often reaches 150-200 kg/ha in this technology.

The most used manner of fertilization with mineral fertilizers is a broadcast application prior to potato planting as a separate practice using automatic or tractor spreaders. A fertilizer lying on the soil surface has no benefit until it is introduced or solved and then transported into the soil. During soil loosening a fertilizer is distributed over the whole loosed soil profile. In conventional soil preparation prior to potato planting the depth of loosening ranges between 100 and 150 mm, in de-stoning technology prior to planting the depth of loosed profile ranges between 160 and 270 mm. The soil is better loosed into the higher depth; however, a fertilizer is distributed into the larger volume of soil and therefore its concentration is reduced, incl. zone surrounding the tubers. Considering soil loosening higher intensity of soil mineralization and nitrification events is exhibited.

A solution, how to prevent nutrient losses and to ensure a sufficient nutrient amount for plants already since early developmental stages of the crop, is utilization of local mineral fertilizer application at planting, particularly nitrogen fertilizers. This manner enables a 25-30% reduction of nitrogen rate compared to classical broadcast application. According to literature in local application of mineral nitrogen fertilizers at potato planting the expected utilization of nitrogen by plants is 45-65%, in broadcasting it is 30-50%. Nutrients are in ready state in soil, loading of the environment is reduced limiting total applied rates of nutrients by up to one third. According to the results from Scotland, on average 13.1% of fertilizers (2-25%) can be saved in local application of a liquid fertilizer and yields of table potatoes are 16.6% higher (10.6-20.6%).

Local application is based on a principle of precise dosage and placement of nutrients into the root zone, what increases a possibility of nutrient utilization from a fertilizer by plants. Nutrient availability is also ensured in a dry period, when a lack of soil moisture could hinder the effective utilization of a fertilizer applied on the soil surface. Nitrogen is gradually released from the incorporated nitrogen fertilizer in the soil and formation of nitrate nitrogen in the soil is similar to application of a slow-acting fertilizer. Better utilization of nutrients is observed.

In addition to the use of solid mineral fertilizers an application of liquid fertilizers is also possible. From technical viewpoint, a local application of fertilizers is performed by adapters on the planter that enable placement of the fertilizer in the optimum distance from tubers. The tank is set on the front three-point tractor hitch. Knives for incorporation of a liquid fertilizer are mounted on the frame of the planter or special included frame between three-point tractor hitch and a planter.

The system of local application of mineral fertilizers at planting has been recently verified in PRI Havlíčkův Brod. Conventional application of broadcasting using of granulated ammonium sulphate with the N rate of 110 kg/ha was compared to the local application of the same N rate and a 30% reduced rate (80 kg N/ha). The same variants were also established with liquid fertilizer DAM 390 (variants are shown in Table 5). Tuber yield was the main studied factor in 2004 and 2005.

Results of yield are given in Figures 3 and 4. The results of 2004 show that the level of yield in the variants with local application of fertilizers at planting with 30% reduced N rate is almost correlated with the level of yield in the variants with the full N rate in broadcasted

Table 5. Variants of a field trial with broadcast and local application of fertilizers.

Broadcast application	**Local application**		**Local application**		**Broadcast application**
Ammonium sulphate*	Ammonium sulphate	Ammonium sulphate	DAM 390**	DAM 390	DAM 390
110 kg N/ha	10 kg N/ha	80 kg N/ha	80 kg N/ha	110 kg N/ha	110 kg N/ha
var. 1	var. 2	var. 3	var. 4	var. 5	var. 6

* ammonium sulphate - SA (21% N), granulated.
** ammonium nitrate + urea - DAM 390 (30%N), liquid.

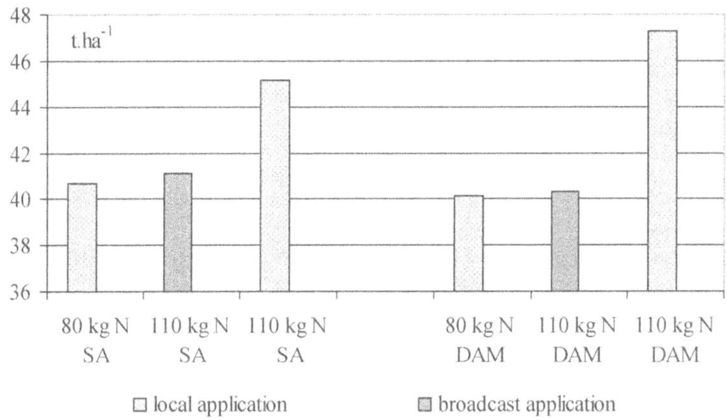

Figure 3. Tuber yield (t/ha) in 2004.

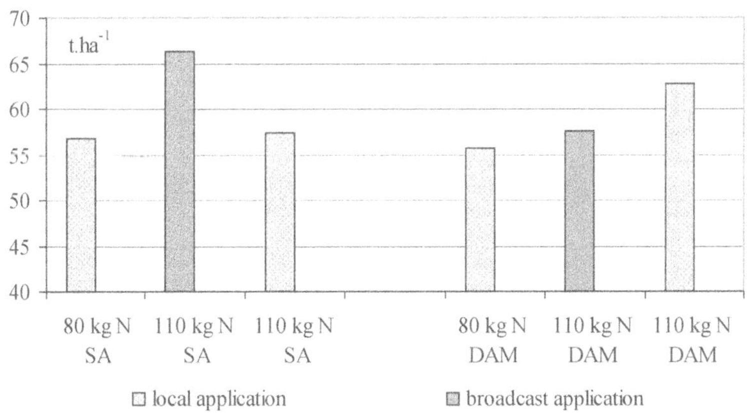

Figure 4. Tuber yield (t/ha) in 2005.

mineral fertilizers. In variants with ammonium sulphate the yield was 9.7% increased in local application of this mineral fertilizer at planting in the rate of 110 kg N/ha compared to broadcasting of the same rate prior to planting. In case of liquid fertilizer DAM 390 a 17% higher tuber yield was observed.

In 2005 local application of N rate 110 kg/ha provided 9,2% higher yields in case of liquid fertilizer DAM 390 application compared to broadcasting prior to planting. After local application of 80 kg N/ha in DAM 390 the yield was 2,5% reduced compared to broadcasting of 110 kg N/ha.

Based on reported results we can conclude that efficacy of locally applied rate of 80 kg N/ha was comparable to broadcasting of 110 kg N/ha.

Conclusion

The de-stoning technology combined with local fertilizer application directly at planting was successful according to the results of the trials and the verification of the technology in agricultural practice. It offers yield increase, improvement of potato quality and reduced environmental loading.

To increase the competitiveness of the Czech potato growing, it is necessary to adopt a series of preventive measures.

- increasing potato yield level in the CR, ensuring yield stability and corresponding quality of production
- ensuring territorial protection of the CR against introduction of quarantine pests and diseases
- support of potato research, breeding and consultancy
- effective promotion and support of potato consumption as a healthy food of high nutritive value
- support of growers to acquire EUREP-GAP certificate to strengthen competitiveness within the single market
- development of conditions comparable to the competition in the EU for Czech growers and potato processors for starch production.

Acknowledgements

This project was supported with means of the Research Plant VÚB MSM6010980701.

References

Andersson, A.J.O. (1997): Praktisk test av stenstranglaggare. Potatis odlaren 1: 41-43.

Costa, J. *et al.*, (1997): Yield, water, use efficiency and nitrogen uptake in potato: influence of drought stress. Potato Research 40: 19-34.

Fér, J., Čepl, J. (1997): Pracovní postupy přípravy půdy a sázení brambor. Bramborářství, 5: č. 3.

Diviš, J., Špína, P. (1998): Vliv odkamenění na fyzikální vlastosti půdy. Bramborářství 6: 6-7

Nazarenko, I.I., Kamarickij, V.M., Vološin, M.M. (1991): Frezernaja obrabotka počvy ulučšajet jeji fizičeskije svojstva. Kartofel i Ovošči, č. 2, s. 7-9.

Curriculum vitae – J. Čepl

Jaroslav Čepl - born on April 23, 1960, Havlíčkův Brod, 1982 – graduation from Agricultural University Prague, Agronomical Faculty, 1983 - assistant, Potato Research Institute Havlíčkův Brod, 1984 – research worker, 1994 - PhD, dissertation: 'Uptake and utilization of nutrients of chosen potato varieties in various nitrogen fertilization' Scientific activities: potato nutrition and fertilization especially in the field of study of nutrient contents and ratios in plant or its parts during vegetation in relation to yields and quality of harvested tubers, potato cultural practices, weed management, methods of estimation of nutrients available in soil, problems of potassium and magnesium in soil, total questions of tuber quality (nitrates, heavy metals, glycoalkaloids), potato growing technologies according to utility types, registration trials with herbicides in potato and cereals. Science organizing activity: member of Plant production department of Czech Academy of Agricultural Sciences, work in Commission of plant production quality and Commission of plant nutrition, a member of scientific committees of agricultural research institutes, advisor in the field of potato growing technologies, a member of European Association for Potato Research. At present, he is focused on the research in the field of quality of potato products, potentials of tuber production of improved user value, mapping of weeds, reduction of *Clavibacter michiganensis* occurrence.

Profile – State Potato Research Institute

Already in October 1921, the idea to found State Potato Research Institute was arisen. It is interesting that this idea was positively accepted by 'potato public' of that time expecting 'enhancement of potato growing, potato industry and economic utilization of potatoes' from this achievement. The State Potato Research Institutes were found in 1923 to create groundwork for research activity. In the history of the institute it is certainly possible to find several important changes, which would gain detailed analysis. The aim of this short essay is not dealing with them in details; however, we have to remember one important date. It is August 1994, when the privatization was done based on the decision of the Ministry of Agriculture and for the first time since its foundation. The Institute discontinued being state owned company and became a private enterprise. The privatization had and still has impact on activity of the Institute, its achievements and to certain extent the character of individual workplaces. The Potato Research Institute still retains its original name and tries to follow the best what was achieved in the past. Research activity remains the main direction and its extent depends on successfulness of applicants in public competition at the Grant Agency of the Czech Republic, Ministry of Education, Youth and Sports of the CR and particularly at the National Agency for Agricultural Research established by the Ministry of Agriculture of the Czech Republic. Besides research and following activities, far greater space must belong to activities, which bring certain sources for institute development. It is mainly providing of registration trials for testing pesticides, post-harvest greenhouse tests of potato seed health state (ELISA), propagation of recovered potato varieties and hybrids via tissue culture technology, tests for resistance to common scab, potato wart disease and potato nematode etc.

Effects of tuber size and date of stem cutting on yield and yield components of Agria potato variety in East Azarbaijan

F. Farahvash and M. Mobasher
Islamic Azad Univ, of Tabriz. Dep. Of Agriculture, P.O. Box 1655 Tabriz, Iran, az2462001@ hotmail. com

Abstract

The aim of this study was to evaluate and select desirable tuber size and the date of stem cutting of the Agria potato variety in semi-cold temperature condition of the East Azarbaijan province in Iran. As regard of qualitative factors (Bacterial and fungous infection) there were significant differences between treatments. The experiment indicated that difference between treatments with evaluation of quantitative factors, a_1b_2 treatment (Whole tuber, with the stem cut off, 20 days before harvesting) and a_1b_3 (The whole tuber, with the stem cut off, 10 days before harvesting) were in the class A, but a_1b_2 treatment in study of quantitative factors, had less fungal; and bacterial infection under semi-cold temperature condition of Tabriz and can be introduced to Azarbaijan province and similar areas.

Keywords: Agria, yield, bacterial infection, fungal infection

Introduction

Potato is the fourth most important food crop worldwide after wheat, rice and maize, cultivated on about 18 million hectares and the yearly production is around 295 million tons (Askew , 2001). Potato is the most important crop after wheat in Iran. It is very sensitive to ecological factors such as temperature, rainfall and photo period.

Potato yield vary markedly from less then 10 t/ha to more than 50 t/ha, depending on climatic condition, irrigation and crop management, also noted factors effects on seed tuber quality.

Potato is one of the rich resources of amino acids, vitamins like B1, B2, C and mineral materials such as phosphorous and potassium. This crop was introduced to Iran about 200 years ago and nowadays it is one of the most important products. The research of this crop started at 1965 in universities and plant research institutes of agriculture ministry of Iran. Various variety of potato imported to Iran from Germany and India. More than 60 species of potatoes have been described in nature. However, cultivated strains of potatoes have been derived from only a handful of species and thus contain low levels of genetic variation. The common potato, Solanum tuberosum, is tetraploid (2n=24) consisting of two subspecies S. tuberosum subsp. tuberosum and S. tuberosum subsp. andigena. Potato is traditionally grown vegetative through seed tubers. This results in continuous accumulation and increase of various tuber-borne diseases in seed tuber and reduction in the crop yield.

Aphid (*Myzus persicae*) is the most important virus vector for potato crop. Amongst potato viruses, leaf roll and virus Y are very serious. The successful cultivation of potato depends on the availability of disease-free seed tubers, soil moistures, low temperature, short day conditions during tuberisation phase resulting in rapid bulking rate. However, some varieties tuberize well under long day conditions also. The potato crop is attacked by some 36 viruses and viruses like organisms (Khurana,1999). Viruses are largely responsible for crop losses along with degeneration of seed stocks (annual losses:30-40% infection causing 25-30% yield loses in India). Viruses are responsible for causing crop losses up to 50% in tuber yield in Indian plains (Pushkarnath,1967). An average 5-10% viral infection hardly affects crop yields but early and severe infection dose cause appreciable yield losses (10-60%).

The aim of this study was to evaluate and select desirable tuber size and the date of stem cutting on the Agria potato variety in semi-cold temperature condition of East Azarbijan province in Iran.

Materials and methods

The experimental farm was located in the agriculture research station of the Islamic Azad University in the south-east of Tabriz, in East Azarbaijan province in the north-west of Iran (Lat 38°5′ N and Long 46°17′ E) at altitude of 1360 m above sea-level. Annual mean temperature is 13.04 °C. The region is classified as cold and semi-arid. Annual precipitation is 232.14mm per unit area. These trials were conducted in the two years (2005-2006) on sandy loam soils.

This experiment conducted on factorial design with 4 replications, 8 treatment and 32 plots. Soil test show followed results:

pH between 8.2 – 8.5, organic carbon below %1 , C/N = 11.72 , CEC ($DM^{(+)}$/Kg) = 6.90 – 18.20.

In this study, the first factor was the size of seed tuber (A) in two levels:
1. Whole tuber seed with 55-60 g (a_1).
2. Whole tuber seed more than 80 g with length cut off (a_2).

The second factor was the date of stem cutting, with four levels (B):
1. Stem cut off 30 days before harvesting (b_1).
2. Stem cut off 20 days before harvesting (b_2).
3. Stem cut off10 days before harvesting (b_3).
4. Stem cutting at harvesting time (b_4).

Each plot was 3 cultivated lines, four meters long and plant distances in rows were 25 cm. Row distances in plots were 75 cm.

Weeds controlled by manual four times during growth period. Prepare the field as for ware potato crop, by adding chemical fertilizers Ammonium phosphate 200 kg per hectare, Potassium sulphate 100 kg per hectare and Urea 100 kg per hectare.

Results and discussion

The variance analysis of the results achieved from the quantitative characteristics in the experiment showed that there were significant variations among mean yield; mean main stems numbers mean of big tubers weight, mean of tubers weight, mean of big tubers numbers, mean tubers numbers. As regard of qualitative factors (bacterial and fungal infection) there were significant differences between treatments (Tables 1, 2, 3, 4 and Figures 1 to 8).

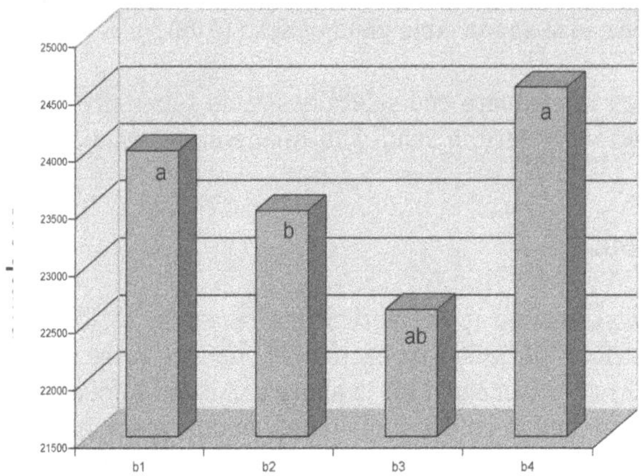

Figure 1. The effects of date of stem cut off on main stem numbers (in two years).

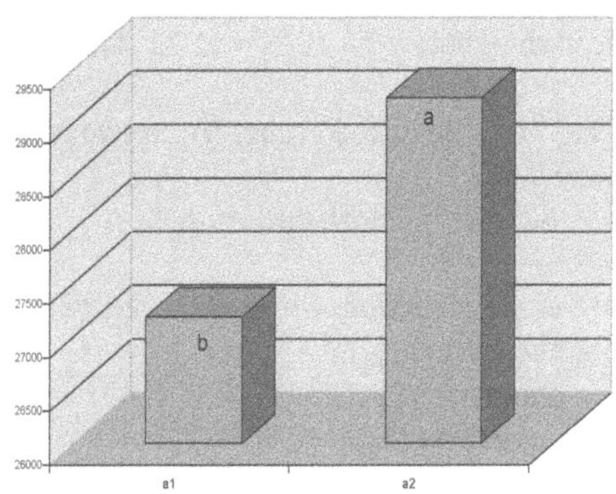

Figure 2. Mean of tuber fungous infection - seed tuber size (in two years).

Table 1. The variance analysis (MS) of studied characteristics (in second year).

SOV (sources of changes)	Degrees of freedom	Yield (ton/ha)	Main stem number	Tuber bacterial infection number	Tuber fungous infection number	Big Tuber number
Replicates (R)	3	60.356 **	25187682.04**	1339723925.61**	1035823350.86**	51583285.61**
Size of tuber seed (A)	1	8.8831 ns	2469753.12 ns	227820521**	115858447.53**	461520.28 ns
Date of stem cut off (B)	3	32.4624*	9006107.12*	41375141.94 ns	3642939.44 ns	107305.19**
Interaction effects (A×B)	3	6.9465 ns	1037407.45 ns	13041865.86 ns	1207616.36 ns	46314.36 ns
Error	21	7.7566	2756482.13	26457678.16	5386075.19	322014.32
CV%		8.25	7.2	16.29	8.38	3.87

Table 2. The variance analysis (MS) of studied characters (in two years).

SOV (sources of variation)	Degrees of freedom	Yield (ton/ha)	Main stem number	Tuber bacterial infection number	Tuber fungous infection number	Big Tuber number
Replicates (R)		58.1467 **	228484372.5**	1043288637.67**	773105992.36**	59038255.32**
Size of tuber seed (A)		13.727 ns	694722.25ns	140176720.14***	66695805.56**	1799957.64*
Date of stem cut off (B)		14.3036*	10950865.27 *	83003557.93**	5789488.72 ns	280311.76 ns
Year		11.56 ns	23365139.06 **	1027963828.51**	17363889 ns	10448672.51**
Interaction effects (A×B)	2	20.0504	2900405.37 ns	9337902.93 ns	669737.72 ns	41746.84 ns
Interaction effects (A×C)		0.2601 ns	1929321 ns	90366412.51*	49780080.24 **	145065.76 ns
Interaction effects (B×C)		19.356*	4004089.6 ns	5554919.47 ns	3499878.75 ns	264091.3 ns
Interaction effects (A×B×C)		5.3438 ns	1616115.29 ns	6920043.39 ns	4431337.5 ns	34248.39 ns
Error	2	6.2884	2840322.53	15962034.45	4657787.2	415726.24
Total	3	-	-	-	-	-
CV%		7.33	7.12	14.5	7.65	4.04

Table 3. The comparison between mean of studied characters with Duncan's test (5%) in different levels of factor A (in two years).

Treatment	Yield (ton/ha)	Tuber fungous infection number	Tuber bacterial infection number	Mean main stem number	Big tuber number (percent)
a_1	33.7 A	27180.1 B	26069 B	23555.1A	33.33 B
a_2	34.63 A	29221.8 A	29028.9 A	23763.5 A	44.15 A

Table 4. The comparison between mean of studied characters with Duncan's test (5%) in different levels of factor B.

Treatment	Yield (ton/ha)	Main stem number	Small tuber number (percent)	Medium tuber number (percent)	Big tuber number (percent)
b_1	32.88C	5.550A	6.05B	5.05A	36.750B
b_2	33.77BC	5.50A	6.20B	5.275A	37.70B
b_3	35.5AB	5.40A	6.75A	5.4125A	39.763A
b4	36.2A	5.39A	6.788A	5.45A	40.750A
L. S. D 5%	0.519	0.432	0.4444	0.5628	1.670

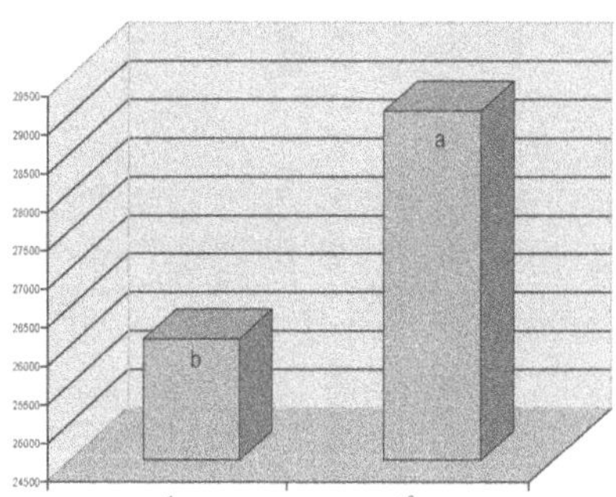

Figure 3. Mean of tuber bacterial infection - seed tuber size (in two years).

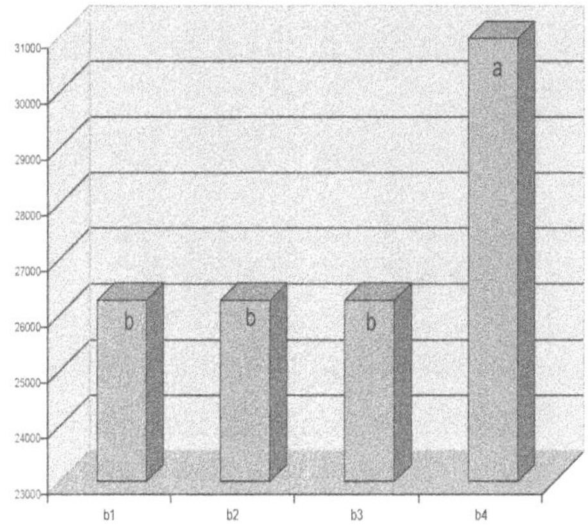

Figure 4. Mean of tuber bacterial infection - date of stem cut off (in two years).

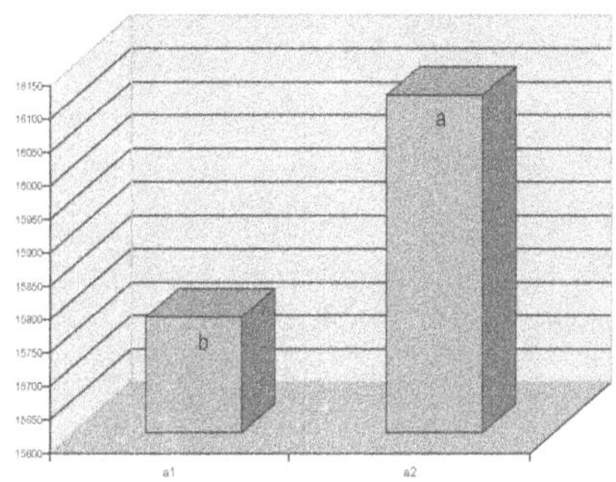

Figure 5. Mean of big tubers - seed tuber size (in two years).

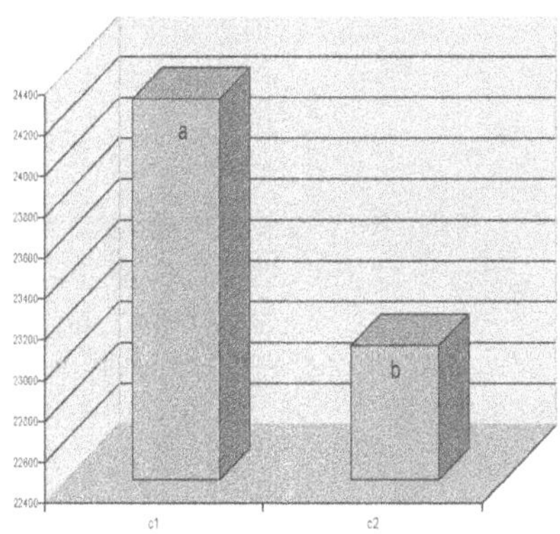

Figure 6. Mean of main stems (in two years).

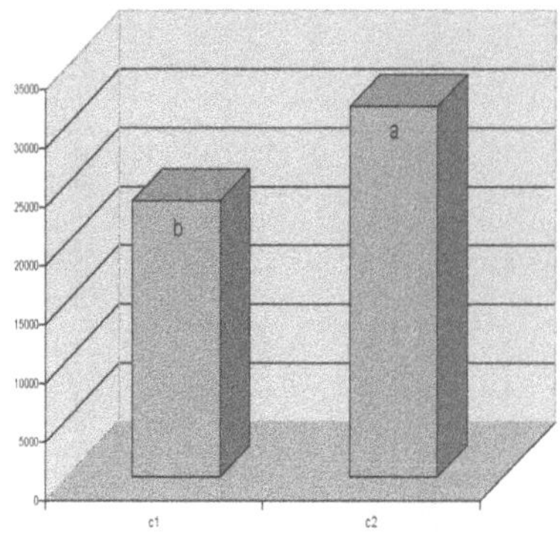

Figure 7. Mean of tuber bacterial infection (in two years).

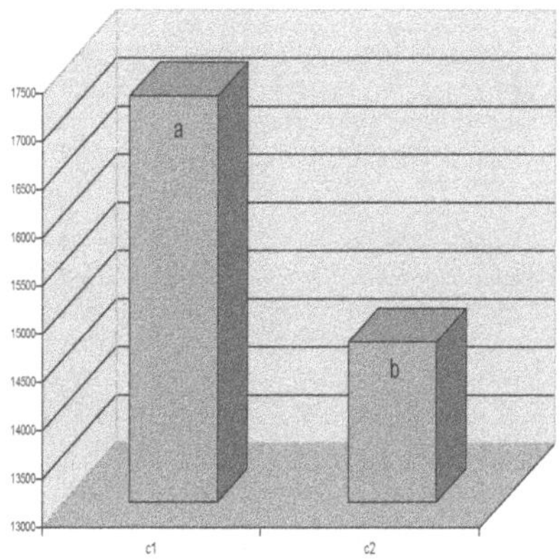

Figure 8. Mean of big size tubers (in two years).

Conclusions

The experiment indicated that difference between treatments with evaluation of quantitative factors, a_1b_2 treatment (Whole tuber, with the stem cut off, 20 days before harvesting) and a_1b_3 (The whole tuber, with the stem cut off, 10 days before harvesting) were in the class A, but a_1b_2 treatment in study of quantitative factors, had less fungous and bacterial infection under semi-cold temperature condition of Tabriz and can be introduced to Azarbaijan province and similar areas. The data reported show that there was a marked relationship between tuber size and time of stem cut off.

Results of this research in two years indicated that for achievement of reasonable efficiency, yield without bacterial and fungal infection, use of seed tuber with 50-60g is economic and use of stem cutting factor is effective for decrease of bacterial and fungous infection. This operation should be done 10-20 days before harvesting.

In addition, the size of seed tuber 50-60g causes to produce more medium tuber numbers. This matter is better in potato production in case of high quality and marketing speciality. Finally in this research in view of qualitative and quantitative characters, a_1b_2, a_1b_3 treatments are suggested for cultivation in cold temperate regions and other similar climate and soil conditions.

References

Allen, E.J. and R.K. Scott. 1980. An analysis of growth of the potato crop. Journal of Agriculture Science, Cambridge 94; 583- 606.

Beukema, H.P. and D.E Vander Zaag. 1979. Potato improvement. Some factors and facts. I.A.C International Agricultural Centre, Wageningen, The Netherland, 222 pp.

Beukema, H.P and D.E Vander Zaag. 1990. Introduction to potato production Purdoc, Wageningen pp 207.

Burton, W.G. 1989. The potato. Longman scientific Technical. New York pp. 742.

Burton, W.G. and A.R. Wilson. 1970. The apparent effect of latitude of the place of cultivation upon the sugar content of potato grown in Great Britain. Potato Research, 13:269-283.

Campbell, M.D., G.S. Campbell, R. Kunkel and R.I. Papendick. 1976. A model describing soil plant water relations for potatoes. American Potato Journal 53:431-441.

Chadha, K.L. 1994. Changing scenario of potato production in the world In Potato: present and future. G.S. Shekhawat *et al.* (eds) Indian Potato Assoc, Shimla, pp 5-11.

Cutter, C.G. 1978. Structure and development of potato plant. In: The potato crop. P.M. Harrris (ed.) Chapman and Hall London -PP. 70-80.

Doorenbos, J. and W.O. Pruitt, 1975. Crop water requirement. FAO. Irrigation and Drainage. PP. 24.

Dwello , R.B., P.J. Hurley and J.J. Pavek. 1983. Photosynthesis and stomatal conductance of potato clones (Solanum tuberosum L). Plant Physiology 72: 172-176.

Feddes, R.A.J. de Graaf , J. Bouma and C.D. Val Loon. 1988. Simulation of water use and production of potatoes as affected by soil compaction. Potato Research 31: 225–239.

Gandar, P.W. and C.B. Tanner, 1976. Leaf growth, tuber growth and water potential. Crop Science. 16:534–538.

Grewal, J.S., R.C. Sharma and S.S. Saini, 1992. Agro techniques for intensive potato cultivation in India. ICAR, New Delhi pp. 126

Harris, P.M. 1978. The potato crop, Chapman and Hall, London PP. 730

Hooker, W.J. 1983. Research for the potato in the year 2000. International Potato Center Lima Peru.

Horton, D. 1987. Potatoes. Westiview Press USA pp. 243

Huaman, Z. 1986. Systematic botany and morphology of potato. Technical Information, Bulletin 6. CIP, Lima Peru. International Potato Center. 1984. Potatoes for the developing world, CIP Lima, Peru.

Manrique, A. 1993. Constraints for potato production in the tropics. J. Plant Nutrition 16:2075-2080.

Moorby, J. and F.L. Milthorpe. 1975. Potato. In L.T. Evans (ed.) Crop physiology. Some case histories, Cambridge University Press. London. PP. 225-238.

Nelson, D.G. 1996. Approaches to controlling numbers of stems tubers and size distribution. Abstracts of Conference Papers and Demonstrations of 13th Conference of the EAPR. pp. 441-442.

Storey. R.M.J. 1992. Tuber quality. In: P.M. Harris (ed.). The potato crop. Chapman and Hall, Lodon. pp. 507-512.

Wiersema. S.G. 1987. Effect of stem density on potato production. Technical Information. Bulletin, CIP, Lima Peru.

Curriculum Vitae – Farhad Farahvash

Farhad Farahvash was born in Tabriz, Iran (1965). He received a Ph. D. degree from the Islamic Azad University (Tehran Sciences and Research) in 2000 based upon his work on Ecophysiological study and salt determination. He is the author of some articles in agricultural sciences journals and he has published a few books (1. Life process of plants; 2. Ecology; 3. Modern technology of vegetables production; 4. Lower plants; 5. Anatomy and activities of plants; 6. Morphology, taxonomy and systematic of plants). He has also supervised some M. Sc. theses on different topics. He attend in some international conferences like, VIII ESA congress (European agriculture in a global context) held in Copenhagen, Denmark July 2004; 3rd Balkan symposium on vegetables and potatoes held in Bursa, Turkey sep 2004, … . His present professional position is member of academic board (Assistant professor of agricultural department) in Islamic Azad university of Tabriz. The topics of his research projects are: Investigation and evaluation of quantitative and qualitative characteristics of true potato seed progenies in East Azarbaijan, Iran; Investigation important and effective factors on true potato seed production of introduced potato varieties in East Azarbaijan, Iran.

Profile – Islamic Azad university of Tabriz

The department of agriculture of the Islamic Azad university of Tabriz was founded in 1982. It consists of two postgraduate courses and 8 study courses undergraduate (Plant diseases, Soil sciences, Environmental sciences, Agricultural extension and education, Horticulture, Plant breeding, Food science and technology). At present over 800 students are studying at agriculture faculty and over 1300 students graduated. Our faculty offers a friendly environment with knowledgeable and efficient staff, up to date facilities and the latest technology. All these elements indicate that the university is fully aware of achieving its role as the key actor in the global agriculture development process and its responsibilities and commitments in materialising the loftiest objectives of higher education. We hope that we can establish relationships with many universities in various countries.

Integrated agronomic practices of early potato production of the Republic of Bashkortostan

A.D. Andrianov, D.A. Andrianov[1] and V.I. Kostin[2]
[1]Bashkir State Agrarian University, 450001, street 50 let Oktyabrya, 34, town Ufa, Republic of Bashkortostan, Russian Federation
[2]Ulyanovsk State Agricultural Academy, 432980, boulevard Novyi Venec, 1, Ulyanovsk Region, town Ulyanovsk, Russian Federation

Abstract

On the basis of generalization and analysis of scientific results of our long-term multifactorial field experiments and laboratory research of early potato culture by utilization of mathematical methods the conception and all the links of science-based integrated agrotechnics are made up, innovative integral agrotechnologies for soil-climatic and organization-economic conditions of the Republic of Bashkortostan are developed.

Keywords: nutrition value, yield and quality of tubers, integral agrotechnologies, irrigation

Introduction

Potato is an essential food product (Anisimov, 2006) and needed by the people of the Republic of Bashkortostan all year round. Even stored under the best conditions, potato loses some quality the longer they are stored. A lot of essential nutrients are lost in the respiration process. Fresh potatoes with high nutrition value should be used for full value ration. Climatic condition of the Republic of Bashkortostan are not suitable for growing potatoes all the year round, but it is quite possible to get adequate amount of fresh potato from June. Nowadays potato growing in the Republic of Bashkortostan has been focused in private and small market farms (97% of all planting area). It is grown on an area of 3,000 hectare at large market farms (Table 1) and on the area of 1000 hectare at peasants farms. Potato consumption in the Republic of Bashkortostan has increased to 180 kg per capita for recent 3 years. At the same time there is a tendency of concentration of potato production. In 6 regions of the Republic of Bashkortostan (Ufimsky, Ilishevsky, Tuimazinsky, Meleuzovsky, Sterlitamaksky, Aurgazinsky) they produce 75% of potatoes and their crop yield are 150-211% higher than the average level in the Republic. In two regions (Chishminsky and Beloretsky) 70% of areas and 85% of yield are concentrated under potato crop. Their productivity is 3-5 times higher than in other regions.

Management of the production process

The authors of the article on the basis of the integrated agrotechniques of early potato used knowledge of natural stable produce process in crop protection for optimizing all its constituents and increasing crop productivity and tuber yield quality. Productivity of early potato is understood as a possibility of early potato to produce maximum output of dry matter per area unit by the concrete date of summer harvest. The constituents of dry matter are starch,

Table 1. Crop Production in the Republic of Bashkortostan.

Index	Year	Types of farms			
		market	collective	private	Total
Harvested area, ha	1990	3,3376	17	76,291	110,000
	1991- 1995	15,058	101	99,989	115,148
	1996-2000	6,784	398	101,546	108,676
	2001-2005	3,244	752	106,736	110,747
	2006	2,422	1,032	111,115	113,569
Gross yield, thousand ton	1990	259.2	0.1	1,120	1,379
	1991- 1995	128.5	0.8	849.8	979.1
	1996-2000	54.2	3.1	987.3	1,045.1
	2001-2005	30.1	9.5	1,221.9	1,261.5
	2006	41.0	15.9	1,348.4	1,404.3
Crop yield, t/ha	1990	7.8	8.0	14.6	12.7
	1991- 1995	8.5	8.6	8.5	8.3
	1996-2000	5.9	8.0	9.7	9.6
	2001-2005	10.1	12.6	11.4	11.4
	2006	16.9	15.4	12.2	12.4

protein, vitamins, mineral salts. It is necessary to study the productivity of early potato in connection with conditions of germination and its individual components. The chief criterion of early potato productivity is tuber yield.

We found out that during the period of intensive accumulation of early potato yield the amount of tubers increased to the end of July and then it didn't change. But the amount of the accumulation in July and early August occurs due to increasing amount and mass of commercial tubers. Two periods can be singled out in the tuber formation process. The first one is the formation of vine, growth of stolons and beginning of tuber formation. The second one is increasing tuber mass due to functioning of overground mass and root system. The data of the Table 2 shows that by the August, 5 early potato may form an average tuber yield of 1458.2 g per hill or with planting density 50 thousand items per hectare the crop yield is 72 t/ha. All known varieties have strong inverse correlation of the tuber amounts and their average mass. At the same time annual cycle between the described components of early potato productivity is typical for them. If there are more tubers per a hill a year then next year average mass of tubers will be larger. Besides it was found out that growing of high plastic commercial tubers gradually were increasing to finish of the harvesting. Thus, yield variety ensures yields of high quality. Different early ripening and medium-early varieties differ response to growing conditions in different periods of vegetation. The indices of early potato productivity change under the influence of outward conditions (Tables 3 and 4). Different varieties change with different degree at different stages of plant growth.

Table 2. Relationship between tuber number and mass in early potato yield (experimental training farm 'Milovsky' BSAU, 1967-1996).

Variety	Connection between tuber amount and mass		Productivity indices during years			
	Correlation coefficient	t-criterion	with maximum tuber amount		with maximum tuber mass	
			tuber number per hill,	average tuber mass, g	tuber number per hill,	average tuber mass, g
Priekulsky early	-0.857±0,.24	6.92	10.2	56.3	16.3	41.4
Iskra	-0.714±165	4.30	11.2	80.8	16.6	55.4
Belorussky early	-0.837±0.142	5.89	10.8	76.8	16.8	48.6
Nevsky	-0.626±0.352	1.78	11.7	77.8	17.8	51.4
Ufimets	-0.749±0.222	3.37	13.2	68.4	15.2	56.4

Table 3. Early potato productivity depending on total effective temperatures during tuber formation (experimental training farm 'Milovsky' BSAU, 1967-1996).

Harvesting time	Tuber crop yield, t/ha		Tuber number per hill		Average tuber mass, g	
	Correlation coefficient	t-criterion	Correlation coefficient	t-criterion	Correlation coefficient	t-criterion
Variety Priekulsky early						
10-07	+0.376	1.20	-0.092	0.28	-0.042	0.12
20-07	+0.302	0.90	-0.012	0.03	+0.340	1.02
30-07	-0.127	0.37	+0.106	0.26	-0.105	0.30
10-08	+0.129	0.36	+0.402	1.21	-0.260	0.74
Variety Iskra						
10-07	+0.642	2.41	+0.438	1.41	+0.668	2.46
20-07	+0.372	1.10	-0.112	0.32	+0.530	1.87
30-07	-0.464	1.48	+0.182	0.52	-0.486	1.54
10-08	-0.524	1.83	+0.213	0.63	-0.322	0.84
Variety Nevsky						
10-07	+0.654	2.07	+0.670	2.97	+0.636	2.64
20-07	+0.672	2.23	+0.352	0.63	+0.657	2.41
30-07	-0.112	0.19	-0.321	0.58	+0.134	0.23
10-08	-0.080	0.14	-0.246	0.45	+0.113	0.20

Table 4. Productivity dependence of early potato on productive moisture store in soil.

Harvesting time	Tuber crop yield, t/ha		Tuber number per hill		Average tuber mass, g	
	Correlation coefficient	t-criterion	Correlation coefficient	t-criterion	Correlation coefficient	t-criterion
Variety Priekulsky early						
10-07	+0.077	0.22	-0.242	0.71	-0.036	0.11
20-07	+0.215	0.62	-0.045	0.12	+0.224	0.65
30-07	-0.507	1.68	+0.005	0.01	-0.352	1.10
10-08	+0.006	0.00	+0.176	0.51	-0.132	0.37
Variety Iskra						
10-07	+0.680	2.91	+0.676	2.18	+0.033	0.09
20-07	+0.534	2.11	-0.052	0.14	+0.142	0.40
30-07	-0.030	0.08	+0.037	0.10	-0.489	1.55
10-08	-0.329	0.98	+0.386	1.18	-0.496	1.56
Variety Nevsky						
10-07	+0.742	4.26	+0.756	3.51	+0.890	8.90
20-07	+0,354	0.64	+0.515	1.37	+0.016	0.03
30-07	-0.717	3.41	-0.719	3.54	+0.316	0.57
10-08	-0.456	1.49	-0.410	1.28	+0.386	0.83

The change of any parameters may result in yield increasing if optimal leaf surface of the crop will be maintained. High productive crops are produced on optimal correlation of architectonic indices of individual plants and crops in general. In order to characterize the duration of photosynthetic work of crops we use the index of photosynthetic potential (PhSP), which is in close positive correlation with tuber yields. The closer to the optimum leaf area (60-70 thousand m^2/ha) the stronger this correlation. The plantings are considered to be good if PhSP is not less than 2 mln m^2 day/ hectare, taking into account 100 days of actual vegetation. Optimum growth of leaf surface and formation of high PhSP can be supplied with proper crop management.

Integrated management of early potato

Many authors (Bolotov (1790), Grachev (1880), Lorh (1948), Lysenko (1951), Yakushkin (1945), Karmanov (1983), Pisarev (1984), Mosin (1978), Kiryuhin (1988), Akatyeva (1984), Shteingard (1989), Meyer (1953), Gladkov (1964) and others) studied yield forming process and tuber quality and agrotechnique of early potato crop in different ecological conditions of growth.

Nowadays different ways of potato agrotechnique improvement are suggested (Allison *et al.*, (2001); Anisimov (2005); Simakov and Anisimov (2006); Stalham and Allen (2004); Zlotikov *et al.*, (2004); Zlotikov *et al.*, (2001); Zlotikov *et al.*, (2000)).

Increasing early potato production in the Republic of Bashkortostan will occur by means of crop yield increase and tuber quality. Applied up to present days chemical- agrotechniques cannot meet the needs of the population with fresh potato. New crop management development represent a constant process. The basis of new proper agrotechnique formation of early potato are farming rules, biochemistry and potato crop physiology, established by practicing long-term multifactorial field experiments.

That's why for every farm a certain level of early potato crop yield must be defined to provide high economical and power production effectiveness. The whole range of peculiarities should be taken into consideration in crop production. (1) Biological peculiarities. To receive an early product the varieties of early tuber formation and active tuber accumulation are most valuable. (2) Soil and climatic conditions of farm area landscapes. In the Republic of Bashkortostan black soils and grey forest soils of heavy and medium loam prevail. (3) Variety differences. In depend of requirements of consumers it is necessary to grow the varieties with different market properties. (4) Mastering of techniques, accelerating plant growth and development and tuber yield accumulation. The main criterion of applying agrotechnology proper is their complexity and differentiation due to soil – climatic and organization – economic conditions, biological peculiarities of early potato culture while conserving soil fertility and ecological safety and obtaining planned yield in calculate date.

The integrated agrotechnology formation scheme of early potato crop in the Republic of Bashkortostan is based on equality and indispensability of the following crop production factors: (1) Ecological conditions and crop growing location. (2) Organizational and economical conditions of producers. (3) Integrated agrotechnique recommendations. (4) Specialized crop rotations. (5) Organic-mineral nutrition and differential plant irrigation. (6) Biologized plant conservation system from diseases and insects. (7) Reclamational different depth layer soil cultivation. (8) Specialized farm machinery. (9) Selection and seed-growing of adapted varieties of early and medium maturity groups. (10) Qualified agrobioenergetic estimation of agrotechnique effectiveness.

To define optimal parameters of early potato in Pre Urals and Southern Urals conditions we prepared and practiced long- term multifactorial field experiments in 1990-2005 in different farm zones of the Republic of Bashkortostan on the medium humus and medium power heavy loam black soils. Forerunners were bare and full fallows, winter rye, legume grasses and vegetable crops. For research super-elite and crop elite Nevsky, Pushkinets, Iskra, Belorussky early, Red Scarlett and others with tuber mass of 60-80g were used. The whole area of plot is 47,6m^2, the experiment was repeated three times. Planting was made with germinated planting material in different times with different planting scheme on the depth 6-7 sm. Final harvesting was made by the complete harvesting method on the first decade of August.

All observations, accounting and analysis were provided by generally accepted methods of Lorkh All-Russian Research Institute for potato farming. Data were mathematically calculated in Windows with the method of dispersal analyses accounted by Dospehov. The main results of our research work were given in previous publications (Andrianov and Andrianov (2006); Andrianov *et al.*, (2005); Andrianov and Andrianov (2004); Andrianov and Andrianov (2003a); Andrianov and Andrianov (2003b); Andrianov and Andrianov (2003c); Andrianov and Andrianov (2002)).

In 1990-1995 we carried out research studying under boghart conditions the effects of planting time, density and sizes on early potato growth and development, formation of tuber yields and its quality in different harvesting time. The scheme of multifactor field experiments included the following variants. A Factor. Planting time under soil temperature on the tuber embedding depth is +3-5 °C (May,1-6), +6-8 °C (May 9-14), +9-11 °C (May, 15-20). B Factor. Planting tuber mass is 40-60g, 61-80g and 81-100g. C Factor. Planting density is 40, 50, 60, 70 thousand tuber/ha. Elite seed material had light sprouting during 28-30 days and seed treatment. Nutrition is 50 t/ha, decomposed manure +N90P120K60 kg active substance per ha. Some results are given in Tables 5 and 6. Early planting of large and middle-sized seed pieces ensured producing the highest crop yield during the final harvesting (August, 5).

The largest yield of dry matter and the highest productivity of photosynthesis were achieved (Table 6) with different planting density of seed pieces with different mass during the first planting. According to the results if you grow early potato first of all you should use seed pieces with mass of 80...100g and 60-70g. There are new data for 2006.

Multifactorial field experiments to determine the optimal regime of drip watering were held in irrigated vegetable crop rotation of the state farm 'Alexeevsky'. These experiences were held to form programme yield of early potato tubers of quality required by concrete summer harvesting time. Irrigation rate makes up from 1376 m^3/ha without fertilization to 1875 m^3/ha with fertilizer used to the planned crop yield of 40 t/ha with pre-watering threshold of 85-90-85% in calculated soil layer.

Drip watering increased vine mass and leaf area to 200-278% relative to rainfed conditions. The largest crop yield (65.9 and 62.2 t/ha) was obtained in the variants of pre-watering threshold of 85-90-85% and 85-85-85% with calculated rate of full mineral fertilizer per 40 t/ha of tubers.

Improvement of growing conditions promoting more intensive mass increase and tuber volume doesn't decrease their starch and ascorbic acid content. So, tuber yield per area unit is considerably higher when watered. Planting water-supplying improvement leads to decrease nitrate content. At the same time it is necessary to notice that in all experimental variants the meaning index was below to the limiting permissible concentration (LPC). Under regulated conditions when potato plants are provided by water and mineral elements, market properties increase significantly.

In total combination of studied signs in field experiments, drip watering variants such as Red Scarlet with pre-watering threshold in calculated soil layer (40-50-60) of the vegetation periods of 85-90-85% and 85-85-85% was recognized the best. Full mineral fertilizers to receive 40t tubers per 1ha were applied.

Multifactor field experiments to study different preparations for biologised protection systems of early potato from Colorado beetles were held in experimental farm 'Milovsky' of the Bashkir State Agrarian University.

Table 5. Crop yield (t/ha) of early potato tubers, variety Belorussky early during the periods of growing season 1990-1991.

Planting date	Seed size, g	Plant density, 1000 per ha	Development phase				
			budding	flowering	Flowering + 10 days	Flowering + 20 days	Harvesting
1	40	40	0.32	9.8	15.5	21.6	28.4
		50	0.35	12.0	1.83	25.0	32.4
		60	0.34	12.7	1.95	26.3	33.7
		70	0.30	13.1	20.0	26.8	33.5
	70	40	0.70	16.0	25.5	34.8	44.0
		50	0.76	18.5	29.0	39.0	48.5
		60	0.71	18.6	29.0	38.4	46.8
		70	0.60	17.2	26.5	30.5	42.8
	90	40	1.00	20.5	32.0	42.3	52.3
		50	1.10	23.2	36.1	47.1	57.0
		60	1.00	21.8	33.6	44.1	52.0
		70	0.88	21.3	32.7	42.2	49.3
2	40	40	0.72	10.2	15.6	23.0	28.1
		50	0.81	12.3	18.6	26.8	32.0
		60	0.84	13.7	20.2	28.7	34.0
		70	0.78	14.0	20.5	28.8	33.5
	70	40	1.70	16.1	25.1	35.7	42.5
		50	2.00	19.0	28.5	40.2	47.0
		60	2.00	19.5	28.7	39.6	45.4
		70	1.72	18.0	26.3	36.3	41.5
	90	40	2.55	20.8	31.2	43.1	50.0
		50	2.75	23.2	34.8	47.2	54.0
		60	2.70	22.7	33.1	44.4	50.0
		70	2.60	22.2	32.7	43.2	48.4
3	40	40	3.40	8.7	13.5	20.0	24.7
		50	3.84	10.3	16.1	23.2	28.1
		60	4.06	11.6	18.0	25.4	30.5
		70	4.18	12.1	18.4	26.0	30.7
	70	40	6.48	14.0	22.0	31.6	38.2
		50	7.14	16.3	25.0	35.3	42.0
		60	7.16	17.1	26.1	36.1	42.1
		70	6.14	15.8	24.0	33.3	38.8
	90	40	8.55	17.7	27.5	38.2	45.0
		50	9.36	20.0	30.4	41.8	48.4
		60	8.62	19.7	30.0	40.6	46.5
		70	8.17	19.6	29.7	40.0	45.4

Table 6. Dry matter biomass, photosynthesis potential and net productivity of early potato photosynthesis of Belorussky early variety, 1990-1991.

Seed size. g	Plant density, 1000 per ha	Dry matter crop yield g/m² Planting date			Photosynthesis potential g/m² d Planting date			Net photosynthesis productivity g/m² d Planting date		
		1	2	3	1	2	3	1	2	3
40	40	975	968	822	189.1	194.6	167.0	5.14	5.00	4.92
	50	1118	1121	963	213.1	219.5	190.6	5.23	5.11	5.03
	60	1178	1190	1048	222.7	230.7	206.5	5.25	5.13	5.03
	70	1179	1191	1067	222.9	230.9	212.3	5.24	5.12	4.96
70	40	1455	1425	1249	245.9	246.2	218.3	5.92	5.78	5.70
	50	1638	1606	1408	268.5	276.1	245.4	6.08	5.80	5.70
	60	1620	1590	1445	264.3	274.6	250.9	6.09	5.75	5.77
	70	1502	1470	1360	247.8	257.6	240.0	6.04	5.68	5.62
90	40	1754	1663	1485	269.6	269.7	246.5	6.48	6.16	6.09
	50	1962	1824	1638	294.0	297.3	269.0	6.63	6.13	6.05
	60	1857	1749	1617	288.3	290.2	266.9	6.40	6.00	6.05
	70	1756	1679	1570	272.5	278.5	259.8	6.40	6.00	6.02

The largest vegetation, assimilation surface area and stem quality both in one potato plant and in planting area unit, quality and tuber masses under the potato plant were marked in the bitoxibacilline used variants.

The largest biological activity (up to 99.8%) was achieved when bitoxibacilline was used. Insects have already died in 3 hours, peaking in 3-5 days after treatment and continues to act during 2-3 weeks. After bitoxibacilline used feeding activity of Colorado beetles are depressed quickly. Bitoxibacilline is a safe and effective component of crop protection. The highest yield (35.8-36.4 t/ha) was achieved when bitoxibacilline was used both in full doze and in reduced doze.

Bitoxibacilline treatment of early potato vegetation increases dry matter content, starch and ascorbic acid considerably in produced fresh tuber yield and increases its crop yield.

The lowest nitrate content is noted in planting protection biologization. Ecologically clean and high quality tuber production in integrated agrotechnique of early potato crop is ensured by biologised plant protection when bitoxibacilline used, germination of planting material, optimal early date plantings and dozes and water-soluble fertilizer correlation.

Conclusion

Thus, the analyses of research carried out and experience of leading potato growers of the Republic of Bashkortostan allows to make the following main conclusions:

1. Soil and climatic conditions of the Republic of Bashkortostan are quite favourable for early potato production in July – first decade of August up to 100 t/ha.
2. Theoretical statements of integrated agrotechnique of early potato un the Republic of Bashkortostan allow to apply different integrated agrotechniques on the different levels to meet the needs of crop plants.
3. It is necessary to establish the Potato Centre of the Republic of Bashkortostan to improve quality of raw potato tubers and potato products, to increase its crop yield and profitableness of potato production complex.

Acknowledgements

Kiryuhin V.P. (deceased), candidate of agricultural sciences, senior scientific worker, the head of Potato Staff, Storage and Processing Department of the Russian Scientific Research Institute of Potato Production; Kiryuhina I.I., candidate of technical sciences, the head of Standardization and Certification Department of the Russian Scientific Research Institute of Starch Products for their sincere and so necessary support and care and exquisite leading in research work.

Nedorezkov V.D., doctor of agricultural the Republic of Bashkortostan, Rector of Bashkir State Agrarian University in 1999-2007 for his all-round help and support in my Doctor thesis work.

The staff of the scientific department of BSAU library with the head Beloborodova O.V., Honoured Worker of Culture of the Republic of Bashkortostan for skilled and timely supplying with current bibliography information and completing all new books.

Guryanov P.V., Honoured machine operator, inventor and rationalizator (experimental farm 'Milovsky' BSAU) for the highest skilled work and responsibility and love to land of Bashkortostan.

The staff of the Foreign Languages Chair of the BSAU with the head Novikova O.N., candidate of philology sciences, assistant professor for their professional translation of the manuscript into English.

References

Allison, M.F., J.H. Fowler and E.J. Allen (2001). Responses of potato (Solanum tuberosum) to potassium fertilizers. The Journal of Agricultural Science 136: 407-426.

Andrianov, A.D. and D.A. Andrianov (2006). Fertilizers and phyto-regulators increase immunity and productivity of the early potato. Potato and vegetables. No. 4: 11-12 (in Russian).

Andrianov, A.D., D.A. Andrianov and Yu. M. Alimbaev (2005). Predecessors and Treatment of early potato. Potato and vegetables. No. 1: 12 (in Russian).

Andrianov, A.D. and D.A. Andrianov (2004). Optimization of the early grown potato water consumption when irrigating. Improvement and water production. No. 3: 22-23 (in Russian).

Andrianov, D.A. and A.D. Andrianov (2003a). The systems of the basis soil tillage andfertilizing in the crop rotation for early potato. Potato and vegetables. No. 1: 12 (in Russian).

Andrianov, D.A. and A.D. Andrianov (2003b). Irrigation of early potato. Potato and vegetables. No. 4: 8-9 (in Russian).

Andrianov, D.A. and Andrianov A.D. (2003c). Technology of growing of early harvest in Bashkortostan republic. Potato and vegetables. No. 8: 19-20 (in Russian).

Andrianov, D.A. and A.D. Andrianov (2002). Soil tillage andmanagement as a basis of higt yield. Potato and vegetables. No. 8: 11-12 (in Russian).

Andrianov, D.A., A.D. Andrianov and A.V. Vorobiyov (2000). The optimal patterns of watering and nutrition of potato crops in Bashkiria. Potato and vegetables. No. 5: 22 (in Russian).

Anisimov, B.V. (2006). Value and role of potato in the healthy human nutrition. Potato and vegetables. No. 4: 9-10 (in Russian).

Anisimov, B.V. (2005). Production of potato and seed potato in Russia. In: Potato in progress. Wagenigen Academic Publishers, the Netherlands, 352-358.

Simakow, E.A. and B.V. Anisimov (2006). Quality improvement and market development of ware and seed potato in Russia. In: Potato developments in a changing Europe. Wageningen Academic Publishers, the Netherlands, 161-174.

Stalham, M.A. and E.J. Allen (2004). Water uptake in the potato (*Solanum tuberosum*) crop. The Journal of Agricultural Science 142: 373-393.

Zlotikov, A.K., S.A. Stefanina , E.V. Kirsanova and Z.I. Glazova (2004). Influence of biopreparation *Albit* on sowing characteristics and yield qualities of buckwheat seeds. In: Advances in Buckwheat Research. Proc. 9th Int. Symp. on Buckwheat, Prague, August 18-22, 2004. Prague. p. 499-500.

Zlotikov, A. K., Y. N. Shapovalova , A. A. Makarov (2001). Association of *Bacillus firmus* E3 and *Klebsiella terrigena* E6 with increased ability for nitrogen fixation. Soil Biology and Biochemistry. 33. p. 1525-1530.

Zlotikov, K. M., T. N . Pustovoitova , A. K . Zlotikov (2000). Metabolites of *Pseudomonas aureofaciens* H16 and *Bacillus megaterium* PC2 increase drought resistance of spring wheat. In: Kulaev I.S. (Ed.) Modern problems of microbial biochemistry and biotechnology. Abstr. Int. Symp., Puschino, June 25-30, 2000. IBPM. Puschino. p. 138-139.

Curriculum vitae – A.D. Andrianov

A.D. Andrianov studied agronomy in the Bashkir Agricultural Institute, which graduated in 1985 and got the speciality of agronomist. During the following 5 years he worked as agronomist at the Bashkir Science Research Institute of Agriculture, the Bashkir project-finding station of chemization. In 1990-1993 he studied at the post-graduate course of Russian Scientific Institute of Potato Production and defended his candidate thesis on early potato topic. He has been working as teaching assistant, senior lecturer and professor assistant at the BSAU since 1994. His research works are devoted to working out of integrated agrotechniques of early potato. He wrote more than 100 scientific works and reports at conferences.

Profile – Bashkir State Agrarian University

Bashkir State Agrarian University is a higher educational institution, which carries out educational and scientific research activities. The University consists of 8 departments, centre of humanitarian education, 58 chairs, scientific technical library, information centre, postgraduate course, scientific experimental complex, refresher course for farmers etc. There are full-time and part-time students at the university. Departments and Specialties: 1. Agronomy Department. 2. Animal Science Department has the specialty of Zootechnics.

3. Veterinary Medicine. 4. Farm Mechanisation Department, specialties of Automobiles and Professional Education agricultural engineering. 5. Farm Electrification and Automation Department, specialty of Energy Supply of Enterprises. 6. Land Management and Forestry Department, specialties of Land Management, Environment-friendly Landscape and Forestry. 7. Food Technologies Department, specialties of Grain Storage and Processing Technologies, Bread, Confectionary and Pasta Technologies, Wine and Brewage Technologies, Milk and Dairy Products Technologies, Meat and Meat Products Technologies, Machines and Apparatus of Food Industries and Food Engineering at Small Business. 8. Economical Department, specialties of Finance and Credit, Accounting, Analysis and Audit, Economy and Management at Enterprises (in agricultural production), Marketing, State Municipal Management and Applied Information in Economics. Directions of Research: 1. Environment-friendly landscape, effective systems of agriculture, adapted crop varieties and hybrids, selection and genetics of field crops. 2. Improvement of pedigree and productive qualities of farm livestock and increase of bee families' productivity. 3. Improvement of diagnosis methods, therapy and prophylactic measures of animal diseases. 4. Improvement of exploitation and maintenance of farm machinery. 5. Effective Usage of energy resources in agriculture. 6. Ecological monitoring of territories and estimation of toxic state of underground water. 7. Rational usage, control and reproduction of land, water and forests. 8. Improvement of economic and financial stock-taking in agriculture. 9. Improvement of processing technologies and increase of quality of agricultural produce.

Prospects of potato growing techniques in wide rows

V.I. Starovoitov[1], N.V. Voronov[2] and O.A. Pavlova[1]
[1]All-Russian Potato Research Institute, 23, Lorkh Street, Kraskovo, 140051, Moscow region, Russia
[2]Ecological Centre of the Russian Academy of Sciences, St. Petersburg, Russia

Abstract

A new technique based on growing high-quality potatoes, including potatoes suitable for making French fries and 'tiny' baked potatoes, where large tubers with a cross-section diameter more than 60-70 mm are used, was developed by scientists of the All-Russian Potato Research Institute. The basic elements of the technique include growing potatoes in wide rows, using bio- organic fertilizers, deep soil treatment before planting (25 cm), using middle and large size seed potatoes and planting density of up to 50 and 20 thousand tubers per hectare. This technique requires building a system of new machines or refurbishment of the old equipment. The proposed potato cultivation technique allows productivity to be raised by 15-30% and power efficiency by 15-25%. It also offers an opportunity to increase the reproduction coefficient 1.2-1.5-fold, obtain twice as many large tubers (over 70 mm), and decrease greatly a number of greenish potatoes by making bigger feeding area and creating better moisture conditions.

Keywords: wide rows, ecological compatibility, large tubers

Introduction

At present, the use of potatoes for French fries and 'tiny' (baked) potatoes is rapidly becoming extensive. The tubers having the largest cross-section diameter of over 60-70 mm suit these purposes. Applied by farms located outside black soil areas, traditional techniques using inter-row distances of 70 cm and planting densities of 50-60 thousand tubers per hectare make it impossible to harvest high-quality tubers of this fraction, as plants are located too close to each other, lacking nutritious substances and moisture.

The Russian food industry currently consumes mostly imported potatoes, whereas native raw material is under-consumed because of the unequal sizes and irregular shape of tubers, their low content of dry matter, large quantities of greenish tubers due to non-availability of specialized technologies. The new technique developed by the All-Russian Potato Research Institute ('Institute') will help master the production of high-quality potatoes outside Russia's black soil areas.

Materials and methods

The correlation between the yield of large tubers, compliance of basic product quality indicators with the food industry's requirements, and reduced costs for production of one tonne of tubers was studied by the Institute in 2002-2004 at the Elite Potatoes Agricultural Firm, Ramenskoye District, Moscow Region. The soils were sod-podzolic, medium-loamy.

The experiment procedure included studying of combinations of the following agricultural techniques:

1. Inter-row distances: 70 and 140 cm.
2. Application rates of the biohumus organic fertilizer: 0.60 and 80 g/plant.
3. Deep soil treatment before planting (25 cm).
4. Weight of seed tubers: medium (50-80 g) and large (100-120 g) fractions.
5. Planting density: 50 thousand/ha (inter-row distance of 70 cm and (110+30) cm) and 20 thousand/ha (inter-row distance of 140 cm).

The multifactorial experiment was started with the method of systematic site placement. Warmed up and sprouted 'Udacha' and 'Zhukovski Ranni' tubers were planted manually. The registered area of the site was 50.4 m^2.

Pre-emergence treatment was given with the help of a KRN-4,2 cultivator equipped with wedge-shaped, lancet, and double lancet arms, followed by rotary cultivation (MTZ-82 + UMVK-1,4M). Two post-emergence mechanic treatments were provided: MTZ-82 + KRN-4,2 equipped with lancet arms and coverers.

Aktara® with an application rate of 0.06 kg/ha was sprayed with an OPSh-2000 sprayer to provide treatment against Colorado beetles. Three treatments against the potato blight were given during the vegetation period: the first treatment used 2 kg/ha Acrobat® during blooming; the second, Ridomil Gold® 14 days after the first treatment; and the third, Polyram® 14 days after the second treatment, 2 kg/ha. All observations, registrations, and analyses were accordance with the Institute's techniques.

Weather conditions: the vegetation period of 2002 was characterized by an acute shortage of moisture in soils and higher air temperatures during the whole vegetation period. 2003 was characterized by lower air temperatures and a large amount of precipitation from planting to emergence, insufficient moisture during tuber formation, and abundant moisture during harvesting. During growth of above-ground parts of plants in 2004, lower temperatures and high moisture content were indicated. A drought was registered during tuber formation.

Results and discussion

Potato plants positively reacted to improved soil environment conditions due to: inter-row distances increased to 140 cm, soil treatment before planting, application of biohumus at a rate of 60-80 g/plant, planting density decreased to 20 thousand pcs/ha of tubers of large seed formation; enhanced biometric indicators (Table 1): number of stems increased by 2.6 pcs, height of stems increased by 10.9 cm, weight of leaves increased by 138 g/plant, assimilation surface of leaves increased by 0.3 m^2. 'Udacha' plants prevailed substantially over 'Zhukovski Ranni' plants.

Weight accumulation of large tubers (> 60 mm) was registered as from the full flowering phase. Data of the best treatments in rows and the reference treatment with an inter-row distance of 70 cm are shown in Figure 1. Here and below, four combined treatments are shown: (1) inter-row distance of 140 cm, planting density of 20 thousand pcs/ha, weight of seed tubers of 100-120 g,

Table 1. Biometric Indicators of Potato Plant Development. According to Inter-Row Distances, Planting Density, and Size of Tubers Planted. Mean Values for 2002-2004 (Bloom Phase).

Inter-row distance, cm	Planting density, thsd pcs/ha	Weight of seed tubers, g	'Udacha' Stems, pcs/ plant	Height, cm	Weight, g/plant	Area of leaves, m^2/plant	'Zhukovski Ranni' Stems, pcs/ plant	Height, cm	Weight, g/plant	Area of leaves, m^2/plant
70	50	50-80	3.4	42.8	251	0.41	2.7	41.7	119	0.21
110+30	50	50-80	3.2	45.7	207	0.32	2.8	41.7	128	0.21
140	20	50-80	3.7	46.4	267	0.46	3.2	42.8	166	0.29
140	20	100-120	6.0	53.7	389	0.71	4.4	46.0	257	0.46

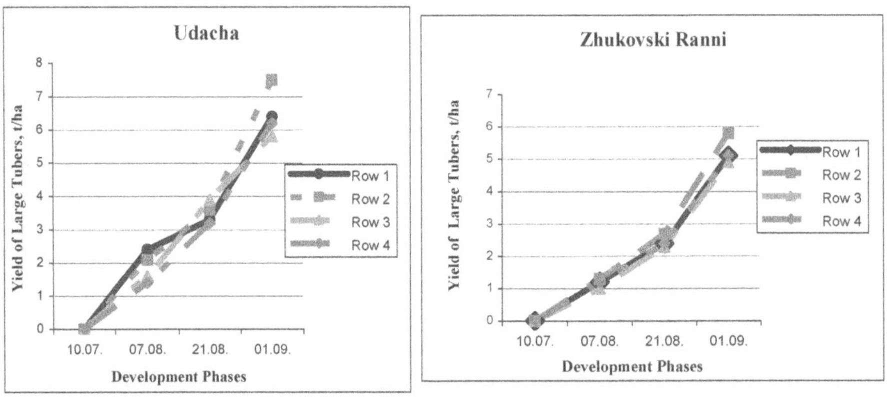

Figure. 1. Changes in Accumulated Tuber Weights (Tubers with the Largest Cross-Section Diameter of over 60 mm), t/ha (2002-2004).

rate of locally applied biohumus of 60-80 g/plant, without deep soil treatment before planting; (20 inter-row distance of 140 cm, planting density of 20 thousand pcs/ha, seed tubers weight of 100-120 g, rate of locally applied biohumus of 60-80 g/plant, with deep soil treatment before planting; (3) inter-row distance of 70 cm, planting density of 50 thousand pcs/ha, weight of seed tubers of 50-80 g, without locally applied biohumus, without deep soil treatment before planting; (4) inter-row distance of 140 cm, planting density of 50 thousand pcs/ha, weight of seed tubers of 50-80 g, rate of locally applied biohumus of 80 g/plant.

The observations of growth and development of potatoes made during the vegetation period revealed an unequal effect of the agricultural combinations applied on yield formation. Tuber formation rates were different and depended on variety features, techniques under study, and weather conditions.

The highest yields were found at the sites with 'Udacha' and 'Zhukovski Ranni' in the coolest and moistest year 2004, especially on the sites planted with large tubes, as large and healthy plants of these treatments reacted more favourably to vegetation conditions.

Weight additions of large tubers begin in late July, and the most extensive growth is observed in the last ten-day period of August, especially on the sites with an inter-row distance of 140 cm, planted with large tubers with the planting density of 20 thousand tubers/ha. Increased biohumus application rates with deep soil treatment before planting favoured a better nutrition inflow to tubers, which led to a more moderate humidification of porous soils. As a result, a considerable increase in the yield of large tubers was registered.

When analyzing the effect of the inter-row distance, one can note that the growth of the crop is more extensive at the beginning of tuber formation with an inter-row distance of 70 cm irrespective of the variety. However, a month after blooming, rowed plants catch up in development. It is especially true with large seed tubers, which gave much higher yield of large tubers.

Such yield structures of potatoes for French fries and 'tiny' baked potatoes are shown in Figure 2. The highest yield of tubers having the largest cross-section diameter of over 70 mm was also attained on the rows with a planting density of 20 thousand/ha, with large tubers (100-120), with a biohumus application rate of 60-80 g/plant (treatments 1 and 2). 'Udacha' gave 3.8-4.3 t/ha; 'Zhukovski Ranni' gave 2.3-3.3 t/ha, which is 1.65 – 1.87-fold ('Udacha') and 1.77 – 2.54-fold ('Zhukovski Ranni') higher than in the case of reference planting (treatment 3) with an inter-row distance of 70 cm (2.3 t/ha for 'Udacha', and 1.3 t/ha for 'Zhukovski Ranni').

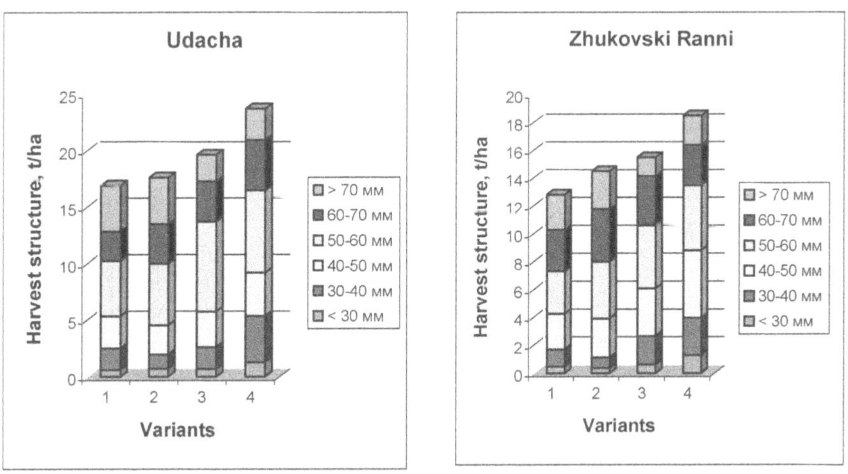

Figure 2. Grades of harvested tubers by cultivation technique, t/ha (2002-2004).

The highest reproduction rate per plant (Figure 3A) was attained on the site with large seed tubers (100-120 g) with a planting density of 20 thousand/ha and a biohumus application rate of 60-80 g/plant (Treatments 1 and 2).

Tuber yield from a site (Figure 3B) rises considerably with a planting density increased to 50 thousand/ha (Treatments 3 and 4). Deep soil treatment before planting increased tuber yield by 10.8 thousand pcs/ha ('Udacha') and by 9.5 thousand pcs/ha ('Zhukovski Ranni').

Harvesting is the most labour-extensive part of potato cultivation. Traction resistance of the combine harvester during potato harvesting is determined by the section area of the tuberiferous soil layer. Rotary ridge-formation makes it possible to obtain the largest volume of a ridge (row) with the following section areas: 818-911 cm^2 for inter-row distances of 70 cm; 1,230-1,250 cm^2 for inter-row distances of 90 cm; 2325-2444 cm^2 for inter-row distances of (110+30) cm (Chistyakov, 2001). In unfavourable conditions, the rowing technique makes combine harvesting easier due to a 38-44 percent reduction in separated soil (Starovoitov, 1993; Banadysev and Yukhnevich, 2000).

The analysis (Figure 4) shows that specific ridge section and, hence, traction resistance of the combine harvester during harvesting on one-line rows is 30-40 percent lower than in the case of the traditional cultivation technique.

Large tubers are damaged very much in the harvesting process due to their weight and a high dropping levels during harvesting. However, harvesting technology, especially related to early potatoes, allows potatoes to be transported directly to consumers or trading networks right from the field. In order to solve this problem, it was proposed that the potato-digger should be upgraded (Patent No. 2208926 'Machine for Harvesting Potato Tubers'). There are also proposals to switch separate machines and machine units to the rowing technology (Patent No. 2199845 'Potato Cultivation Technique', Patent No. 2231942 'Potato Cultivation Device').

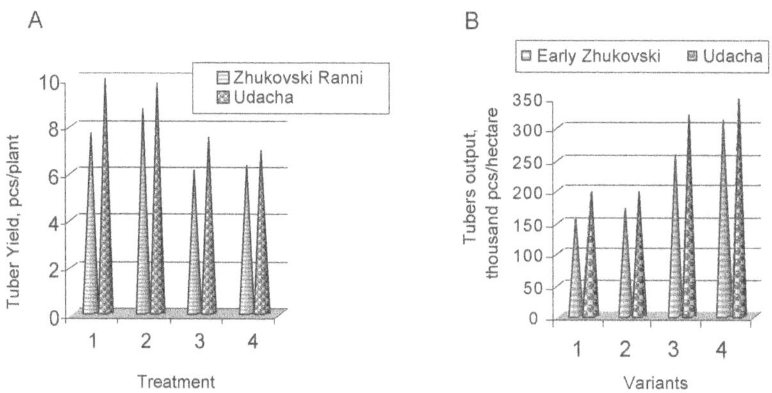

Figure 3. Tuber yield by cultivation technique: (A) from one plant, (B) from one hectare (2002-2004).

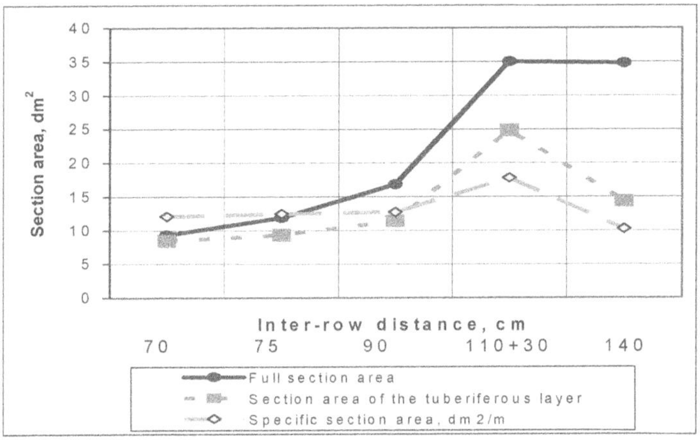

Figure 4. Dependence of Ridge/Row section on inter-row distance during potato cultivation.

Scientific research results have undergone production tests at the same time as being implemented by Tavra LLC, Krasnoufimsk District, Sverdlovsk Region, as of 2004, and by Elite Potatoes Agricultural Firm, Ramenskoye District, Moscow Region, as of 2005.

Conclusions

1. Planting of large-sized seed potatoes (100-120 g) produced better results with lower temperatures and higher moisture content during the vegetation period.
2. In order to grow potatoes suitable for French fries and 'tiny' baked potatoes with large yield of high-quality large tubers, one should apply a row technique using large seed potatoes (100-120 g), a planting density of 20 thousand/ha, a biohumus application rate of 60-80 g/plant, and deep soil treatment before planting.
3. The highest reproduction rate per plant (Figure 3A) was attained on the site with large seed tubers (100-120 g) with a planting density of 20 thousand/ha and a biohumus application rate of 60-80 g/plant.
4. Tuber yield from a site considerably rises with a planting density increased to 20 thousand plants per hectare.
5. Deep soil treatment before planting increased tuber yield by 10.8 thousand pcs/ha ('Udacha') and by 9.5 thousand pcs/ha ('Zhukovski Ranni').

References

Banadysev, S.A. and Yukhnevich, M.I. (2000). Application Pattern for Modern Potato Cultivation Technologies. Collection of Scientific Works, Issue 10, Minsk, Merlit: 230-241.

Chistyakov A.V. (2001). Identifying Efficient Techniques for Care for Potato Plantings with Different Inter-Row Distances on Sod-Podzolic Loamy soil. Author's Dissertation Abstract, Candidate of Science (Agriculture), Moscow, All-Russian Potato Research Institute, 21.

Methods Applied in Studies of Cultivated Potatoes (1967). Moscow, All-Russian Potato Research Institute, 263.

Starovoitov, V.I. (1993). To Master Technologies, Considering Specific Conditions. Potatoes and Vegetables, No 2: 10-11.

Crop protection

Effective and efficient approach to potato pathogen sensitive detection and identification

S.K. Zavriev[1], D. Ryazantsev[1,2], D. Abramov[2], T. Koshkina[2]*
[1]Shemyakin and Ovchinnikov Institute of Bioorganic Chemistry, Russian Academy of Sciences Ul. Miklukho-Maklaya, 16/10, 117997 GSP, Moscow V-437, Russia
[2]AgroDiagnostica Ltd' Kashirskoe sh. 23-5, 115478 Moscow, Russia

Abstract

Various pathogens of viral, bacterial, and fungal origin as well as pests strongly affect potatoes, the most widespread agricultural plant in Russia. The present paper deals with the development and characteristics of the FLASH-based diagnostic kits, which provide fast and accurate detection of main potato pathogens in soil and plant samples with the minimum contamination risk of the working area by PCR products.

Keywords: phytopathogen, PCR, molecular diagnostics, FLASH

Introduction

The extension of viral and bacterial infections in agro- and biocenosis is rapidly increasing. Potatoes, as a widespread agricultural plant in this country, are drastically contaminated by various fungal, bacterial, and viral pathogens and pests. Infectious diseases are characterized by their activity spread; moreover, a tendency of their action is growing in major potato regions. A dramatically dangerous situation is observed due to a sharp increase in viral and viroidal contamination of potato variety (grown and sold), e.g. potato virus Y – PVY, potato leafroll virus – PLRV, potato spindle tuber viroid – PSTV, etc. (Anisimov, 2004). Of significance is the control of dangerous quarantine potato pathogens and pests, such as potato Andean latent virus, pathogen of bacterial brown rot, and potato cyst Nematodes.

Nowadays about 40 phytopathogenic viruses are known, which have been identified on potato plants in various countries and regions with different climate. According to present data the most widespread pathogenic potato viruses are: the potato leafroll virus, as well as viruses Y, X, S, and M. The following potato viruses appear to be less spread and harmful: virus A, aucuba mosaic- and mop top viruses (though the latter has found a broader extension in the northern areas of Europe), rattle tobacco, etc. The crop losses from these pathogens can reach over 50%. In the past few years in many regions of Russia potato spindle tuber viroid (PSTV) was observed on different potato plants, the crop yield in this case decreased up to 80%.

Today the development of unique, high quality and reproductive potato seed growing is an important task. To solve this problem, virus control approaches deserve their wide application (Anisimov, 2004; Gnutova, 2005).

Among the tools for detecting phytopathogens, including potato infections, the PCR method plays the key role; it is broadly used in medicine for clinical diagnostics due to its simplicity, sensitivity, and high specificity [1, 3, 4]. Using the immunoenzyme assay, minimum 0,1 ng of specific protein of a pathogen is necessary for detection. In case of the PCR method around 10 copies of DNA or cDNA target gene have proved enough. The method promotes to identify any type by a specific nucleotide sequence of its genome, thus providing detection of pathogens or pests (Lopez *et al.*, 2003; Cullen *et al.*, 2000; Gnutova, 2005; Anisimov, 2004).

This paper presents our PCR-based diagnostic kits for main potato pathogens, i.e. bacterial – potato ring rot (*Clavibacter michiganensis* subsp. *sepedonicus*), brown rot (*Ralstonia solanacearum*), black stem and soft spindle rot (*Erwinia carotovora* subsp. *carotovora*); viral - potato leafroll virus, viruses M and S, virus X and aucuba mosaic virus, viruses Y and A, Andean latent virus, mop top virus, as well as spindle tuber viroid; fungal - potato phytophthora (*Phytophthora infestans*), and pests - pallid *(Globodera pallida)* and golden *(Globodera rostochiensis)* potato cyst forming nematodes.

PCR-based diagnostic kits for potato pathogens

Healthy and pathogen-infected potato plants have been rendered to the Lorkh Potato Research Institute, RAAS and to the Research Institute of Phytopathology, RAAS.

To synthesize DNA from micro-organism cultures, a kit 'Sample GC' was used, while in case of DNA from different plant tissues, we used 'Sample CTAB'. The DNA synthesis of Nematodes from individual cysts was performed by the 'Sample GC'. Use was made of the kit 'Sample NK' for isolating total RNA from potato tissues (all the kits manufactured by 'AgroDiagnostica Ltd' (Russia). The reverse transcription reactions with the use of the random primer $d(N)_6$ and the PCR involved the appropriate kits and reagents by the same company. The PCR ran in the thermocycler 'Tercyk', while the PCR-detector 'Gene' (ZAO 'RPC DNA-Technology') was applied for fluorescence detection. To enhance the PCR efficiency, all the kits use a 'hot start' to decrease the risk of non-specific amplification product formation. The 'hot start' is achieved through separation of the PCR-mixture components with the paraffin layer to prevent their mixing before the reaction. Standard values of the specific amplification product (150-300 nucleotide pairs) and the internal control (560 nucleotide pairs) allow use of agarose gel-electrophoresis as an alternative way to detect amplification products. Every system includes a universal internal control (IC) promoting to check the reaction accuracy and the inhibitor absence. The template amount for the IC is taken so that not to inhibit amplification of the specific fragment. Therefore, the IC band can be absent at high concentrations of the specific template. All the PCR-test systems are designed using original primers and fluorescence-labelled DNA probes.

Based on the multiple alignment of the nucleotide sequence of different genes for each pathogen we searched several primer pairs. The most efficient was selected after a set of experiments on sensitivity and specificity.

Interestingly, such an obvious advantage of the PCR method as its high sensitivity leads a probable increase in errors and incorrect results. These errors occur due to contamination with amplicons (amplification products), which appear in the environment when detecting the PCR results while applying the amplification products from the tube to the corresponding agarose gel well. Even a small amount of these fragments in the amplification tube can entail false-positive results. Of significant interest are the modifications of the PCR method, which allow for the reaction results even with the tube closed. One of the variants is FLASH (Fluorescent Amplification-based Specific Hybridization) method, viz. specific fluorescent hybridization upon amplification. The key feature of the FLASH method is the using of oligonucleotide hybridization probes labelled with fluorophore and 'dark' quencher molecules. The probes are added to the reaction mixture together with the primers and other components. Upon the formation of specific amplicons the Taq polymerase destroys the probe and the fluorophore is released from the quencher. Thus, the fluorescence level appears to be proportional to the number of specific PCR products. The scheme for the PCR study in the 'FLASH' format covers three stages. The first two: the sample preparation and amplification do not differ from the conventional PCR process. The third stage – detection of amplification products, is different in principle. The test tubes are placed from the thermocycler into the PCR detector, then without opening them fluorescence is measured. Data obtained are kept in computer database. Thus, a major PCR problem - the risk of contamination with amplicons, can be solved, besides, detection of the results, their interpretation and archiving are essentially simpler. Moreover, no special room for electrophoresis is necessary. The PCR analysis and data recording take only a few minutes (Elov *et al.*, 2005; Laptinov, 2005).

Results

At present, we have developed the FLASH PCR-test systems for the above organisms. The systems are designed so that two reactions can simultaneously run in one test-tube, i.e. amplification of a pathogen DNA genome fragment and of IC DNA. The IC system promotes to verify adequacy of the PCR and the inhibitor presence in the DNA sample.

Figures 1-3 outline data on the analysis of potato samples for availability of RNA leafroll virus, DNA ring rot, as well as of cysts of potato Nematodes for the specific DNA sequence of pallid Nematode, with further detection of amplification products in the FLASH format by 'Gene' detector and for control and data comparison by electrophoresis.

Figure 4 presents data on the potato sample testing for phytophthora DNA. The given system helps distinguish *Ph. infestans* from other *phytophthora* types spread in Russia (*Phytophthora cactorum, Ph. cambivora, Ph. capsici, Ph. cinnamomi, Ph. erytroseptica, Ph. fragariae, Ph. nicotianae* var. parasitica) and many other pathogens. Noteworthy, this approach provides detecting a pathogen in soil as well as in different plant tissues without having distinct symptoms of contamination.

Thus, we have developed simple and potent diagnostic kits to reveal main potato pathogens and covering the majority of viruses as well as several bacteria, fungi, nematodes, including quarantine organisms. Moreover, the test systems in the FLASH format have been designed for 14 pathogens (all major potato viruses and PSTV, *Clavibacter michiganensis* subsp. *sepedonicus*, *Ralstonia solanacearum*, as well as cyst forming Nematodes); the kits promote fast analysis and data recording after testing. Furthermore, the diagnostic kits for the rest of the pathogens are available only in the electrophoretic format.

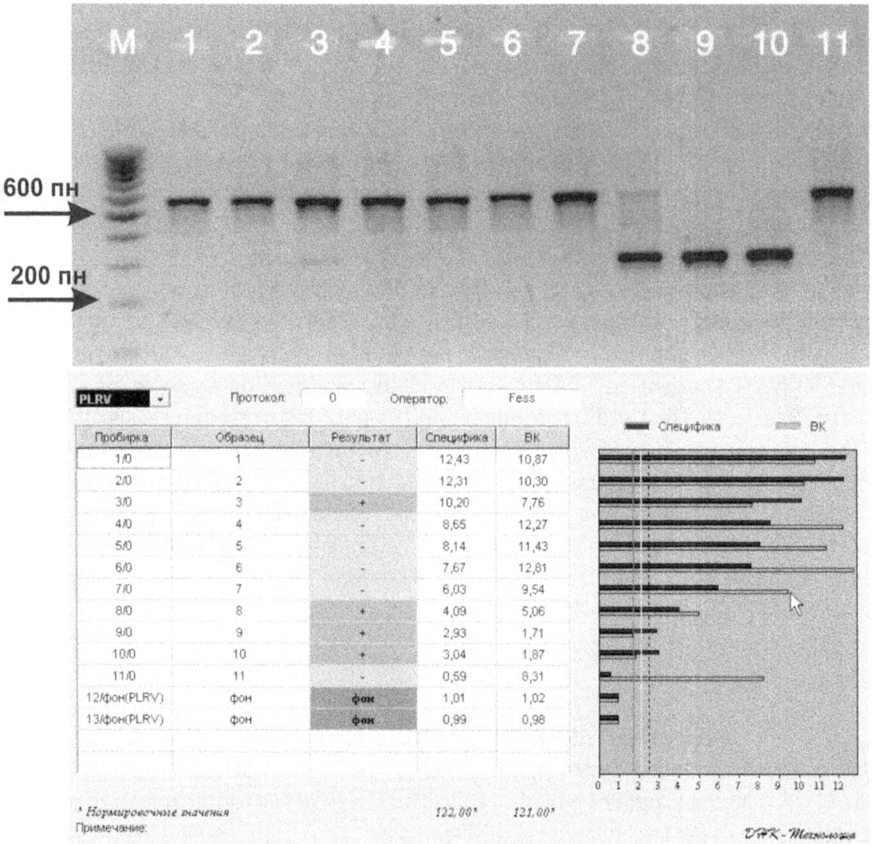

Пробирка	Образец	Результат	Специфика	ВК
1/0	1	-	12,43	10,87
2/0	2	-	12,31	10,30
3/0	3	+	10,20	7,76
4/0	4	-	8,65	12,27
5/0	5	-	8,14	11,43
6/0	6	-	7,67	12,81
7/0	7	-	6,03	9,54
8/0	8	+	4,09	5,06
9/0	9	+	2,93	1,71
10/0	10	+	3,04	1,87
11/0	11	-	0,59	8,31
12/фон(PLRV)	фон	фон	1,01	1,02
13/фон(PLRV)	фон	фон	0,99	0,98
			122,00*	121,00*

Figure 1. Data on the potato samples analysis for the leafroll virus. From 1 to 10 are the samples under analysis, 11 – negative control sample; '+' – the sample containing the cDNA sequence typical of the potato leafroll virus. '-' - no DNA sequence was revealed typical of the potato leafroll virus.

All the diagnostic kits have been tested in and approved by the Potato Research Institute, RAAS, Research Institute of Plant Quarantine, Ministry of Agriculture of Russia, and several potato seeds production centres.

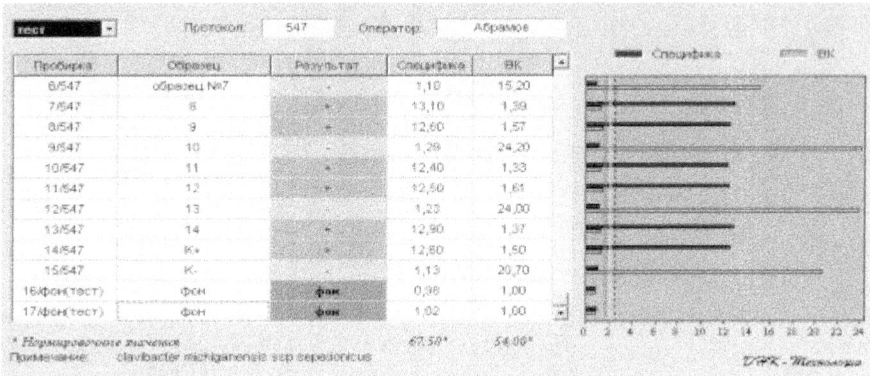

Figure 2. Data on the potato samples analysis for the ring rot virus Clavibacter michiganensis *subsp.* sepedonicus. *Sample 7 -* Ralstonia solanacearum, *samples 10, 13 - healthy plant, samples 8, 9, 11, 12, 14 -* Clavibacter michiganensis *subsp.* sepedonicus, *K+, K- positive and negative amplification controls; '+' - the sample displayed the DNA sequence typical of* Clavibacter michiganensis *subsp.* Sepedonicus *'-' - no DNA sequence was revealed typical of* Clavibacter michiganensis *subsp.* Sepedonicu.

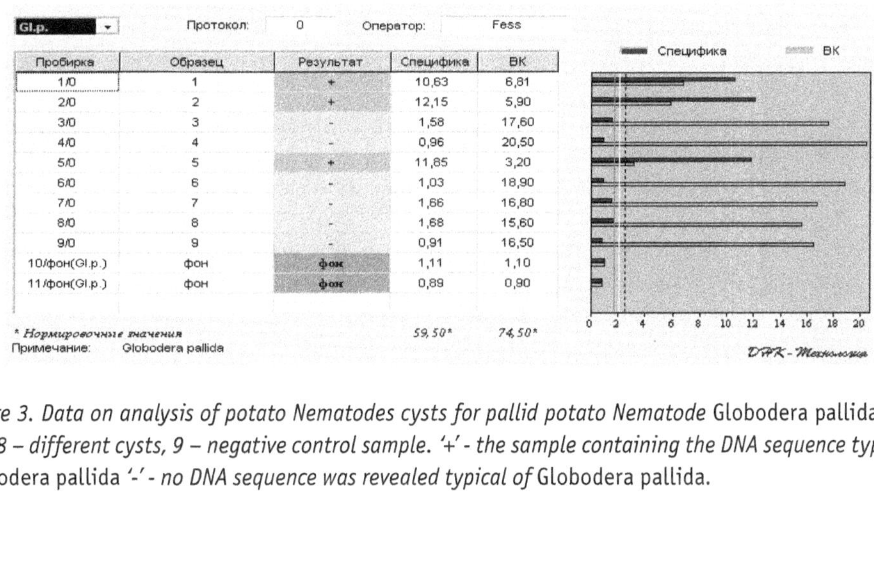

Figure 3. Data on analysis of potato Nematodes cysts for pallid potato Nematode Globodera pallida. *From 1 to 8 – different cysts, 9 – negative control sample. '+' - the sample containing the DNA sequence typical of* Globodera pallida *'-' - no DNA sequence was revealed typical of* Globodera pallida.

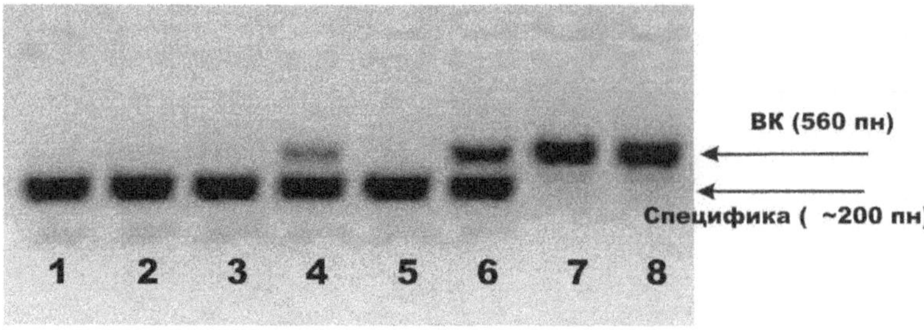

Figure 4. The PCR-based detection of Oomycete Phytophthora infestans. *1 -* Ph. infestans, *race A1, mycelium, 2 -* Ph. infestans, *race A2, mycelium, 3 -* Ph. infestans, *brown spot on the tuber, 4 -* Ph. infestans, *mycelium from the tuber, 5 -* Ph. infestans, *healthy area of the tuber, 6 -* Ph. infestans, *bud from the contaminated tuber, 7 -* Ph. nicotiana, *washing off the medium, 8 – healthy plant.*

References

Cullen D. W.; Lees A. K.; Toth I. K.; Bell K. S.; Duncan J. M., 2000, Detection and quantification of fungal and bacterial potato pathogens in plants and soil. Bull. OEPP., –Oxford. –V. 30. –N. 3/4. –P. 485-488.

Marıa M. Lopez, Edson Bertolini, Antonio Olmos, Paola Caruso, Marıa Teresa Gorris, Pablo Llop, Ramon Penyalver, Mariano Cambra, 2003, Innovative tools for detection of plant pathogenic viruses and bacteria/ In: *Int Microbiol.*, –V. 6. –P. 233-243.

Anisimov B. V., 2004. Plant pathogen viruses and their control in potato seed-growing. (Practical handbook). FGNY 'Rosinformagrotech', p. 1-15.

Gnutova R. V., 2005. Strategy of the study of potato infecting viruses in the Far East region. Agricultural biology (rus), No.1. –p. 46-54.

Yelov A. A., Fedorov N. A., ZHiburt E. B., Kofiadi I. A., Trofimov D. Yu., 2005, Quantitative fluorescence-hybridization PCR-based on domestic equipment and test systems. Healthcare and medical technique. No.2, p. 10.

Laptinov I. A., 2005, Non electrophoresis PCR technology. Lab diagnostic in Russia. Annual reference book. World of Medicine Ed. 'Tchelovek'. 2004/2005, p. 162-163.

Curriculum vitae – Sergei Zavriev

Sergei K. Zavriev studied at the Tbilisi Sate University biophysics and got his master degree in 1971. His PhD work in biophysics related to the study of nucleic acids confirmations, was done at the Institute of Molecular Biology of the Russian Academy of Sciences, Moscow, where in 1975 he receives a PhD degree in Biology. The degree of the Doctor of Sciences in Biology was received in 1985 for the works done on the T-odd phages and the studies of DNA dependent RNA polymerases. During 1980 -1982 S. Zavriev is a researcher in molecular biology at the Shemyakin Institute of Bioorganic Chemistry, USSR Academy of Sciences, Moscow. In the period 1982-2005 he is the Deputy Director for Research and Head of the Laboratory of Molecular Virology of the Institute of Agricultural Biotechnology of the Academy of Agricultural Sciences, Moscow. Starting from May 2005 prof. S. Zavriev is the Head of the Dept. of molecular diagnostic, Shemyakin-Ovchinnikov Institute of Bioorganic Chemistry of the Russian Academy of Sciences.

In 2005 he was elected as a Correspondence member of the Russian Academy of Agricultural Sciences.

The research activity during last 10 years was targeted on investigations of plant virus genome structure and expression, cloning and expression of virus specific genes and their application for functional studies, producing antibodies against different recombinant antigens including allergens, development of technologies for PCR and RT-PCR based detection of DNA- and RNA containing pathogens and producing correspondent diagnostic kits. Prof. S. Zavriev is an author of more than 130 papers and patents.

Prof. S. Zavriev was working as a visiting professor of the Department of Plant pathology in the University North Carolina, USA (1993). He had International grants from USA Academy of Sciences (1993), INCO-Copernicus (1998), INTAS (2000, 2002), FP6 (2007) and several Russian Federation grants. Prof. S. Zavriev collaborates for last 14 years with 'Agdia inc' (Elkhart, IN USA) on PCR-based plant pathogen diagnostics, with the Nexxtec GmbH (Leverkusen, Germany), Agrobiogen GmbH (Munich, Germany) and other companies on development of nucleic acids purification systems and PCR technologies for pathogen and GMO detection. He is a Council for Ecology, The Government of Moscow, Russia; member of the Organizer committee of the Moscow International Biotechnology Congress, of the International Society for Pest Information (ISPI), International Virus Taxonomy Committee (*Flexiviridae* team). From 2005 S. Zavriev is a consultant at the newly founded company 'AgroDiagnostica Ltd'' Russia.

Late blight of potato in Russia

S.N. Elansky[1,3], Yu.T. Dyakov[1], D.I. Milyutina[1], V.P. Apryshko[1], M.A. Pobedinskaya[1], A.V. Filippov[2], B.E. Kozlovsky[2], M.A. Kuznetsova[2], A.N. Rogozhin[2] and N.V. Statsyuk[2]
[1]Lomonosov Moscow State University, Biological Faculty, Moscow, 119992 Russia;
[2]All-Russian Research Institute of Phytopathology of Russian Academy of Agricultural Sciences, Bolshie Vyazemy, 143050 Russia;
[3] Research-and-production company 'AgroKlimatMash', Moscow, 119331, box 31, Russia.

Abstract

Two types of *Phytophthora infestans* populations have been registered in Russia: (1) Siberia and Far East populations, characterized by a very low genotypic variability, and (2) greatly varying European populations. The possible reasons for the low variability of Siberian and Far East populations of *P. infestans* are the use of a local seed material for the potato and tomato reproduction and weather conditions, that do not allow the soil overwintering of oospores. On the contrary, European populations are influenced by the active exchange of the seed material and a permanent import of potato tubers and tomato fruits from abroad. In this paper we discuss the role of different mechanisms (mutations, migrations, sexual and parasexual recombinations) in the variability of *P. infestans* populations in Russia and also the influence of fungicide application and an intraspecific specialization on the structure of these populations.

Keywords: *P. infestans*, variability, tomato, local seed, mating type ratio, fungicides

Introduction

Phytophthora infestans (Mont.) de Bary, the causal agent of late blight of potato and tomato, attracts the fixed attention of scientists from different countries for more than 150 years. In the middle of nineteenth century, when this oomycete suddenly appeared in Europe, it caused the late blight epidemic, engraved in the memory of many generations. Even now this pathogen is often called as 'Irish Famine Fungus'. Almost 100 years after the first epidemics scientists found in Mexico some wild potato species, resistant to late blight and developed the techniques of their crossing with the cultivated potato (Muller, 1935), that resulted in the creation of new cultivars, resistant to late blight (Pushkarev, 1937). However, shortly after the start of the commercial growing of such cultivars, there have been revealed some virulent races of *P. infestans*, able to strike these resistant cultivars, so the new resistance genes, that have been introduced into cultivated potato from the wild Mexican potato, quickly lost their efficacy. Such failure in the use of the monogenic (vertical) resistance made potato breeders to seek more complex ways of the use of a non-specific polygenic (horizontal) resistance.

The same problems arose in the case of chemical protection of potato. The significant differences between the oomycetes and the true fungi, including their chemical composition, ultra structure, and metabolism, cause many systemic fungicides, used for the protection of plants against the fungal diseases, to be inefficient in the case of the late blight. So, the chemical protection of potato against the late blight was based on the multiple spraying with contact fungicides. The development of phenylamide fungicides, toxic for oomycetes and systemically distributed in plants, became a revolution in the field of the chemical protection of potato against the late blight. However, the wide use of these preparations soon resulted in the accumulation of phenylamide-resistant strains in *P. infestans* populations (Davidse *et al.*, 1981), that complicated the use of plant protection techniques. In view of the aforesaid, one can easily understand such great attention of many scientists in the study of the structure of *P. infestans* populations, mechanisms of changes in the number of different races and their genetic structure, and genetic mechanisms, providing the variability of this pathogen.

The structure of *P. infestans* populations in Russia

In the absence of a recombination or the insignificant contribution of this process into the population structure, the population consists of several clones, very rarely exchanging their genetic material. In such case the study of individual gene frequencies will be less informative, than the study of frequencies of genotypes, that have the common origin (clonal lineages) and differ only in the point mutations. The RFLP-RG57 method, developed in the early nineties, significantly accelerated population studies of the late blight agent and its clonal lineages. Along with this technique, scientists also use the mating type analysis, DNA fingerprinting, allozyme analysis for peptidase and glucose 6-phosphate isomerase, and the determination of the mitochondrial DNA (mtDNA) haplotype.

The US-1 clonal lineage was the predominant lineage in all *P. infestans* populations up to the late eighties, when it was replaced by other clones and finally vanished from Europe and North America. In Russia (Moscow and Ekaterinburg regions) this genotype was not registered since 1993 (Fry *et al.*, 1992).

In Asiatic Russia, i.e. on the huge territory lying between Ekaterinburg and Sakhalin, the predominated clonal lineage is SIB-1, that can be characterized by the A1 mating type and IIa mtDNA haplotype. In 1993 this genotype was registered in the Moscow region, and some field populations consisted mainly of the metalaxyl-resistant strains of this clonal lineage. Since 1993 this genotype is not widespread. In 1997-1998 the SIB-1 lineage was widespread beyond the Ural Mountains, except the Khabarovsk Territory and the Jewish Autonomous Region, where the SIB-2 genotype predominated. The spatial separation of clones with different mating types excludes the sexual reproduction in the Siberian and Far East populations of *P. infestans*. In contrast to Siberia, the population of the Moscow region is represented by many clonal lineages, and almost every isolate has its unique multilocus genotype (Elansky *et al.*, 2001). One can not explain such diversity only by the fact of the 'import' of new strains through the infected seed material, imported from different parts of the world. Since both mating types are presented in this population, then such diversity can be also conditioned by the recombination process. For example, it seems that BC-2, BC-3, and BC-4 genotypes arose in British Columbia because of a hybridization between BC-1 and US-6 clonal lineages (Goodwin *et al.*, 1995). It is also possible, that the *P. infestans* populations of the Moscow region also contain some hybrid strains. For example, MO-4, MO-8, and MO-11 strains, heterozygous in the PEP locus, can be the hybrids between the MO-8 strain, which is homozygous in the one allele of the PEP locus and has A1 mating type, and MO-12, MO-21, and MO-22 strains, respectively, which have A2 mating type and are homozygous in another allele of the same locus. In this case there is a tendency to the strengthening of the role of the sexual process in the current *P. infestans* populations, and the value of information, obtained by the clonal lineage analysis, will reduce, as it was shown for the populations from the Moscow region.

Role of different mechanisms in the variability of *P. infestans* populations

Mutations

Since mutations represent a stochastic process with a low frequency, their generation in any locus depends on the mutation frequency for this locus and the magnitude of population. A study of the mutation frequency of different *P. infestans* strains usually involves a determination of the number of colonies grown on the selective nutrient medium after a treatment with chemical or physical mutagens. A high frequency of mutations, determining a metalaxyl resistance, can be one of the reasons of the accumulation of resistant strains in the nature. However, one-step mutations usually cause some genomic disbalance, and, along with the providing of a new property, reduce the total adaptation potential of the organism. The most of laboratory mutations provide the reduced aggressiveness and are not registered in the native populations. For example, the correlation coefficient between the level of phenylamide resistance of mutant *P. infestans* strains and their growth rate in artificial medium averages (-0.62). At the same time, correlation coefficient between the resistance to fungicides and the aggressiveness of these strains on potato leaves averages (-0.65) (Derevyagina *et al.*, 1993), that testifies the low adaptation level of mutant strains. Any mutations, affecting the level of dimethomorph resistance, were also accompanied by a sharp reduction in the strain viability (Bagirova *et al.*, 2001).

The size of population plays the critical role in the appearance of spontaneous mutations. In the case of the moderate infection of a potato field (35 spots per one plant), each hectare of the field generates 8×10^{12} spores a day (Dyakov and Suprun, 1984). Apparently, such populations contain all mutations, permitted by the exchange type for each locus. Even any rare mutation with the frequency equal to 10^{-9} will be acquired by a thousand organisms from the billions, that live on the one hectare of this field. Therefore, it is quite possible that any resistant and competitive strains can be selected among the many thousands organisms, obtained any resistance mutation.

Migrations

Today we know two main types of migration of *P. infestans*: a short-distance migration (within the bounds of the same or neighbouring fields) via the spreading of zoosporangia by air flows or rain drops, and a long-distance migration via the seed material or transported tomato fruits. The first way provides the nidus expansion, and the second one allows the appearance of new remote niduses. The spread of infection with potato tubers or tomato fruits facilitates the appearance and development of the disease in new places and also represents the main source of the genetic diversity of populations. Agricultural enterprises of the Moscow region grow potato cultivars obtained from different regions of Russia and Western Europe. Tomato fruits are usually brought from the southern regions of Russia (Astrakhan region, the Krasnodar Territory, and North Caucasus). Tomato seeds, which also can serve as the source of infection (Rubin *et al.*, 2001), are brought from the Southern Russia, China, European and other countries.

Apparently, this is one of the reasons of such high genetic diversity of *P. infestans* populations in the Moscow region.

Agricultural enterprises, located on the huge territory of Asiatic Russia (from Ekaterinburg to Sakhalin and Vladivostok), grow potato mainly from the local seed material; tomato is not widespread on this territory and also is grown from the local seed material. The genetic diversity of *P. infestans* populations on this territory is minimal; until now only three genotypes have been detected, and one of them (SIB-1) was registered in 79% of isolates (Elansky *et al.*, 2001). However, within Sib 1 and Sib 2 clonal lineages there are some differences between the tested isolates in the level of their metalaxyl resistance and the number of virulence genes. Such differences can be caused by any mutation processes.

Sexual recombination

To provide the contribution of a sexual recombination into the variability of a population, the presence of two mating types at a ratio of 1 : 1 and the initial variability of this population are necessary.

Almost all *P. infestans* populations from the European part of Russia contain strains of both mating types, but their proportion varies greatly in different populations and even in the same population in different years.

There are some indirect data, testifying the existence of sexual reproduction in certain years and in certain regions of Russia:

- The study of *P. infestans* populations from the Moscow region showed that the total genetic diversity of 13 populations, where the A2 mating type was registered for less than 10% isolates, was 0.08 (calculated for three isozyme loci); for 14 populations, where the share of this mating type exceeded 30%, the genetic diversity was twice as much (0.15; Elansky *et al.*, 1999). So, the higher the sexual process probability, the greater the genetic diversity of the population.
- The relation between the mating type ratio and the intensity of oospore formation was registered for the *P. infestans* populations of Israel (Cohen *et al.*, 1997) and Holland (Flier *et al.*, 2004). Our studies showed that for populations, containing 62, 17,9, and 6% of isolates of A2 mating type, the presence of oospores on examined potato leaflets with two or more lesions was registered for 78, 50, 30, and 15% of leaflets, respectively. Samples with two or more lesions contained oospores more often, than those with only one lesion (32 and 14% from the total number of samples, respectively; Apryshko *et al.*, 2004).
- The distribution of DNA fingerprint and isozyme loci in some populations from North Caucasus corresponds to the Hardy-Weinberg equation (Ahmatkhanova *et al.*, 2004), that can be evidence of a significant contribution of the sexual recombination into the population variability. The study of *P. infestans* populations from other regions of Russia did not show any correspondence to the Hardy-Weinberg distribution, but revealed the presence of a nonequilibrium linkage, testifying the prevalence of a clonal reproduction (Elansky *et al.*, 1999).

Parasexual recombination and mitotic gene conversion

The joint growing of *P. infestans* strains with mutations, affecting their resistance to different growth inhibitors, caused the appearance of isolates resistant to both inhibitors (Shattock and Shaw, 1975; Dyakov and Kuzovnikova, 1974; Kulish and Dyakov, 1979). Strains, resistant to both growth inhibitors, appeared as a result of a mycelium heterocaryotization; they were split in the case of their reproduction by uninuclear zoospores (Judelson and Ge Yang, 1998) and were not split in the monozoosporic progeny, since they have diploid (tetraploid, because the initial isolates were diploid) nuclei (Kulish and Dyakov, 1979). The segregation frequency of heterozygous diploids was very low because of the haploidization, nondisjunction of chromosomes, and mitotic crossingover (Poedinok *et al.*, 1982). The frequency of these processes can be increased through the certain impacts on heterozygous diploids (UV-treatment of germinating spores, mycelium treatment with *p*-phthorphenylalanine, etc.).

In spite of the unexpectedly high frequency of generation of heterocaryons and heterozygous diploids (tens of percents), this process occurs only in the case of the joint growing of mutant strains, originated from the same one. Because of a vegetative incompatibility, the heterocaryotization does not occur (or occurs with very low frequency) between differing native isolates (Poedinok and Dyakov, 1981; Anikina *et al.*, 1997; Cherepennikova-Anikina *et al.*, 2002). So, the role of the parasexual recombination can be reduced to the intraclonal recombination in heterozygous nuclei and the transition of certain genes into the homozygous state without any sexual process. This process can be epidemiologically important for the strains with recessive or semi dominant mutations, affecting their resistance to fungicides. The transition of such mutation into the homozygous form due to the parasexual process increases the resistance of its carrier (Dolgova and Dyakov, 1986).

Use of fungicides

The use of fungicides not only decreases the magnitude of population of phytopathogenic fungi (i.e. influences on the quantitative characteristic of their populations), but also can change the frequencies of certain genotypes (i.e. influences on the quality composition of the populations). The changes in the aggressiveness, virulence, and resistance to fungicides are among the most important parameters of the populations, affected by fungicides. The level of such influence is determined mainly by the type of the fungicide used. All fungicides can be generally divided into three types: *poly-site*, *oligo-site*, and *single-site* ones.

The *poly-site* group includes many contact fungicides. The resistance to such preparations (if it is really possible) is controlled by a large number of weakly expressing genes. Such properties determine the lack of any significant changes in the resistance of a population after the fungicidal treatment (though several experiments with different fungi, including *P. infestans*, demonstrated some increase in this resistance). The population, surviving after the treatment with contact fungicides, consists of two groups of strains: (1) strains, survived on the non-treated plant surfaces (since they did not contact with the fungicide, then their aggressiveness and resistance remain unchanged), (2) strains, contacted with a fungicide at a sub lethal concentration. The resistance of this part of the population does not change, but the general adaptation level and its parasitic component, i.e. an aggressiveness, decrease due to a partial damaging effect of a fungicide (even in the sub lethal concentrations) on the fungal metabolism (Derevyagina and Dyakov, 1990). Thus, even survived part of the population, that was in contact with the fungicide, has a low aggressiveness and can not generate any epidemic. Therefore, any careful treatment, reducing the frequency of that part of population, which will not contact with a fungicide, is a necessary condition of the successful protection activity.

The resistance of *P. infestans* to the *oligo-site* fungicides (some translaminar and systemic fungicides, for example, dimethomorph) is controlled by several additive genes. Each gene mutation results in a certain increase in resistance, and the total resistance is determined by the summation of such mutations. Therefore, such increase in resistance will be stepwise. For example, the dimethomorph resistance is polygenic and additive. The one-step mutation with frequency equal to 6,3 x 10^{-7} slightly increase the resistance. Every additive mutation reduces the target size and, therefore, the frequency of any subsequent mutations (Bagirova *et al.*, 2001). The improvement of the average resistance of *P. infestans* after the multiple treatment with the oligo-site fungicide occurs gradually. The rate of this process is determined by at least three factors: (1) mutation frequency for resistance genes, (2) resistance coefficient (the ratio between the lethal doses for resistance and sensitive strains), and (3) effect of the resistance gene mutations on the adaptation ability of the organism.

The frequency of each subsequent mutation is lower, then for the previous one, so the process is of damped character (Bagirova *et al.*, 2001). However, in the case of any recombination (sexual or parasexual) processes, there is a possibility of the union of different parental mutations in the hybrid strain and the acceleration of the above-mentioned process. Therefore, panmictic populations acquire the heightened resistance more quickly, than agamic ones; the agamic populations, which do not have any vegetation incompatibility barriers, acquire this feature more quickly, than populations, separated by such barriers. As a result, the presence of strains with different mating types in a population accelerates the acquisition of the resistance to oligo-site fungicides.

The second and third factors do not promote any fast accumulation of dimethomorph-resistant strains in the population. Each subsequent mutation doubles the resistance, and, at the same time, reduces both growth rate on artificial medium and, especially, aggressiveness (Bagirova *et al.*, 2001; Stem, Kirk, 2004). Probably, this fact can explain almost complete lack of resistant strains among the native *P. infestans* strains, including those, that have been collected from potato fields, treated with dimethomorph (Derevyagina *et al.*, 1999, Elansky *et al.*, 2007).

A population, treated with any oligo-site fungicide, also consists of two groups of strains. Strains from the first group did not come into contact with the fungicide and, therefore, did not change their initial characteristics (if this group contains resistant strains, then they will not be accumulated because of the higher aggressiveness and competitive ability of sensitive strains). The second group includes strains, affected with the sub lethal concentrations of the fungicide. The accumulation of resistant strains is possible in the second group, since in this case they have some advantages in comparison with the sensitive strains. Therefore, using oligo-site fungicides, one should remember, that the high concentration of the fungicide (several times more than the sub lethal one) is more important factor, than the careful treatment, since in the case of the stepwise mutagenesis, the initial resistance of mutant strains is low.

Finally, mutations, affecting the resistance to the *mono-site fungicides*, are highly expressive, i.e. a single mutation can determine a high resistance level (up to the total lack of sensitivity). Therefore, the resistance of populations increases very quickly. Phenylamides, including the widespread Ridomil (the acting substance is methalaxyl) and Ridomil-Gold (the acting substance is mefenoxam), can serve as examples of such fungicides. Mutations, causing the resistance to these preparations, have a high frequency (6.9×10^{-6}), and the resistance level of such mutants is very high and exceeds that of a sensitive strain in one thousand times or even more (Derevyagina *et al.*, 1993). Though the growth rate and aggressiveness of the resistant mutants reduce (because of the lack of aggressive sensitive strains), the magnitude of resistant population quickly increases along with its aggressiveness. Thus, after several years of fungicide application, the aggressiveness of resistant strains can achieve the level of the sensitive ones and even exceed it (Derevyagina, Dyakov, 1990). The aggressiveness grows more quickly, if the sexual recombination is possible in the given population.

The study of metalaxyl resistance of *P. infestans* strains, isolated in 1988-2004 in different regions of Russia, showed that resistant strains are usually isolated from the infected leaves and stems of potato and tomato and also from potato tubers. However, the frequency of resistant strains, isolated from tomato fruits, was very low, even in the case when the leaves of the same plants were infected with resistant strains. It seems that the sensitive strains have any advantages regarding the process of the fruit infection.

After 1993-1994 we registered the reduction in the frequency of resistant strains in most of the *P. infestans* populations. It was possibly caused by some changes in the structure of potato growing in Russia: in the beginning of the nineties it was concentrated mainly in small private gardens. The use of expensive fungicides, including metalaxyl, was practically ceased. According to our data, metalaxyl-sensitive *P. infestans* strains predominated on such private gardens, so we can consider these small gardens as the global reserve of sensitive isolates.

After 1994 the share of metalaxyl-resistant strains in almost all studied populations of the European part of Russia did not exceed 30%. A population usually contains a small number of resistant isolates even if metalaxyl-based fungicides did not used for a long period of time. In the case of the treatment of a potato field with such fungicides, the number of resistant strains sharply increased (Elansky *et al.*, 2007).

Intraspecific specialization

P. infestans can cause epidemics on two commercial crops - potato and tomato. In the case of potato the epidemics began shortly after this oomycete had spread to the new regions. The infection of tomato was also registered shortly after the appearance of the infection, but the first epidemics befell only one hundred years later - in the middle of XX century. In Russia the late blight of tomato was registered in XIX century. 'During a long time scientists did not pay attention to this disease, since it did not cause any significant crop losses, but in the sixties and seventies the late blight epidemics on tomato were registered in the Soviet Union, mainly in the Lower Volga, Ukraine, North Caucasus, and Moldavia...' (Balashova, 1979). Since then the late blight of tomato was registered every year and spread over the whole territory of the industrial and private vegetable cultivation, causing huge economical damage.

In 1993 the study of population markers of strains, isolated from the neighbouring potato and tomato fields, showed that about ¼ of isolates, collected from tomato leaves, were transferred from the potato field (Dolgova *et al.*, 1997). In theory one can suppose that the divergence of these populations from different hosts will increase and result in the development of intraspecific specialized forms (f. sp. potato and f. sp. tomato), especially since oospores can survive in plant residues (Drenth *et al.*, 1995; Bagirova and Dyakov, 1998) and tomato seeds (Rubin *et al.*, 2001) and, therefore, tomato plants have a potato-independent source for the spring revival of *P. infestans*.

However, the real events did not correspond to this hypothesis. The wintering of oospores allowed the parasite to avoid the most narrow stage of its life cycle, i.e. a soil vegetation stage, when its parasitic abilities are reduced and then gradually recovered in summer during outbreak. Primary zoosporangia and zoospores, germinating from oospores, are highly active, especially if oospores were parthenogenetical generated under the influence of pheromones of the strain with the opposite mating type. Thus, the infection material, grown from tomato seeds, infected with *P. infestans* oospores, is highly pathogenic for both tomato and potato. These changes resulted in the new population rearrangement, expressed through the following important epidemiological changes:

- Infected tomato sprouts became the important source of the primary infection of potato (Filippov *et al.*, 2004, Ivanyuk, personal communication).
- Epidemics on potato fields occurred one month earlier than usual (in June).
- The percentage of T1 race on the potato fields increased, though earlier it was very low (Ulanova *et al.*, 2003).
- The virulence of strains, isolated from tomato leaves and tested on the appropriate differentiator cultivars of potato, came up with that of the strains, isolated from potato; the aggressiveness of 'tomato' strains exceeded that of 'potato' strains on both tomato and potato plants (Lavrova *et al.*, 2003; Ulanova *et al.*, 2003).

Thus, the convergence of populations occurred instead of the divergence, i.e. the united population, characterizing by the high aggressiveness and virulence regarding to both tomato and potato, has developed on these host plants.

Conclusion

Two types of *Phytophthora infestans* populations have been registered in Russia: (1) Siberia and Far East populations, characterized by a very low genotypic variability, and (2) greatly varying European populations. The possible reasons for the low variability of Siberian and Far East populations are the use of the local seed material for potato and tomato reproduction and also the weather conditions, that do not allow the soil wintering of oospores. On the contrary, European populations are influenced by an active exchange of the seed material and a permanent import of potato tubers and tomato fruits from abroad.

One of the most urgent questions for the potato and tomato growers concerns the future changes in *P. infestans* populations. In our opinion, the population structure of this pathogen will change in the following directions:

- Increase in the genetic variety of all Russian populations, caused by the global warming, intensive exchange of the seed material, and import of tubers from other regions by processing enterprises.
- Increase in the aggressiveness of strains, caused by the sexual recombination and the increase in the genetic variety.
- Increase in the percentage of metalaxyl-resistant strains in *P. infestans* populations, caused by a predomination of large agricultural enterprises, practicing intensive chemical treatments, in the potato-growing sector of economics. Such global reserve of sensitive strains as the small private gardens is gradually vanishing. The mass appearance of strains resistant to oligo- and poly-site fungicides is highly unlikely.
- Disappearance of the differences between *P. infestans* populations, developing on potato and tomato plants, and the prevalence of strains with a high aggressiveness towards both hosts.

Taking into account these possible changes, we can recommend growers to use cultivars with the heightened level of the non-specific late blight resistance, to control the quality of the seed material, to observe the anti-resistant strategy, to use the forecast of the disease development during the application of chemical treatments, and to provide a spatial separation of potato and tomato fields.

Acknowledgements

The authors are grateful to graduate and undergraduate students and employees of the Department of Mycology and Algology of the Biological Faculty of Moscow State University and All-Russian Research Institute of Phytopathology, who studied *P. infestans* populations in different years: T.A. Kuzovnikova, V.B. Kulish, N.L. Poedinok, Ya.V. Petrunina, A.V. Dolgova, V.A. Terekhova, L.M. Suprun, M.K. Derevyagina, I.N. Rybakova, S.F. Bagirova, A.N. Smirnov, A.S. Kravtsov, F.H. Ahmatkhanova, E.I. Lavrova, T.I. Ulanova, I.N. Kozlovskaya, E.N. Morozova. Since 2007 this project was supported by International Science and Technology Centre (project No. 3440).

References

Lavrova, O.I., Elansky Ahmatkhanova, F.X., Dyakov, Yu.T., Petrunina, Ya.V., Pobedinskaya, M.A., Elansky, S.N., Kozlovskaya, I.N., Kozlovsky, B.E., Morozova, E.V., and Smirnov, A.N. (2004) Populations of *Phytophthora infestans* on the North Caucasus. Mikologia i Phitopathologia 38(3): 71-78.

Apryshko, V.P., Petrunina, Ya.V., Pobedinskaya, M.A., and Elansky, S.N. (2004) Oospores of *Phytophthora infestans* in the natural late blight focuses of the Moscow region in 2003. In: proceedings of the 'Micology and Algology - 2004' conference, M.: Prometei, 21-22.

Bagirova, S.F., An Zsan Li, Dolgova, A.V., Elansky, S.N., Shaw, D.C., and Dyakov, Yu.,T. (2001) Mutants of *Phytophthora infestans* resistant to dimethomorph fungicide. J. Rus. Phytopathol. Soc. 2: 19-24.

Bagirova, S.F. and Dyakov, Yu.T. (1998) Role of oospores in the renewal of the primary late blight infection of tomato. Selskohozyaistvennaya biologia 3: 67-71.

Balashova, N.N, (1979) Resistance of Lycopersicon genus to *Phythopthora infestans* and methods of its use in the tomato breeding. Kishinev, Shtinitsa.

Cherepennikova-Anikin, M.I., Savenkova, L.V., Dolgova, A.V., Shaw, D.S., and Dyakov, Yu.T. (2002) Vegetative incompatibility of *Phytophthora infestans*. J. Russ. Phytopathol. Soc. 3: 19-32.

Cohen Y., Farkash S., Reshit Z., and Baider A. (1997) Oospore production of *Phytophthora infestans* in potato and tomato leaves. Phytopathology 87: 191-196.

Davidse, L.C., Looijen, D., Turnensteen, L.I., and Vanderwal, D. (1981) Occurrence of metalaxyl resistant strains of *Phytophthora infestans* in Dutch potato fields. Neth. J. Plant Pathol. 7(2): 65-68.

Derevyagina, M.K., Dolgova, A.V., and Dyakov, Yu.T. (1993) Mutants of *Phytophthora unfestans* resistant to phenylamide fungicides. Mikologia i Phytopatologia 27(3): 57-61.

Derevyagina, M.K. and Dyakov, Yu.T. (1990) Influence of systemic and contact fungicides on the infection of potato leaves with the late blight. Mikologia i Phytopatologia 24: 252-256.

Derevyagina, M.K., Elansky, S.N., and Dyakov, Yu.T. (1999) Resistance of *Phytophthora infestans* to dimethomorph fungicide. Mikologia i Phytopatologia 33(3): 208-213.

Dolgova, A.V. and Dyakov, Yu.T. (1986) Genetics of the resistance of *Phytophthora infestans* to the metalaxyl fungicide. Genetika 22(10): 2423-2429.

Dolgova, A.V., Smirnov, A.N., Maleeva, Yu.V., Bagirova, S.F., Kozlovskaya, I.N., Kolesnikov, A.A., Shaw, D.S., and Dyakov, Yu.T. (1997) Comparison of *Phytophthora infestans* populations isolated from potato and tomato plants in the Moscow region. In: 'Problemy optimizatsii fitosanitarnogo sostoyaniya rastenievodstva' [Problems of optimization of the phytosanitary state of the plant breeding]. St.-Petersburg, 348-357.

Drenth A., Janssen E.M., and Govers, F. (1995) Formation and survival of oospores of *Phytophthora infestans* under natural conditions. Plant Pathol. 44: 86 -94.

Dyakov, Yu.T. and Kuzovnikova, T.A. (1974) Heterocaryosis in *Phytophthora infestans*. II. Genetic studies. Mikologia i Phytopatologia 8(2): 81-89.

Dyakov, Yu.T. and Suprun, L.M. (1984) Probabilistic approach to the calculation of the frequencies of virulence genes and its use in the population analysis of *Phytophthora infestans*. Selskohozyaistvennaya biologia 3: 111 - 118.

Elansky, S.N., Apryshko, V.P., Milyutina, D.I., and Kozlovsky, B.E. (2007) Resistance of Russian *Phytophthora infestans* strains to Metalaxyl and Dimethomorph fungicides. Vest. Mosk. Univ. 1 (in press).

Elansky S.N., Smirnov A.N., Dolgova A.V., and Dyakov Yu.T. (1999) Populations of *Phytophthora infestans* in the Moscow region. 1. Systems of reproduction. Mikologia i Phytopatologia 33(5): 346-352.

Elansky S., Smirnov A., Dyakov Yu., Dolgova A., Filippov A., Kozovsky B., Kozlovskaya I., Russo P., Smart C., and Fry W. (2001) Genotypic analysis of Russian isolates of *Phytophthora infestans* from the Moscow Region, Siberia and Far Eeast. J. Phytopathology 149: 605 - 611.

Filippov, A.V., Gurevich, B.I., Kuznetsova, M.A., Rogozhin, A.N., Spiglazova, S.Yu., Kravtsov, A.S., Smetanina, T.I., and Smirnov, A.N. (2004) Horizontal resistance of potato leaflets to the late blight and the aggressiveness of *Phytophthora infestans* isolates from different geographical regions. Mikologia i Phytopatologia 38(5): 74-87.

Flier, W.G., Kessel, G.J.T., and Shepers H.T.A.M. (2004) The impact of oospores of *Phytophthora infestans* on late blight epidemics. Plant dreeding and seed science 50: 5-13.

Fry, W.E., Goodwin, S.P., Matuszak, J.M., Spielman, L.J., and Milgroom M. G. (1992) Population genetics and interconinental migrations of *Phytopthora infestans*. Ann. Rev. Phytopathol. 30: 107-129.

Goodwin, S.B., Sujkowski, L.S., Dyer, A.T., Fry, B.A., and Fry, W.E. (1995) Direct detection of gene flow and probably sexual reproduction of *Phytophthora infestans* in northern North America. Phytopathology 85: 473-479.

Judelson, H.S. and Ge Yang (1998) Recombination pathways in *Phytophthora infestans:* polyploidy resulting from aberrant sexual development and zoospore-mediated heterokaryosis. Mycol. Res. 102: 1245-1253.

Kulish, V.B. and Dyakov, Yu.T. (1979) Isolation of diploid strains from the mixed cultures of two physiological races of *Phytophthora infestans*. Dokl. Akad. Nauk 244(3): 435-438.

Lavrova, O.I., Elasky, S.N., and Dyakov, Yu.T. (2003) Selection of *Phytophthora infestans* strains in asexual generations. J. Rus. Phytopathol. Soc. 4: 1-7.

Muller K.O. (1935) Uber den augenbicklichen Stand unserer Kenntis zur biologischen Spezialisierung des Krautfauleerrggers der Kartoffel (*Phytophthora infestans*). Der Zuchter 7.

Poedinok, N.L., Dolgova, A.V., and Dyakov, Yu.T. (1982) Parasexual process of phytopathogenic fungus *Phytophthora infestans*. Genetika 18: 1423-1428.

Poedinok, N.L. and Dyakov, Yu.T. (1981) Discovery of the vegetative incompatibility in the phytopathogenic fungus *Phytophthora infestans*. Mikologia i Phytopatologia 15(4): 275-279.

Pushkarev, I.I. (1937) A novel potato cultivar 8670 resistant to late blight. M., Selhozgiz.

Rubin, E., Baider, A., and Cohen, Y. (2001) *Phytophthora infestans* produced oospores in fruits and seeds of tomato. Phytopathology 91: 1074-1080.

Shattock, R.C. and Shaw, D.S. (1976) Novel phenotypes of *Phytophthora infestans* from mixed cultures of antibiotic resistant mutants. Transact. British Mycol. Soc. 67: 201-206.

Stem, J.M. and Kirk, W.W. (2004) The generation and quantification of resistance to dimethomorph in *Phytophthora infestans*. Plant Disease 88: 930-934.

Ulanova, T.I., Elansky, S.N., Filippov, A.V., Dyakov, Yu.T., Apryshko, V.P., Kozlovsky, B.E., Smirnov, A.N., and Coffey, M.D. (2003) Resistance to late blight of some promising lines of *Lycopersicon hirsutum*. J. Rus. Phytopathol. Soc. 4: 9-15.

Curriculum Vitae – Sergey N. Elansky

Sergey N. Elansky studied biology and physics at Moscow M. V. Lomonosov State University (MSU) and received his MSc in 1994. After this he studied as a post-graduate student in the department of mycology and algology of MSU and received his PhD in 1998. The thesis was entitled: 'Populations of *Phytophthora infestans* in Russia'. He then specialized in potato diseases and worked in the department of fungal potato diseases of All-Russian Research Institute of Phytopathology for 6 years. In 2006 Dr. Elansky joined the department of mycology and Algology of MSU, where he works now.

The main object of his scientific interest is the phytopathogenic oomycete *Phytophthora infestans*. The main tasks are: comparison analysis of strains and populations on different host-plants and in different regions, the role of oospores and sexual process in genetic structure of *P. infestans* populations, mechanisms of changes of aggressiveness during the vegetation seasons and winter storage, monitoring of fungicide-resistant strains in field populations and search for new marker features for population research.

S.N. Elansky is author of over 60 scientific paper and co-author or co-editor of several books. He is the member of the 'Council of fungal diseases of plants' of the Russian Agricultural Academy, member of the editorial board of the journal 'Potato and vegetables'. In 2004 he organized the site http://www. kartofel. org, oriented to Russian potato farmers. Dr. Elansky also organized and manages the company 'AgroKlimatMash' that elaborates and constructs microprocessor units and soft for ventilation and microclimate control systems for agricultural objects such as potato and vegetable storages.

Company profile – AgroKlimatMash

Research-and-production company 'AgroKlimatMash' elaborates and constructs microprocessor units and soft for ventilation and microclimate control systems of agricultural objects such as potato and vegetable storages. Company offers electric equipment for new and reconstructed potato and vegetable storages. Our specialists can mount, repair, and service of ventilation and microclimate control systems. Company offers completely equipped storages based on fast-build steel constructions with unsupported spans.

We are open for collaboration. All information about the company and its production are available online: http://www. kartofel. org. E-mail: kartofelorg@yahoo. com, phone: (7906) 056-58-61.

High-sensitive technologies for molecular diagnostics of potato virus and viroid infections

Yu.F. Drygin[*1], *S.N. Chirkov*[2], *O.A. Kondakova*[2], *R.A. Zinovkin*[2], *P.A. Ivanov*[2], *A.N. Blintsov*[2], *E.S. Gavryushina*[3], *A.V. Zherdev*[4], *N.A. Byzova*[4], *B.B. Dzantiev*[4] *and J.G. Atabekov*[1,2]
[1]*A.N. Belozersky Institute of Physico-Chemical Biology, Moscow State University, 119992 Moscow, Russian Federation*
[2]*Faculty of Biology , Moscow State University, 119992 Moscow, Russian Federation*
[3]*Faculty of Bioengineering and Bioinformatics, Moscow State University, 119992 Moscow, Russian Federation*
[4]*A.N. Bach Institute of Biochemistry,* Russian Academy of Sciences, *119071 Moscow, Russian Federation*

Abstract

Here we report the development of various diagnostic methods for potato virus and viroid infections including ELISA and immunochromatography on test-strips, combined technologies of DNA probing and dot ELISA as well as RT-PCR-DNA probe-dot ELISA. Immunostrips for detection of potato viruses X and Y (PVX, PVY) were prepared using colloidal gold particle–polyclonal rabbit antibodies conjugates. Clear visual results can be achieved with 5 min incubations detecting as low virus concentrations as 0.2 microgram per ml. Non-toxic and water-soluble agent was found out for viroid, virus and cellular RNA extraction from plant cells and viruses. Comparison of the developed procedure and SDS-phenol method for RNA extraction demonstrated identical yield and purity of RNA preparations. Several recombinant plasmids with insertions of cDNAs of potato viruses A, M, X, S, Y, PLRV and PMTV were constructed. To obtain DNA probes, the recombinant plasmids with respective cDNA insertions were chemically labelled with chlordiethylenetriamineplatinum chloride, (dien)Pt chloride. Primary antibodies against (dien)platinated DNA and secondary immunoglobulins conjugated with peroxidase were employed in DNA probe-dot ELISA detection of the target pathogen. Sensitivity of the recombinant plasmid detection by DNA probe-dot ELISA method with chemiluminescent substrate was 0.1 pg and less than 1 pg for virus RNAs. Individual molecules of PSTVd and potato virus RNAs can be found by combined RT-PCR-DNA probe-dot ELISA technique.

Keywords: ELISA, immunostrips, RNA isolation, DNA probe, RT-PCR-DNA probe-dot ELISA

Introduction

One of major reasons for the chronically low potato crops in the Russian Federation is a widespread use of seed stocks infected with viruses, viroid and bacteria. Quality control of microplants, potato nuclear stock production, and following reproduction of seeds has low has low efficiency due to the use of visual testing as a main screening technique. Formation of a modern market for potato seeds generated a need for radical improvement in seed quality control and its certification. Officially existing system of certification for the primary seed stock assumes mandatory examination of the potato seeds by qualified personnel in special diagnostic centres (laboratories) to discover the economically important harmful infection agents (Anisimov *et al.*, 2005). Thus, the development of the virus-free potato cultivation depends directly on obligatory employment of modern efficient methods for testing for viral and viroid infections. Molecular diagnostics of viral and viroid diseases is a corner stone in the modern seed-growing, pathogen identification, sanitation of seed stocks and quarantine service.

Current methods of molecular diagnostics of pest infections are based on the high specificity of the molecular recognition of antigens by antibodies and target polynucleotide with nucleic acid-based probe. High sensitivity in the pest detection is achieved by enzyme catalysis in ELISA or by the amplification of the specific molecular interaction signal up to million-fold by polymerase chain reaction.

Molecular diagnosis of potato virus and viroid infections

Currently several industry-scale technologies for molecular diagnostics of potato virus and viroid infections are being used. In potato growing countries, including Russia, the most practiced method is ELISA technology (Clark and Adams, 1977).

ELISA

The method demonstrates high specificity and sensitivity in virus detection, with detection limits being as low as 1 ng/ml. More importantly, ELISA is high throughput assay, it allows dozens or even hundreds of samples to be analyzed for in days by qualified technician. Also immunoreagents for ELISA, diagnostic kits and specialized equipment are readily commercially available.

Generally for detection the conjugates of antibodies with alkaline phosphatase or peroxidase are used in ELISA. In Moscow State University modified sandwich ELISA called 'Pyrotest' was developed, it employs secondary antibodies that are conjugated with *E. coli* inorganic pyrophosphatase (Mizenina *et al.*, 1991; Surguchova *et al.*, 1998; Bobkova *et al.*, 2003). 5-10 ng/ml of potato viruses can be readily detected with 'Pyrotest' in the plant extract. Concentration

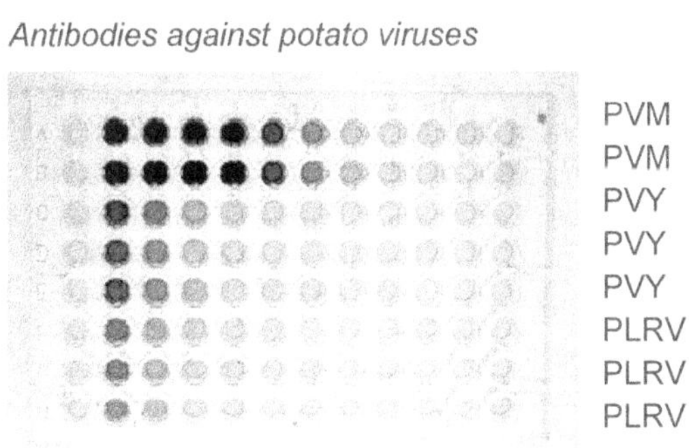

Figure 1. Leaf sap of the potato variety 'Nevsky' was tested. Mixed infection with PVX, PVY and PLRV (potato leaf roll virus) was revealed with 'Pyrotest'.

of most potato viruses in infected potato is higher than that, therefore 'Pyrotest' can be used as a testing method to ensure virus-free potato growing. It should be emphasized that ELISA of infected samples shows 7-20 times higher optical density than uninfected samples. Thus direct unambiguous visual diagnostics is possible. 'Pyrotest' sets for detection of PVX, PVY, PVS, PVM, PVA and PLRV infections were worked out (Chirkov *et al.*, 2000) and are now commercially available via 'Immunotech', a small company incorporated in MSU.

Immunochromatography method

High sensitivity testing technologies for virus infections are employed in special laboratories where control of regulated pest infections is carried out by professional staff which uses cumbersome methods and employing expensive equipment, where microplants, potato nuclear stock, etc., are the main subject for examination. Alternative approach called immunochromatography on test-strips (immunostrips) is developing for analysis of field plants. This approach is simple, has good sensitivity, lower cost and is accessible for common use in plant pathology. There are several variations of this method. Commonly, immunostrip is divided for three zones: start zone, analytical and control one. In analytical zone antibodies against viral antigen are fixed as a band on membrane filter strip. Control zone contains a band of fixed antispecies immunoglobulins. Approximately in the middle of the start zone the conjugates of the primary antibodies with colloidal gold are applied. The end of the start zone is dipped in a testing plant extract resulting in viral antigen migration with the front of ascending liquid along the strip. The complex formed by the interaction of viral antigen with gold-labelled antibodies continues to migrate until it reaches fixed virus-specific antibodies in analytical zone. Concentrated conjugate forms purple stained band (Figure 2). Excess of the conjugate moves through analytical zone and is halted by antispecies specific antibodies producing another stained band. In the case of uninfected plant testing, the only purple coloured band appears in the control zone. Thus, presence of virus in the plant extract and its concentration is reflected by the number and intensity of the coloured bands. In addition to its simplicity the immunochromatography on test-strips is fast. Bands are formed after 5-10 min chromatography.

In collaboration with colleagues from the Bach Institute of Biochemistry we developed an immunostrip analysis for the detection of PVX, PVY and other viral infections at virus concentration of 0.2 microgram per ml (Figure 2).

DNA probe-dot ELISA

30 years ago ELISA was complemented with diagnostic technologies that use molecular hybridization of nucleic acids. Immunoassays have definite limitations: i) pure virus isolates are necessary to raise monospecific antibodies: ii) essential quantities of the pure virus are needed to immunize animals; iii) large scale of antibodies purification has to be carried out for serial testing, and finally potato viroid infection can not be detected by ELISA.

During the 90-ies decade of the last century Russian Federation had s serious loss in potato crop because of potato spindle tuber viroid (PSTVd) disease that has been spread over many

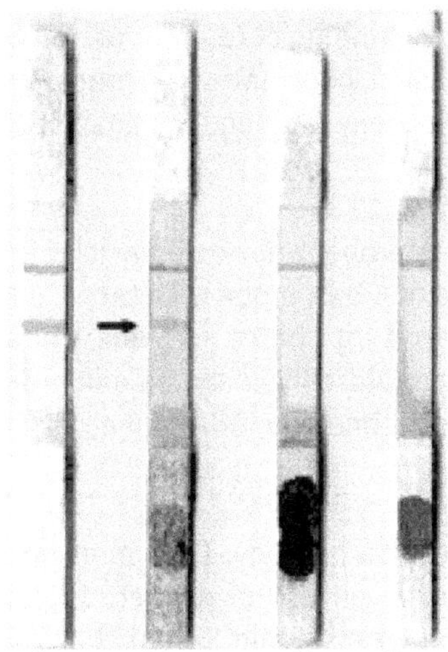

Figure 2. Detection of PVY infection by immunochromatography method on a test-strip. A – purified PVY preparation, 5 microgram/ml. B -crude sap of PVY infected potato variety 'Nevsky'. C - crude sap of healthy potato. D - crude sap of PVX infected potato variety 'Nevsky'. An arrow points stained band in analytical zone.

country regions (Mozhaeva *et al.*, 2002). PSTVd infection spreads systemically. According to EPPO standards PSTVd is a quarantine pest.

PSTVd is a small (about 360 nucleotides) circular single-stranded RNA that has rod-like stable secondary structure (Semancik, 1987). This is non-encapsidated free pathogenic RNA that does not code any protein, therefore its can not be targeted by traditional immunological methods.

Viroid infection can be only diagnosed by technologies based on nucleic acid molecular hybridization. There are multiple technologies of molecular diagnostics for pest infection detection based on the principle of complementary pairing of nucleic acid heterocyclic bases. Neither virus isolation and purification nor immunization of animals with virus specific antigens is necessary for these technologies. Moreover, both sensitivity and specificity of diagnostics obtained by the DNA probe-dot ELISA combined technology are superior to ELISA.

The DNA probe consists of labelled recombinant DNA containing DNA (cDNA, usually 300 or more nucleotides in size) complementary to the target pathogen nucleic acid. Recombinant DNA's are manufactured by genetic engineering methods. Usually they are cloned as bacterial plasmid DNA's and amplified in bacteria. This is cDNA, which is responsible for the specificity of the pest recognizing by the DNA probe.

Recombinant plasmids containing PVX, PVY, PLRV, PVA, PVM, PVS, PMTV (potato mop top virus), PAMV (potato aucuba mosaic virus) and PSTVd cDNAs (all being most economically important pests in Russian Federation) were constructed by virologists of Moscow State University. We used (dien)Pt (chlordiethylentriaminoplatinum chloride) to label recombinant DNA (Kiseleva *et al.*, 1994) because of quantitative binding of (dien)Pt with DNA with negligible effect on the specificity of DNA probe hybridization, good antigenic properties of (dien)Pt-DNA modified by (dien)Pt haptene, and the fact that immunoglobulins against (dien)Pt-DNA are universal and recognize any dien-platinated DNA.

To detect hybridized (dien)Pt-DNA probe, we use sandwich ELISA, where primary antibodies are targeted toward hybrid of (dien)Pt-DNA probe with target phytopathogen nucleic acid, and secondary antibodies are conjugated with horse peroxidase or alkaline phosphatase. Because analyzed sample is fixed on nitrocellulose filter and all diagnostic reactions are carried out in heterogeneous phase we call this technology 'DNA probe-dot ELISA'. The schematics of it are presented in Figure 3.

DNA-probe-dot ELISA is a multistep technology. It requires isolation of total RNA from crude plant cellular extract. This depends on use of dangerous and very toxic substances, such as guanidinium thiocyonate (Chomczynski and Sacchi, 1987), phenol, chloroform as well as detergents. We developed a novel method for virus and cellular RNA isolation using a water-soluble non-toxic chaotropic agent (NTCA, patent pending).

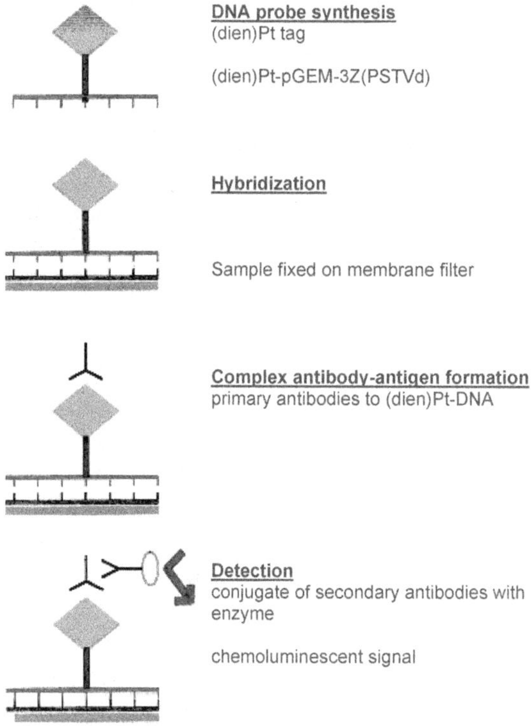

Figure 3. Scheme of DNA probe-dot ELISA technology for PSTVd infection testing with (dien)Pt-pGEM-3Z(PSTVd).

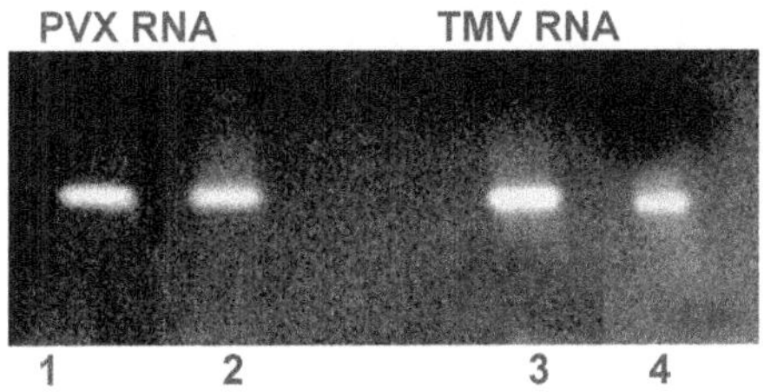

Figure 4. Agarose electrophoresis of virus RNA preparations obtained by deproteinization of PVX and TMV (tobacco mosaic virus) with NTCA (1 and 3) and SDS-phenol (2 and 4).

The resulting yield and purity of RNA were identical to those produced by SDS-phenol method (Figure 4). Our method is particularly appropriate for extensive use in diagnostics due to its simplicity, accessibility and biosafety.

We compared efficiency of PSTVd extraction and detection by the DNA probe-dot ELISA technology in total RNA preparations isolated by SDS-phenol (A), guanidinium thiocyonate (Chomczynski и. Sacchi, 1987) with following sorption of viroid RNA on silica (Malinowski, 1996) and NTCA (C) methods. Equal aliquotes of RNA preparations were spotted onto nitrocellulose membrane and fixed with UV irradiation. As Figure 5 shows, all isolation methods produced comparable quantities and quality of the PSTVd preparations.

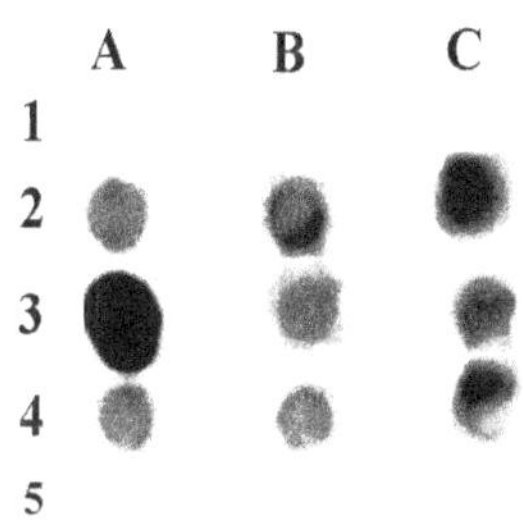

Figure 5. Detection of PSTVd with (dien)Pt-DNA probe in viroid RNA preparations obtained from infected potato (lines 2 - 4) by three different methods (columns A – C) from leaf tissue:
Lines: 1 and 5 - healthy plants;
Columns: A – by SDS phenol;
B – according to (Malinowski, 1996);
C – with NTCA.

Testing of PSTVd, PVX and PVY infections of potato sort varieties received from Lorch Potato Institute collection with respective (dien)Pt-DNA probes is presented in Figures. 6-8.

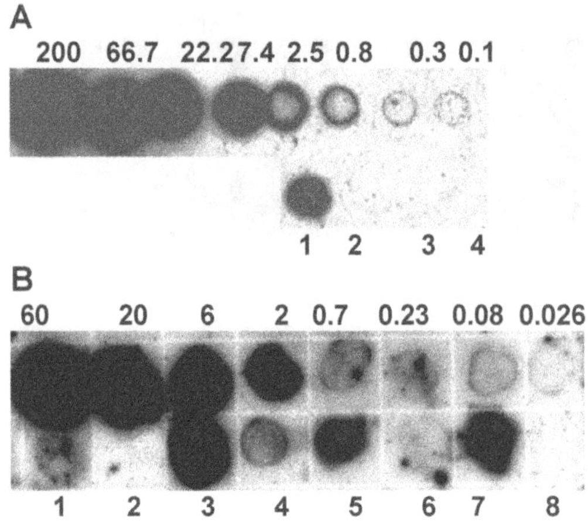

Figure 6. Diagnostics of PSTVd infection with (dien)Pt-pGEM-3Z(PSTVd) probe. Total RNA preparation was spotted on nitrocellulose membrane and fixed by UV-light radiation.
Upper line: A and B – sensitivity of pGEM-3Z(PSTVd) detection, picograms.
Lower line: detection of PSTVd in tube potato microplants and leaf tissue of tomatoes.
A – 1 – (positive control), 1 ng of the total RNA preparation from infected potato was applied onto membrane filter; 2 – 4 (negative control), the same as in 1 but from uninfected potato.
B – total RNA preparations from tube microplants of potato varieties: Jukovsky early (1), Udacha (2), Nevsky (3), from PSTVd infected (4-7) and healthy (8) tomatoes.

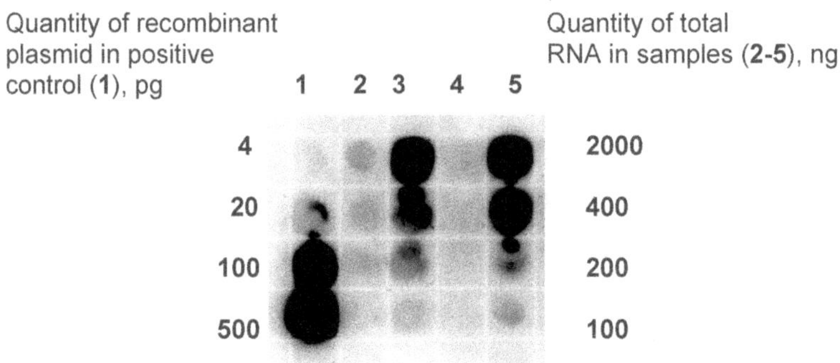

Figure 7. Testing of PVX infection by DNA probe-dot ELISA technology.
1 – quantity (pg) of recombinant plasmid pUC-19 with PVX cDNA insertion applied.
2-5 –total RNA preparations (2, 4) isolated from healthy (according to ELISA) and PVX infected (3, 5) potato (var. Nevsky) by phenol deproteinization (4, 5) and with NTCA (2, 3).

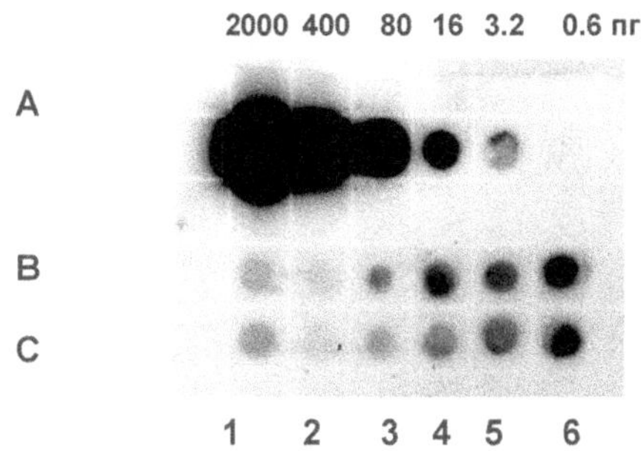

Figure 8. Determination of PVY by DNA probe-dot ELISA technology in total RNA preparations from greenhouse potato var. Jukovsy early (1,2) and Nicotiana tabacum var. Samsun infected with PVY (3-6).
A – positive control, recombinant plasmid with PVY cDNA insertion spotted and fixed on membrane filter, pg.
1 B, C –infected potato.
2 B, C – healthy potato.
Total RNA was isolated with NTCA from 100 (3), 200 (1, 2, 4), 400 (5) and 800 (6) mg of fresh leaf tissue.

RT-PCR technology

The alternative to DNA probe technology is the polymerase chain reaction that is able to amplify target nucleic acid. This powerful technology is widely used in medicine diagnostics. It is possible to reach the limit of sensitivity in the pathogen detection with PCR. It seems natural to use this technique while compiling the growing Central Collection of potato varieties of Russian Federation (Simakov *et al.*, 2004). Limitations for the use of PCR for disease testing are related to the frequent occurrence of multiple products of amplification resulting in the necessity to prove their specificity or to tune conditions of chain reaction for each virus genome, and higher cost of analysis.

In a model PCR experiment we examined sensitivity of PVX detection by RT-PCR using purified virus RNA as a matrix. We found that one femtogram of virus RNA in the sample can be detected with a pair of primers PVX5755+/PVX6105- by electrophoresis of RT-PCR products in agarose gel using ethidium bromide staining (Figure 9). In our experiment it corresponded to 300 molecules of PVX RNA for RT reaction and to 30 molecules for RT-PCR.

RT-PCR-DNA probe-dot ELISA technology

Sensitivity of RT-PCR technology and, what is more importantly, its specificity can be increased if the products of amplification are analyzed with DNA-probe instead of electrophoresis with ethidium bromide. RT-PCR product was spotted on nitrocellulose membrane, fixed by UV-light and hybridized with DNA probe specific to PVX genome. Sensitivity of PVX RNA detection increased more than one order of magnitude (see Figure 10) up to less than 0.1 fg or 0.05 attomoles of PVX RNA at concentration 0.25×10^{-18} M.

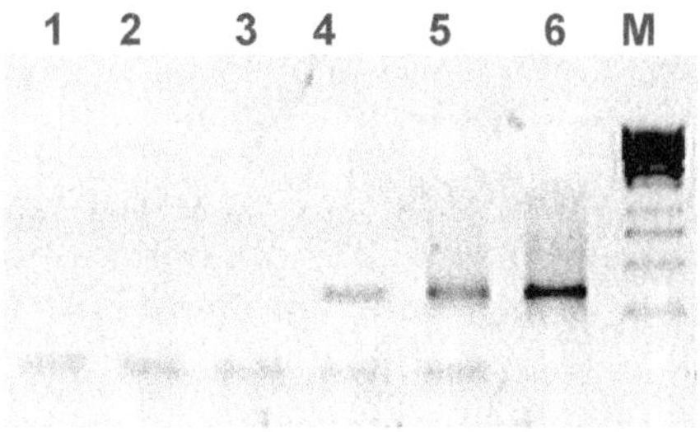

Figure 9. Analysis products of PVX RNA amplification by RT-PCR after electrophoresis in 2% agarose gel with ethidium bromide.
1 negative control of RT reaction (no RNA added);
2 0.03 fg;
3 0.1 fg;
4 1 fg;
5 10 fg;
6 100 fg of PVX RNA were added to RT reaction cocktail;
M molecular mass markers (GeneRuler 1kb Ladder).

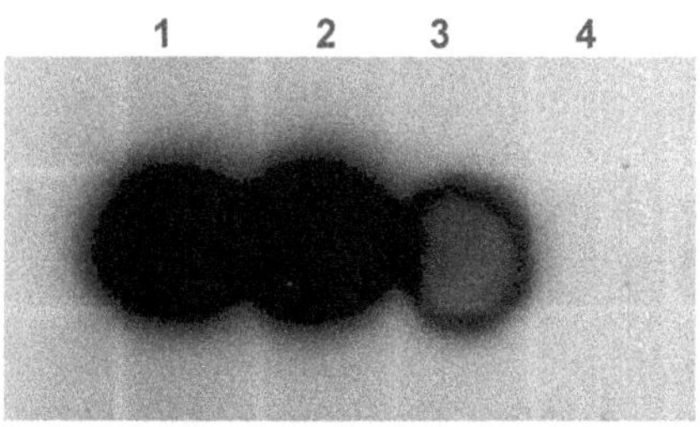

Figure 10. Combined RT-PCR-DNA probe-dot ELISA technology increases sensitivity of PVX RNA detection.
1 – 10 fg of PVX RNA, 2 – 1 fg; 3 – 0.1 fg of PVX RNA added to reaction cocktail of RT reaction, 4 – negative control of RT-reaction, no RNA added.

This value corresponds to 30 molecules of RNA in RT reaction and to 3 molecules of cDNA in PCR reaction mixture. Thus, combined RT-PCR-DNA probe-dot ELISA technology makes it possible to detect individual molecules of a viral RNA.

In our experience the sensitivity of virus RNA detection in total RNA preparations isolated from infected cells drops down 10 times almost, nevertheless we use methods good enough to detect one molecule of the pathogen RNA in the total RNA preparation isolated from several milligrams of potato leaf tissue.

We compared sensitivities of DNA probe-dot ELISA, RT-PCR and RT-PCR-DNA probe-dot ELISA technologies for potato sort varieties infected with PSTVd (Figures 11 and 12).

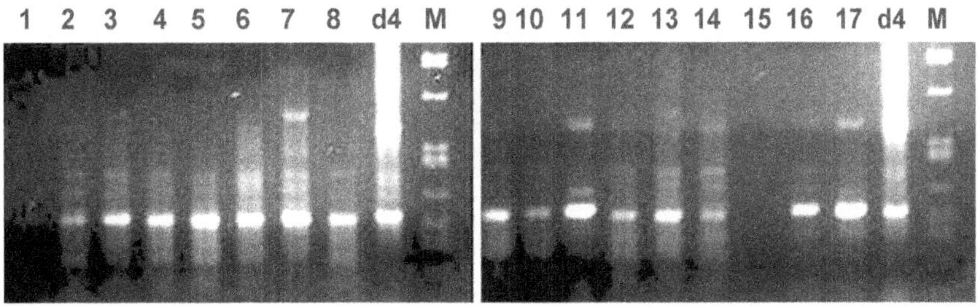

Figure 11. RT-PCR technology for testing of the total RNA preparations isolated from potato tube microplants and leaf tissue of tomatoes presumably infected with PSTVd.
To amplify PSTVd DNA we used primer 'GAAAAAAGAAGGCGGCTCGGAGGA – rev' for reverse transcription and pair of primers 'GAAAAAAGAAGGCGGCTCGGAGGA – rev' and ' GCCGACAGGAGTAATTC – forward' for PCR. RNA was extracted from 30 mg of plant leaf tissue. Amplified DNA was analyzed by 1% agarose electrophoresis in presence of ethidium bromide. d4 – positive control, recombinant plasmid pGEM-3Z(PSTVd)$_n$; then RNA preparations from potato varieties: Udacha (1), Golubizna (2), Il'innsky (3), Pransa (4), Logovskoy (5), Bryansky (6), Skoroplodny (7), Jukovsky early (8), Bryanskaya novizna (9), Osen (10), Rojdestvensky (11)/, Peterburgsky (12), Elizaveta (13) and RNA preparations from leaf tissue of tomatoes (14-17).

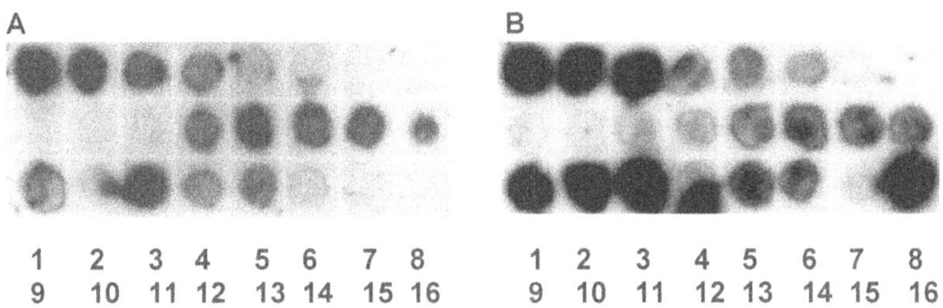

Figure 12. Comparision of sensitivities for PSTVd infection tested by RT-PCR (Figure 9), DNA-probe-dot ELISA (Figure 10A) and RT-PCR-DNA probe-dot ELISA (Figure 10B) in preparations of total RNA npenapamoв isolated from collection potato and laboratory tomatoes. In case of the DNA-probe-dot ELISA diagnostics, one microliter of each RNA preparation (from 60 microliters of each RNA preparation) was deposited on nitrocellulose filter and fixed; for RT-PCR-DNA probe-dot ELISA technology a probe corresponding to 1/500 of microliter was amplified.
Upper line in Figures 12A and 12B present a positive control and consist of serial dilutions of pGEM-3Z(PSTVd) in a range: 200; 66,7; 22,2; 7,4; 2,5; 0,8; 0,3; 0,1 pg.
Second line Figure 12A and 12B) presents results of testing for samples 1-8 and third line – for samples – 9-15 that correspond to numbers of samples in Figure 9. Sample 16 in Figure 10A is negative control (RNA preparation from var. Udacha) and in 10B - positive control, RNA preparation from severely infected tomato.

Comparative study of testing technologies demonstrated that powerful RT-PCR-DNA probe-dot ELISA technology revealed PSTVd infection in samples 15, 10, and 14 that were previously undetected.

Our experience indicates that potato testing should be separated in two steps. First screening of samples should be carried out by the low-cost throughput DNA-probe-dot ELISA technology (to exclude heavily infected samples) later followed by more cumbersome and expensive technologies.

Conclusions

Development of technologies for industrial-scale molecular diagnostics of virus and viroid infections approached its limits in sensitivity, being able to detect individual nucleic acid molecules in small plant organism.

Future trends in development of these technologies will be determined by technical simplification of each step, beginning with preparation of the object under examination and ending with registration of data obtained.

Acknowledgements

Authors are grateful to our colleagues from Loch Potato Institute and Institute of Phytopathology of Agricultural Academy of Russian Federation for their help with the study.

The reseach presented in this paper was supported by Russian Foundation for Basic Research for grant N 06-04-08290.

References

Anisimov B.V., A.I. Uskov, E.A. Simakov *et al.*(2005). The procedure of field inspections and post-harvested potato seeds quality control (*in Russian)*. IKAR-Press, Moscow.

Bobkova A.F., A.N. Blintsov and S.N. Chirkov (2003). Pyrotest as a new das-ELISA test for potato viruses. Agrarnaya Rossia (*in Russian)*. 3: 23 – 27

Chirkov S.N., V.K Novikov, A.F. Bobkova and J. G. Atabekov (2000). Pyrotest is improved ELISA-test of potato viruses. Plant Protection and Quarantine (*in Russian)*. 12: 15 -16.

Chirkov S.N., A.P. Osipov and J.G. Atabekov (2002). The 'Pyrotest' kits for laboratory diagnostics of potato viruses. Potato and Vegetables (*in Russian)*. 1:29.

Chomczynski, P. and N. Sacchi (1987). Single-step method of RNA isolation by acid guanidinium thiocyonate-phenol-chloroform extraction. Analytical Biochemistry 162: 156–159.

Clark, M.F. and A.N Adams (1977). Characteristics of the microplate method of enzyme-linked immunosorbent assay for the detection of plant viruses. Journal General Virology 34: 475–483.

Drygin Yu.F., S.M. Musin, O.A. Kondakova *et al.* (1996). Infectiousness of standardized potato sort samples from Russia Federation collections with potato spindle tuber viroid. Proceedings Russian Agriculture Academy (*in Russian*) 6: 24-25.

Kiseleva V.I., M.F. Turchinsky, T.B. Kolesnik and A.M. Poverenny (1994). The application of chlorodiethylentriaminplatinum in hybridization analysis of nucleic acid specific sequences. Bioorganic Chemistry (*in Russian*) 20:14-20.

Kondakova O.A. and Yu.F. Drygin (1999) The diagnostics of potato viroid disease with (dien)Pt-DNA probe. Biotechnology (*in Russian*) 4: 83-90.

Malinowski, T. (1996). In: H.-W Dehne, G. Adam. *et al.* (Eds.) Developments in Plant Pathology. vol. 11, Kluwer Acad. Publishers, 445-448.

Mizenina, O.A., O.V. Borisova, V.K. Novikov, O.A. Evtushenko, A.A. Baykov and J.G Atabekov (1991). Inorganic pyrophosphatase from E. coli as a label for the detection of plant viruses by ELISA. Journal Phytopathology 133: 278-288.

Mozhaeva, K.A., N.V. Girsova, N.V. and T.B. Kastalyeva (2002). Main causes of Potato Spindle Tuber Viroid distribution in seed potato in Russia in 1990s. Journal Russian Phytopathology Society 3: 39-42.

Semancik, J.S. Viroids and Viroidlike Pathogens (1987). (Semancik, J. S., Ed.). CRC Press, Boca Raton, FL 127-160.

Simakov E.A. (2004) On the improvement of scientific basis of potato seeds production in Russia. In Proceedings of Russian Potato Research Institute, Moscow.

Surguchova N., S. Chirkov and J. Atabekov (1998) ELISA-test bei Nepoviren mit einem neuen Enzym - einer anorganischen Pyrophosphatase aus *Escherichia coli*. Arch. Phytopath. Pflanz. 31: 535-541.

Curriculum vitae – Yuri F. Drygin

Dr. Yuri F. Drygin has graduated from Lomonosov Moscow State University in 1965 with MS in Chemistry. As a part of his research he studies the chemistry and biochemistry of nucleoproteins. He later moved onto working in the Interfaculty Laboratory of Molecular Biology and Bioorganic Chemistry at Moscow State University (later renamed as Belozersky Institute of Physico-Chemical Biology) as a senior technician. Dr. Drygin has received his PhD degree from the Lomonosov Moscow State University in 1971 and Dr. Sci. (Chemistry) degree in 1989 as a result of his studies of Tight Complexes of DNA and Residual Polypeptides, as well as Natural Covalent Complexes of Nucleic Acids and Proteins. In 1975/76 Dr. Drygin worked as a researcher in IREX Program in laboratory of James C. Wang in the Berkeley University of California (UCB). On his return to Moscow State University, the main interest of his work was concentrated on the structural-functional relationship of animal and plant viruses. Currently Dr. Drygin holds the position of Head of Laboratory of Nucleic Acid-Protein Interactions in this Institute, he is the author of over 60 peer-reviewed manuscripts and multiple columns in conference papers. Since 1989 he collaborates with Department of Virology in Lomonosov MSU (working directly with the head of Department of Plant Virus Biochemistry, Professor Joseph Atabekov) to develop molecular diagnostic technologies for the plant viral and viroid infections using ELISA or molecular arrays. Dr. Drygin played an important role in obtaining the certification for molecular diagnostics of viroid and viral infections of collection potato samples by Ministry of Agriculture for diagnostic team of Lomonosov MSU together with diagnostics company Immunotech (incorporated in MSU). The diagnostics effort led by Dr. Drygin was awarded by the medal of All-Russian Exhibition of National Economy.

New way of isolation of potato cyst nematodes from soil tests

R.D. Korzhuk and P.A. Melnyck
Ukrainian scientific-research station of plants' quarantine,UAAS, v. Boyani, Novoselitza district, Chernivtsy region, Ukraine, ndskr@myrealbox. com

Abstract

The results of efficiency of application of a new method of nematode cyst isolation from soil tests are stated. Application of a way of nematode cyst isolation from soil samples with the help of the device 'Ruta', the principle of which is based on washing of samples with submission of water from below through a three-way tap, is based on cysts' ability to emerge on a surface of water, as it enables maximum precision (with lower losses) and without significant expenses of time to conduct the cysts' isolation.

Keywords: nematodes, soil test, method, degree of infection, device

Introduction

Potato nematodes are quarantine organisms which are represented by two kinds: potato golden cyst nematode (PGCN) – *Globodera rostochiensis* (Woll), and white - *Globodera pallida* (St). (PPCN), that belong to the genus *Globodera* (Gus'kova A. A. *et al.*, 1982). Areas of distribution of both species of pests are growing because of expansion of international economic connections and non-following of quarantine instructions.

Golden and white potato nematodes are parasitizing on solanaceous plants such as potato and tomatoes. They parasite mainly on roots of plants and pass their life cycle in soil and roots of the host plant (Gus'kova *et al.*, 1982; Dekker *et al.*, 1972).

Inside cysts during several years the eggs with larvae which develop in them are kept, and their quantity changes from several individuals (depending on soil type, degree of its infection, their age) up to several hundreds (Dekker *et al.*, 1972; Kiryanova *et al.*, 1963). Larvae in cysts can keep their viability for more than 10 years (Yefremenko *et al.*, 1988).

The potential danger of potato nematode is that it's almost impossible to detect it at the initial stage of site infection. By existing means its detection is possible only after the accumulation of cysts in soil. It is considered, that potato nematodes damageability can be shown already with soil infection at a level of 500-1000 larvae in 100 cm^3 of soil (Dekker *et al.*, 1972; Yefremenko *et al.*, 1988).

The distribution of cyst nematodes can result in significant economic losses and put under threat the development of potato crops in many regions of the Ukraine.

Among the quarantine measures primarily there is the nematode's locus. Special attention should be paid to monitoring of sites near farmsteads. In areas, where the locus is detected, the annual selection of soil tests is carried out during 3 years; at their absence, once in 5 years. Annually it is necessary to check the fields where potato is grown, the seeds which are delivered from distribution areas; and also research and selection centres, nurseries, seed-production areas (Yefremenko *et al.*, 1988; Podgayetskiy *et al.*, 1995).

Depending on presence and the degree of nematode infestation of soil areas free from infection; are distinguished with the partial (limited) distribution of the pest - less than 50% of the population areas; wide distributed - within the infected area more than 50% of the population areas; the areas with aggressive pathotypes or other species of a nematode (if such ones are detected).

The degree of soil contamination is established on a scale: high (more than 25 cysts in 100 cm^3 of soil), average (up to 25 cysts in 100 cm^3 of the soil), low (less 5 cysts in 100 cm^3 of soil).

For quarantine service the important task is the duly detection and determination of locus borders of potato nematode with the purpose of their further localization and systematic liquidation. The performance is possible under condition of improvement of known or development of new more effective selection methods, allocation and identification of plant nematodes.

Purpose of research: selection and improvement of effective rapid methods of nematode cyst isolation in soil tests.

Methods

Density of nematode population is determined during potato planting (initial number) and after harvest or in another period of year, except winter, when the soil is frozen by selection of soil tests. Selection of tests for the analysis on globoderal presence is carried out by certain methods (Dekker *et al.*, 1972; Yefremenko *et al.*, 1988; Melnyck *et al.*, 1998, Shesteperov *et al.*, 1995). All of them are based on selection of primary 50 collections, each in volume 5 cm^3 of soil on the regularly designed grid (conditional equal square, rectangular, triangular, depending on the form of a site) with the further association in (middle) average test. On the area of 1 hectare 4 average soil tests are selected, 250 cm^3 each.

With the purpose of improvement of methods of potato nematodes detection in crop protection we tested the different route schemes (diagonal, triangle, envelope) on selection of soil tests with different quantity of collections for every 5 m of a route. The most exact was the selection under the circuit of an envelope, 20 collections on the area 25 m^2. The soil tests are selected by the drill on 5 cm^3 for creation of the united test 400 cm^3 from the area of 0.01 hectare. Accordingly on every 0.01 ha 4 registration grounds are made, 20 collections each. So a single soil sample is made.

The middle samples are poured in one little bag, to which a label is detached, on which the place, time and area of a site are pointed and sent direct for the laboratory analysis (Yefremenko *et al.*, 1988).

The following stage in work is realization of the laboratory analysis of the selected soil samples for establishing degree of infection of nematodes' cysts (Yefremenko *et al.*, 1988; Korzhuk *et al.*, 2005; Melnyck *et al.*, 1998).In practice the isolation of invasive cyst potato nematodes from soil was conducted by express-method, called 'Ruta' device, developed by us, the principle of work of differs from the previous methods (Korzhuk *et al.*, 2005; Korzhuk *et al.*, 2005). For more exact establishment of invasive loadings it's necessary to increase flotation. In case of realization of the analysis of damp soil samples it can be achieved, having replaced water by the sated solution of kitchen salt (Dekker *et al.*, 1972).

Method of washing with the help of the device 'Ruta'

The given method of washing with application of device 'Ruta' patented by USRSPQ (Figure 1) (Korzhuk *et al.*, 2005), as well as a number of other methods, is based on ability *cyst nematodes'* to emerge on a water surface (Kiryanova *et al.*, 1963).

The cyst nematodes' isolation by a rapid method is conducted from soil samples previously dried at a temperature not exceeding 40 °C. For the analysis from a sample-select 100 cm^3 of soil, which is sifted through a sieve with a diameter of grid apertures 2-3 mm. From the system of water supply (at its absence - from the tank) submit water from below through water supply stack and turn of the three-way tap (in condition of water supply) to working capacity on 1/4 its volume. Soil button is put in working capacity, already on a quarter filled by water, and

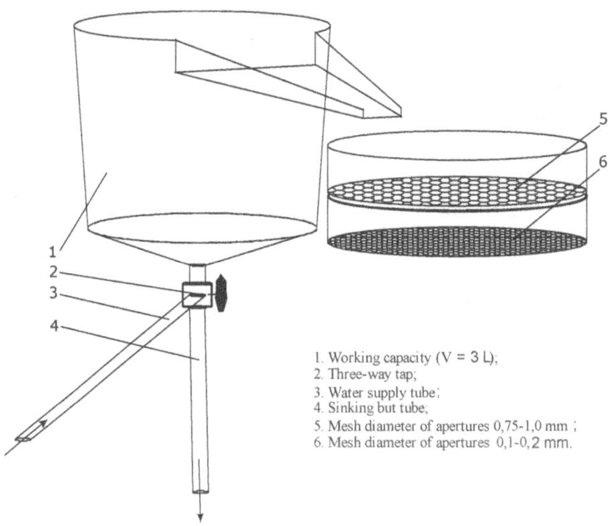

Figure 1. Schematic image of the device 'Ruta'.

is slightly stirred by a glass rod. Continue submission of water further. Water, which moves through the three-way tap of the device 'Ruta' from below under pressure, creates effect 'slight swirl ', therefore the soil is mixed up. Gradually volume of water is increased, filling all working capacity (V=3l). On water surface the fine fossils and cysts emerge which are pushed out by water on to sinker throat of the device. A sinker throat is a slot, which is cut in the wall edge of working capacity of the device, on 2-3 cm beneath its surface. Further, on a throat, water with cysts and other rests gets to the top sieve with a diameter of apertures of 0.75-1.0 mm, and from this - on the bottom sieve with a diameter of apertures of 0.1-0.2 mm. Sieves are placed one above another and keep on timbering, which is placed below working capacity on distance of sinker throat. Having convinced, that on a surface of water in working capacity already there are no plant rests and cysts, lead on water surface with a glass rod near walls of capacity, so that particles and cysts stuck to walls have washed off on sinker throat. The plant rests and cysts, passing through a sieve, linger over their surface: large ones on the top sieve, and fine rests and cysts on bottom.

Water through a sieve oozes to a wash-stand (or bucket). Submission of water in capacity stops, turning the three-way top.

At turn of the three-way top in other condition the fulfilled mixture on a stack is deduced in to a special sewer drain (or bucket). Capacity is rinsed by small amount of water (from a glass) or several turns of the tap in a way of submission of water from system of water supply. Further, the meshes are removed. The top mesh is release from the plant rests and rinsed, and bottom one the rinse is examined through lens. If it is necessary to select the cysts for the further analyses, then a filtrate from a sieve is washed off in capacity (on 1 l). Walls of a mesh are well rinsed by small amount of water, so that all the rest of cysts have got in same capacity. Assert 2-3 min and merge 1/3 volumes of water in capacity on the filter in a funnel. It is possible at once to examine filters or to number and to postpone to the moment of possible analysis on cardboard or glass plants. The output of cysts of potato golden nematodes with 100 cm^3 of soil sample from total available in test in this case made 94.9% Efficiency of application of a new way is supported also with that the expenses of water for realization of 1-st analysis are1.5 times less than in the prototypes, and economy of work time is almost 2 times (Table 1).

The offered method of washing on the device 'Ruta' with the subsequent nematode cyst isolation can be simplified both in time and in methods of conducting, if during operations to put between the mesh sieve with a diameter of apertures of 0.75-1.0 mm and bottom one with a diameter of apertures of 0.1-0.2 mm a calico fabric. In this case, washed off large –sized plant rests will linger over the top mesh, and the small rests and cysts at filtering water get at once on a fabric. After washing soil samples, mashes are separated, and the filtrate on a fabric is examined with the help of lens or under binocular. The fabric filters are very convenient, as the filtering through them goes faster. Besides they can be maintained, leaving up to the assay (before the assay watersoaked), and also used several times.

With the help of the device 'Ruta', the assay of tubers is possible for cyst nematodes' isolation. Usage of the device is convenient for examination of soil and potato tubers in natural conditions.

Table 1. Isolation of potato cyst golden nematode by different methods from soil sample (100 cm3).

No. n/n	Treatment		Flotational method		Isolation by the device Fenuic		Isolation by the device 'Ruta'	
			Quantity of cysts	of % of isolation	Quantity of cysts	of % of isolation	Quantity of cysts	of % of a isolation
1	soil test selected from the natural locus of golden potato nematode in vil. Yasinya, Zakarpattya region		269		271		285	
2	soil test selected from the natural locus of golden potato nematode in t. Rahivof the Zakarpattya region		176		168		182	
3	artificial	10 cysts/100 cm^3	8.3	83.0	8.5	85.0	9.3	93.0
	background	25 cysts/100 cm^3	22.6	90.4	21.3	85.2	24.0	96.0
		50 cysts/100 cm^3	46.2	92.4	45.6	91.2	47.8	95.6
4	average efficiency of method			88.6		87.1		94.9

There are also other methods, which give more or less exact quantitative estimation at isolation of nematodes: sifting on Boynake (through three meshes set), method of paper streaks (Bur). Vilne method centrifugation on Remuth method (in centrifuges at 4500 turns round 1 min), with the help of the Fenuic device (Dekker *et al.*, 1972; Korzhuk *et al.*, 2005; Marks *et al.*, 1998). Cysts isolation by water is spent also with the help of a number of modern cyst extractors by flotation principle (Kiryanova *et al.*, 1963).

After cyst isolation a degree of soil infection is defined by recalculation of amount of viable larvae and eggs in cysts selected from 100 cm^3 of a sample (Yefremenko *et al.*, 1988; Korzhuk *et al.*, 2005). The importance of establishment of invasive loading of soil by globoderas takes place at planning measures of control and definition of their efficiency. The establishment of a degree of soil infection has the important practical significance at creation of skeleton maps of nematodes' dispersal in a country (Melnyck *et al.*, 1998; Shesteperov *et al.*, 1995).

After the definition of a quantitative structure of cysts in soil it is necessary to take into account their qualitative condition, i.e. viability. The viability of eggs and larvae in cysts can be established by different methods: by definition of larvae's activity; on a condition of their internal structure by microscope; study of ability of larvae to hatch under action of root dopants of a potato or stimulation by chemical stimuli; by a method of staining (Dekker *et al.*, 1972; Kiryanova *et al.*, 1963; Marks *et al.*, 1998).

Conclusions

1. By the employing USRPQS in 2004, a new way of nematode cyst isolation from soil tests with the help of the device 'Ruta' (declaration patent No. 7248 from 15-06-05) is developed and patented. The task is achieved by offering a method of nematode cyst isolation the device 'Ruta' as the means of isolation .
2. Application of a way of nematode cyst isolation from soil samples with the help of the device 'Ruta', the principle of which is based on washing of samples with submission of water from below through three-way tap, is based on cysts' ability to emerge on a surface of water, as enables maximum precisely (with smaller losses) and without significant expenses of time to conduct the cysts' isolation.
3. At cyst isolation from soil tests with use of the device 'Ruta' the efficiency of a way made 94.9% of an cysts exit from presence of total in sample, at expenses of water 1.5 times less than at the prototypes, and economy of working time almost in 2 times.
4. The invention concerns phytopathology and can be used as a rapid method in quarantine of plants for isolation of cyst potato nematodes from soil samples at inspection.

Acknowledgements

The Proceedings were agreed by the scientific Board of UkrSRSPQ on December 7, 2006, Protocol No.5.

References

Gus'kova A.A., Makovskaya S.A. Methods of differentiation of potato cystogenous nematodes' species and pathotypes. – L., VIZR, 1982. – 15p. (in Russian Гуськова А.А., Маковская С.А. Методика дифференциации видов и патотипов цистообразующих нематод картофеля. – Л., ВИЗР, 1982. – 15с.).

Dekker H. Plant nematodes and their management and control.– M., Kolos, 1972. – 443p. (in Russian Деккер Х. Нематоды растений и борьба с ними. – М., Колос, 1972. – 443 с.).

Yefremenko T.S., Borovikova A.N., Dudick O.R., Gus'kova A.A., Makovskaya S.A. Instructions on detection of golden and pale potato nematode and management options. – M., VO 'Agropromizdat', 1988 – 45p. (in Russian Ефременко Т.С., Боровикова А.Н., Дудик О.Р., Гуськова А.А., Маковская С.А. Инструкция по выявлению золотистой и бледной картофельной нематод и мерам борьбы с ними. – М., ВО 'Агропромиздат', 1988. – 45 с.).

Korzhuk R.D., Mats'kiv T.I., Melnyck P.O., Solomiychuk M. P. Method of extraction of phytohelminths' cysts out of soil tests. Declarative patent No. 7248 from 15-06-2005. (in Russian Коржук Р.Д., Мацьків Т.І., Мельник П.О., Соломійчук М. П. Спосіб виділення цист фітогельмінтів із грунтових проб. Деклараційний патент No. 7248 від 15-06-05).

Korzhuk R.D., Melnyck P.O., Pruntsev S.Ye., Ostryck I.M., Babych A.G. Methodological recommendations on detection of potato cystogenous nematodes. – Chernivtsy, 2005. – 43p. (in Russian Коржук Р.Д., Мельник П.О., Прунцев С.Є., Острик І.М., Бабич А.Г. Методичні рекомендації з виявлення картопляних цистоутворюючих нематод. – Чернівці, 2005 р. – 43 с.).

Kiryanova E.S. Research methods of nematodes, agricultural crops, soil and insects. M., L.: A.N. USSR, 1963. – 33p. (in Russian Кирьянова Е.С. Методы исследования нематод, сельскохозяйственных растений, почвы и насекомых. – М., Л.: А.Н. СССР, 1963. – 33 с.).

Melnyck P.O., Malakhanova Ye.L., Ivanchyck S.G., Homyak V.V. Methods of Ukraine territory passportization on presence of potato wart and potato cystogenous nematode loci. Information bulletin VPRS POBB, No.32. – Stavropol, 1998. – p. 101-102. (in Russian Мельник П. О., Малаханова Є. Л., Іванчик С. Г., Хом'як В. В. Методика паспортизації території України на наявність вогнищ раку картоплі та картопляної цистоутворюючої нематоди. Інформаційний бюл. ВПРС ПОББ, No.32. – Ставрополь, 1998. – С 101-102.).

Podgayetskiy A.A., Miroshnyck G.G. Potato cyst nematodes, management and control (sanitary analysis and recommendations). – K., 1995. – p. 10 – 19. (in Russian Подгаєцький А. А., Мірошник Г. Г. Цистоутворюючі нематоди картоплі та боротьба з ними (стан аналіз та рекомендації). – К., 1995. – С. 10 –19.).

Shesteperov A. A., Savotykov Yu.F., Quarantine phytohelminthoses. Book 1, M., Kolos, 1995. – 418p. (in Russian Шестеперов А. А., Савотиков Ю. Ф. Карантинные фитогельментозы. Книга 1, М., Колос, 1995. – 418 с.).

Marks R.I., Brodie B.B. Potato Cyst Nematodes, Biology, Distribution and Control. CAB International, 1998. – 410 p.

Dissemination of potato virus Y (PVY) isolates under field conditions

R. Schwärzel[1], P. Gugerli[1], W. Reust[1], T. Hebeisen[2]
[1]Swiss Federal Research Station for Plant Production, Agroscope Changins-Wädenswil ACW, case postale 1012, CH-1260 Nyon, Switzerland
[2]Swiss Federal Research Station for Plant Production, Agroscope Reckenholz-Tänikon ART, CH-8046 Zurich, Switzerland

Abstract

In the last few years, many new isolates of PVY (potyvirus Y) appeared in the potato fields in Switzerland. New recombinations have been revealed by PCR methods (Rigotti and Gugerli, 2004). The dissemination rate and the sensitivity of the cultivars changed fundamentally. Switzerland has intensive trade exchange of potato seed with different European areas and countries. These exchanges accelerate the dissemination of the different predominant PVY types. Early nineties, first PVY^{NT} potato tuber ringspot disease symptoms on new breeding variety Lido have been observed in Switzerland. In the beginning of this millenary, PVY^{Wilga} has been detected. The dissemination of six different PVY isolates on six different cultivars has been studied under field conditions. The rectangular small potato plots, 800 seed tubers each, were separated by twelve meters bands of oat. This cereal was sown at same time than potatoes but since oat grew before potatoes, it could help reducing attractiveness for wing aphides and to move between plots at the beginning of the growing season. In each variety plot, 4% of a single PVY isolate infected potato plants were inserted. For the control of the dissemination of PVY by wing aphids from one plot to another, PVY^{O} and PVY^{N} antibody ELISA tests have been conducted separately. Under these field conditions, PVY ring necrosis symptoms on tubers have been detected in promising new varieties. Dissemination of different PVY isolates from one plot to another, isolated with oat bands, was very low in 2005 and 2006. The cultivar Lady Christl was resistant to all PVY isolate infections; Bintje and Charlotte were sensitive to PVY and very sensitive to PVY^{NT} and PVY^{Wilga} isolates. Charlotte was very sensitive to PVY^{N} isolates in 2006. Spread of PVY^{O} and PVY^{N} was lower then PVY^{NT} and PVY^{Wilga} isolates. Ring necroses on tubers were mostly observed on Nicola and rarely on Bintje, Desiree or Charlotte.
Some new PVY isolates were responsible for PVY ring necroses on tubers for sensitive varieties like Erntestolz, Hermes, Nicola, Ditta and Gabriela. The sensitivity of cultivars to PVY is variable and differs according to the necroses tuber-sensitivity. The different PVY isolates affect the yield of Nicola more or less intensively.

Keywords: potyvirus Y (PVY), PVY isolate, ELISA test, variety assessment

Introduction

Switzerland is a small mountainous country with 7 million inhabitants and a large mountain area. Arable land represents only one quarter of the total surface (40,000 km^2). Around 30 European potato varieties are cultivated on 13,000 ha with an average yield of 38 to 42t/ha.

The seed production covers about 1700 ha. Located in the centre of Europe, Switzerland is not a member of the European Union.

In Switzerland the main potato production area is located on the Plateau, a long piece of land, 400 to 800 m above sea level between the Jura (a low-altitude mountain range) in the North and the Alps in the South. Aphid population, as virus vector, appears early in spring at the same time as potato plant emergence. In Switzerland, the aphid development is very favourable, much more then in the northern part of Europe with constant cool, windy areas. For economical and rational evolution reasons, the seed production in Switzerland has been transferred from isolated high altitude above sea level areas to high-density potato growing areas on the Swiss plateau. The renewal rate of seed potatoes is high with about 70%. Usually, 10 to 15% of the seeds, mostly basic seed category, are imported from Holland, Germany, or Bretagne (France), and exceptionally from North of France, Belgium or Austria. Because of the regular potato import of variable origin and the fast dissemination rate of the virus, Switzerland is characterized as a collector area of many types of potyvirus. Soft viruses such as PVS, PVX, PVA and PVM, but also severs viruses such as PLRV and many types of potyvirus Y isolates.

Since 1995, 95% of the severe virus infections have been caused by PVY. In the last few years, many new isolates of PVY (potyvirus Y) were characterized on potato fields in Switzerland. New recombinations have been revealed by PCR methods (Rigotti and Gugerli, 2004). The dissemination rapidity of the old PVY isolates and the sensitivity of the cultivars changed fundamentally. One of the reasons is that Switzerland has regular intensive trade exchange of potato seed with different European areas and countries. These exchanges accelerate the dissemination of the diverse predominant PVY types originally found in different parts of Europe. Early nineties, the first PVY^{NT} potato tuber ringspot disease symptoms on the new breeding variety Lido have been observed. The real dissemination of ringspot types started in 2000 with the cultivars Nicola, Erntestolz, Ditta and Hermes. At the beginning of this millenary, PVY^{Wilga} has been detected. Since then, we have been testing the sensitivity of new cultivars. The effect of different isolates of PVY on the yield of the cultivar Nicola was determined.

Method

The dissemination of six different PVY isolates on six potato cultivars (Bintje, Charlotte, Desiree, Lady Christl, Marlen, Nicola) have been carried out under field conditions. Sixteen plots of 200 m^2 containing each a total of 800 healthy seed tubers of 6 cultivars were separated by twelve meters bands of oat to reduce the attractiveness for wing aphides to move between plots at the beginning of the growing season. In every plot and for every cultivar two rows of 50 tubers each were grown. In each plot 4% of PVY infected plants of the cultivar Nicola were planted on the 15th and on the 35th plant-position of every 50 plant rows of a single PVY isolate and harvested one week before the end of the trial. PVY isolates were produced under greenhouse conditions (each PVY isolate in a separate compartment) one year before the field trial. The infection with the different PVY isolates was made manually by abrasion of 5 leafs when the potato plants reached 30-50 cm high. In the control plot Nicola plants were treated with water. From each cultivar and plot, 200 tubers were harvested. The dormancy of these tubers was then broken by Rindite. Five weeks later, control of the dissemination of PVY by wing aphids between plots was done conducting PVY^{O} and PVY^{N} antibody ELISA tests separately on each tuber. A second sample of 200 tubers was also controlled visually for ring spot symptoms just after harvest and 2 months after stocking.

Using similar oats band isolation design, two other experiments were carried out. First, the yield of healthy Nicola plants was compared with Nicola plants infected by different PVY isolates. Secondly, PVY ring spot necrosis sensitivity on tubers was studied for all promising varieties present in the Swiss official cultivar field trials. Since five years, this study has been part of the variety assessment in Switzerland. In each no infected variety plot, 4% PVY^{NT} infected potato plants (Nicola and Erntestolz) were inserted. These infected plants were harvested one week before the other plants in the plots. The yielded tubers of promising varieties are controlled visually.

Results and discussion

The cultivar Lady Christl was surprisingly highly resistant to all PVY isolate infections; Bintje and Charlotte were sensitive to PVY and very sensitive to PVY^{NT} and PVY^{Wilga} isolates. Charlotte was very sensitive to PVY^{N} isolates in 2006. Spread of PVY^{0803} and PVY^{N605} was lower than PVY^{NT} and PVY^{Wilga} isolates (Figure 1). Formerly, Nicola and Desiree were reputed to be well resistant to PVY, but today, these cultivars are sensitive to the new isolates of NT (necrotic) or PVY^{Wilga}.. The isolates of PVY^{0803} and PVY^{NT854} are very phytotoxic for Nicola and Desiree. Some of the infected plants died before harvest and did not serve as a virus source for the healthy plants during the growing season. The cultivation of vigorous infected plants with the different isolates of PVY is very challenging. The small PVY^{0803} infected plants where rapidly covered by the healthy vigorous plants. After the harvest, ring necroses on tubers were mostly observed on Nicola and rarely on Bintje, Desiree or Charlotte.

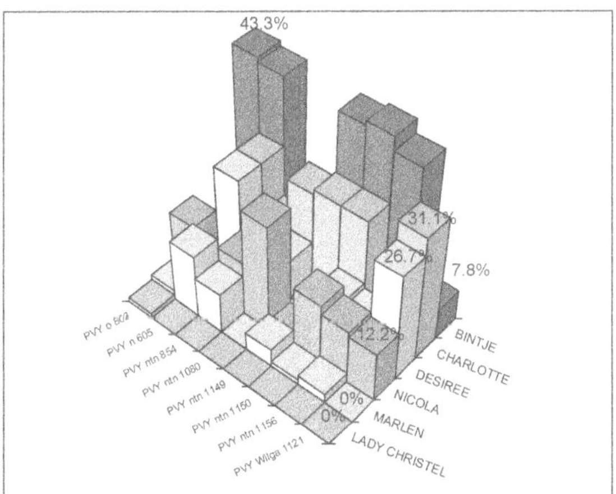

Figure 1. Results of ELISA test on tubers growing under field conditions. The six cultivars have different sensitivities to the Potyvirus Y isolates. Lady Christl was surprisingly high resistant. While Nicola and Desiree were reputed to have good PVY resistances, these cultivars are now sensitive to new PVY isolates. Charlotte and Bintje are very sensitive to PVY.

Under field conditions, the spread of diverse PVY isolates between plots was very low in 2005 and 2006 (Figure 2). The 12 meters isolation bands of well-established oat between the plots were sufficient to prevent isolates transfer via wing aphids.

A complementary study has shown that different PVY isolates affect the yield of Nicola more or less intensively (Figure 3). Though plants were infected with the isolate PVY^{N605}, they did not show strong virus symptoms on leaves and yield represented 85% of the control with healthy plants. The isolate $PVYNT^{854}$ however, had a strong impact on the yield performance, reducing it to less than 20% of normal yield. This trial shows that not only the degree of sensitiveness but also the performance of the plants is affected differently by the diverse isolates of PVY.

A further symptom evaluation on the cultivar assessment denotes that cultivars with high sensitivity to PVY dissemination are not necessarily sensitive to ring necrosis on the tubers (Figure 4). s Varieties like Erntestolz, Hermes, Nicola, Ditta and Gabriella that are sensitive to

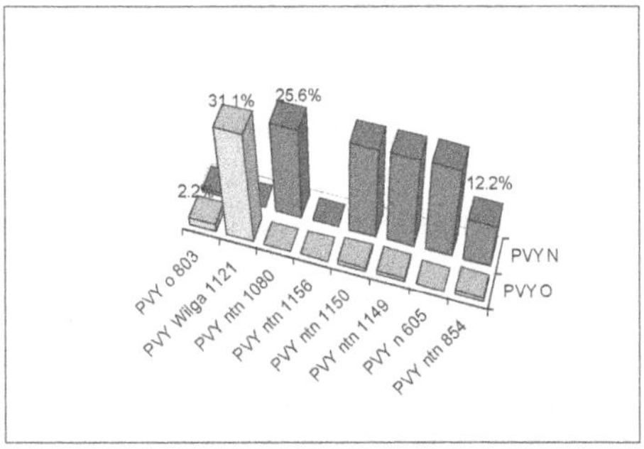

Figure 2. The ELISA test enables to distinguish PVY-o and –n isolates. ELISA test results of the trials under field conditions show that contaminations from PVY^N plots to PVY^O plots are very low. The separation band of oat between the different plots was sufficient. The contamination level of the new PVY isolates was higher for the Wilga type or for the NT isolates.

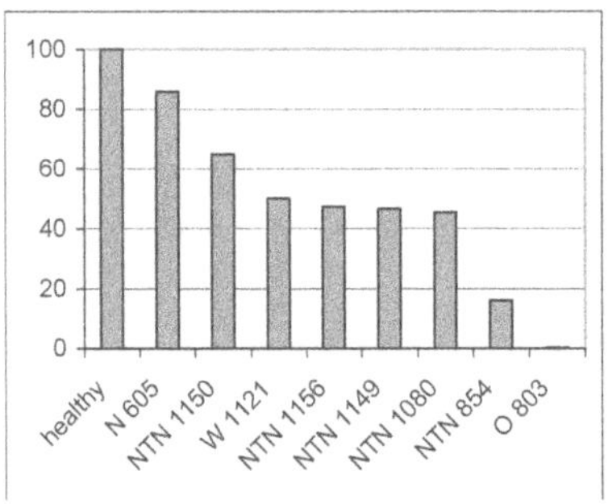

Figure 3. The yield of different isolates of PVY under field conditions was observed in another trial. The yield of the cultivar Nicola was not much affected by the isolate PVY N605. Most of NT isolates gave only half yield and 80% of the tubers showed ring necroses. The isolates PVYNT854 and PVYO803 were very phototoxic for Nicola and many infected plants died before harvest.

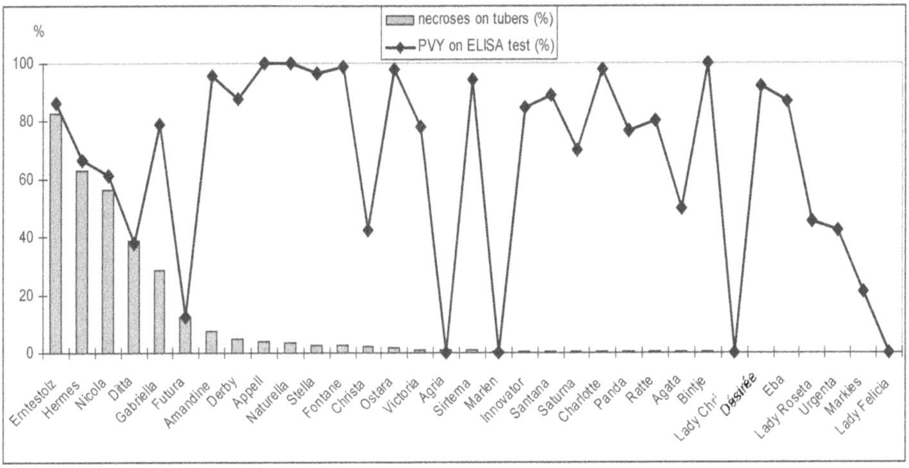

Figure 4. Cultivars with high sensitivity to PVY dissemination are not necessarily sensitive to ring necrosis symptoms on the tubers. These results were obtained in Changins with very high vector pressure and many infected plants of different isolates of PVY NT.

ring necroses on tubers should be avoided. Recent results show, that Verdi, Derby, Arielle and Amandine are sensitive to this disease as well (results not shown), since ring-necroses have been observed on tubers.

Cultivars showing an important sensitiveness to PVY should then not be selected as recommended cultivars for Switzerland. We have shown in our study that the sensitivity of the cultivars tested is variable and differs from the necroses tuber-sensitivity. The large variability of the virus PVY requires further large studies to answer the questions raised for the diverse cultivars.

Reference

Rigotti S. and Gugerli P., 2004. Quantification of Potyvirus PVY^N and PVY^{NT} isolates in Swiss potatoes by one-step diplex RT-PCR. Abstract of the presentation EAPR Virology Section meeting Rennes, France.

Curriculum vitae – R. Schwärzel

Mr. Ruedi Schwärzel from Switzerland – born in 1960 on a small mountain farm. After an apprenticeship in farming and training in viticulture, he accomplished a 4-year study at the Agricultural Technical High School of Zollikofen (Bern), where he obtained an engineer degree in agronomy in 1983. During the next six years he was first responsible for the field production of vegetables in the Cornu Company at Yens, before, for one year, leading the activities of a grapevine nursery.

Since 1989 Mr. Schwärzel has been professionally engaged at the Swiss research station for plant production, currently called Agroscope Changins-Wädenswil ACW; for the first two years as a technical collaborator in protein crops and potato research area, before taking the responsibility for the seed certification. Since 1997 he has been in charge of the potato seed production and field trials in ACW . His duties involves trials on plant susceptibility to various diseases and ravages, using in-vitro generated plants and pre-basic seeds as well as haulm-killing techniques. He also evaluates the relevance of novel seed storage methods for Switzerland.

Profile – Agroscope Changins-Wädenswil ACW, Swiss research station for plant production

Objectives: to maintain and improve agricultural products yield and quality, to diversify the production and to adapt it to the market requirements, to assure long-term soil protection and fertility, to develop environmentally friendly and economically sustainable plant production methods. The research group 'field crops' studies all technical aspects related to food production and conservation. Studies are conducted in small grains for human and animal nutrition, corn, potatoes, sugar beet, pulses, oil crops and various other industrial crops. Soil management, crop rotation, plant nutrition, control of pests and diseases as well as varietal testing are also studied in this group. The 'plant protection' group improves strategies and general knowledge regarding the choice of the production system for this specific topic, as well as monitoring and diagnostic tools to enhance the development of pest prevention methodology in all plant crops. Activities and duties of other research groups are related to grassland management, arboriculture, horticulture, viticulture and oenology. Breeders are involved in the development of new wheat, triticale and soybean varieties, as well as new cultivars of grapes, vegetables, fruits, aromatic and medicinal plants. ACW is composed of several research centres, located in the German-, Italian- and French-speaking parts of Switzerland employing around 250 staff.

Methods of definition of resistance to potato wart *Synchytrium endobioticum* (Schilb). Perc.

A.G. Zelya and P.A. Melnyck
The Ukrainian scientific-research station of plants' quarantine,UAAS, v. Boyani, Novoselitza district, Chernivtsy region,Ukraine, ndskr@myrealbox. com

Abstract

Here, the results of research on the definition of resistance of potato to wart *Synchytrium endobioticum* (Schilb) Perc. are shown by a direct method of potato seedlings' infection by zoospores of fresh wart outgrowths, by protein extraction from wart outgrowths and definition of resistance by a method of infra-red spectroscopy, which is more regularly used to diagnose resistance of potato to wart.

Keywords: potato wart, resistance, infection, zoospores, protein extraction, infra-red spectroscopy.

Introduction

During the years of independence agriculture of Ukraine has changed essentially. Now the land is in property of citizens of Ukraine. Therefore, along with large producers, there are many collective and individual farms. Thus there is a large movement of goods between different countries, an intensive migration of people, that increases the probability of penetration and distribution of quarantine organisms.

To duly reveal a disease agent and define its specification, it is necessary to have reliable and fast methods of diagnostics and identification of quarantine organisms (Zelya *et al.*, 2001; Kiray *et al.*, 1974; Melnyck *et al.*, 2003).

Methods

The Ukrainian research station of plants quarantine has developed methods of diagnostics of a potato wart agent *Synchytrium endobioticum* (Schilb). Perc. (Melnyck *et al.*, 2003).

For this purpose tuber samples of varieties and hybrids of a potato are planted in substrates; when seedlings reach1-2 mm, we infect with summer zoosporangia from fresh wart outgrowths with exposition during 24 hours at 11-13 °C. For this purpose on the tuber top of potato around seedling part fix a paper ring with the help of a mixture of paraffin and Vaseline (1:1). We pour the distilled water in a ring, put slices of fresh wart outgrowth containing summer zoosporangia of the wart agent. Samples are left in a climatic chamber with an increased humidity and temperature 11-13 °C. Time of exposition - 24 hours. After that delete slices of wart outgrowths with paper rings from used tubers and leave samples of a potato at temperatures 17-18 °C for 20 days, then determine the reaction of potato seedlings on pathogen infection. For this purpose they are analyzed under a binocular microscope and the degree of infestation in these gradations is determined:

1. necrotized tissue of a seedling, individual soruses;
2. scattered soruses (if more than 5), dense soruses without seedling deformation;
3. dense soruses with seedling deformation, wart outgrowth (Saltykova *et al.*, 1982; Saltykova *et al.*, 1979).

Another method is based on usage of an albumin (protein) extract. For this purpose take wart outgrowths. Weigh 10 g, which then are minced and grinded with 5 ml of sodium-phosphate buffer containing 12,5 g of glucose, 1% of ascorbic acid, 1% potassium metabisulphite and 0,018% magnesium-chlorine-2. Reaction pH is leaded up to 7,4 with the help of1 n of caustic soda. The homogenate is wrung out through a kapron grid and centrifuged 25 minutes at 12000 rev/min. Thus the proteins are isolated in the supernatant, and the rests dropped out in a pellet. Supernatant 7ml is cleared through a column with sefadex G - 25, counterbalanced by the buffer, which contains 4,66 g of trissaccaride; 0,59 EDTA (caustic)soda and 0,3 g of boric acid on 1000 ml of water; pH of the buffer - 8,6 (Jacovleva *et al.*, 1973).

After the beginning of protein isolation we select 2-3 ml and incubate in a refrigerator at 4-6 °C. In the obtained protein extract diluted 1:15 add pure waterless glycerol and several crystals of neutral red stain. The drops of protein are put by a thin glass rod on 8-12 leaf lobules of each potato breed or hybrid. Leaves for application of protein extract are taken of an average or bottom circle of the well-developed healthy plants in a budding phase. The survey is conducted on the 5th-6th day after protein extract application.

The leaf reaction on the protein extract at susceptible varieties and hybrids of potato is expressed in strong yellowing, browning and dry lobules with protein drops put on them and lobules or whole leaf abscission (3d degree). Weak yellowing of separate leaf lobules with applicated proteins tells about comparative resistance of potato samples (2nd degree). At resistant plants in a zone of protein application a local dieback of tissue is observed (1st degree) (Jacovleva *et al.*, 1973). Enough exact method, which is used in some control investigations is the method with usage of infra-red spectroscopy (Vilkova *et al.*, 1986).

In this case, from the infected potato samples take samples which then place in ditches and analyze on the infra-red analyzer IFA-61. At the analysis use length of waves 1510 nm. With the purpose of increasing the accuracy of the analysis for a result take the mean from three levels obtained at turn of ditches on 90°. A degree of resistance of potato to the wart agent is determined by Equation 1:

$$Y = C \times OD \times (W) \tag{1}$$

Where:

C: constant of the graduated equation;
OD: optical density;
W: analytical length of a wave (1510 nm);
Y: degree of potato seedlings' resistance to the wart agent (%) (Zelya *et al.*, 2001).

Mathematical data processing was conducted by a method of dispersive analysis by Dospehov (Dospechov *et al.*, 1973).

Results

As a result of the conducted investigations on infection of 48 samples of potato by summer zoospores of the wart agent the first degree of resistance (necrotized tissue) – was obtained by 13 samples, second degree of resistance (trace sources, dense soruses without deformation of potato seedling) – was obtained by 7 samples of a potato, third group of resistance (dense soruses with deformation of a seedling, wart outgrowth) was obtained by 28 samples of a potato (Table 1).
As a result of infection of samples of potato with the help of protein extract the first group of wart-resistance (local tissue dieback) was not obtained by any hybrid of potato; the second group of resistance (weak leaf lobules' yellowing of potato) - was obtained by 4 samples, the third degree (browning and drying of potato leaves) - was obtained by 44 samples of potato. In some cases the reaction of plants after protein extract isolation from wart outgrowths did not coincide with reaction of plants to direct infection by wart agent's zoospores. (Table 1).

At diagnostics of potato wart by infra-red spectroscopy, not enough material is used by infection (1 g against 10 g, which is used in other methods.). The offered method of diagnostics allows to determine the disease agent in a short period - 21 days (at other used methods – during vegetative period of potato - 3 months).

Table 1. Degree of resistance to wart of a selection of material of a, determined by different methods (2001 - 2005).

No. n/n	Varieties, hybrids of a potato	Degree of resistance		
		Direct way of infection	Infection by an albmin extract	Way of infra-red spectroscopy, % (M±m)
1	P94 27 – 4	3	3	13.6±0.73
2	P94 27 – 6	1	3	99.2±0.56
3	P94 27 – 11	1	2	92.4±0.56
4	P94 27 – 12	3	3	17.6±0.77
5	P94 27 – 21	3	3	14.9±0.54
6	P94 27 – 22	3	3	17.6±0.77
7	P94 27 – 26	3	3	24.4±0.66
8	P94 27 – 39	1	3	88.3±0.53
9	P94 27 – 46	1	3	96.4±0.54
10	P94 27 – 49	2	3	62.5±0.49
11	P94 27 – 52	1	3	91.1±0.57
12	P95 49/3	1	2	93.8±0.56

Table 1. Continued.

No. n/n	Varieties, hybrids of a potato	Degree of resistance		
		Direct way of infection	Infection by an albmin extract	Way of infra-red spectroscopy, % (M±m)
13	P95 49/5	3	3	21.7±0.72
14	P95 49/21	2	3	66.7±0.43
15	P95 49/25	1	3	93.8±0.54
16	P95 49/28	1	3	97.8±0.56
17	P95 49/35	3	3	16.3±0.74
18	P95 49/48	3	3	16.3±0.74
19	P95 49/53	2	3	58.4±0.42
20	P95 49/54	3	3	17.7±0.77
21	P95 8 – 12	3	3	14.9±0.69
22	P95 8 – 15	2	3	50.2±0.46
23	P95 8 – 22	2	3	54.3±0.43
24	P95 8/33	3	3	13.6±0.73
25	P95 9 – 1	3	3	17.7±0.76
26	P95 9 – 6	3	3	19.0±0.66
27	P95 9 – 7	2	3	62.5±0.49
28	P95 9 – 10	2	3	61.2±0.46
29	P95 11/16	1	3	86.9±0.52
30	P95 11/22	1	3	85.6±0.51
31	P95 11/26	3	3	16.3±0.76
32	P95 11/28	3	3	21.7±0.72
33	P95 11/30	3	3	25.8±0.53
34	P95 11/37	3	3	23.1±0,76
35	P95 11/43	3	3	20.4±0,74
36	P95 11/46	3	3	21.7±0,72
37	P95 11/47	3	3	23.1±0,76
38	P95 11/57	1	3	97.8±0.57
39	P95 11/68	3	3	19.0±0.66
40	P95 11/70	3	3	17.7±0.76
41	P95 11/75	3	3	13.6±0.73
42	Solami	1	2	99.2±0.57
43	Polessky pink	3	3	13.6±0.73
44	Radich	1	2	93.7±0.56
45	Post - 86	3	3	17.7±0.76
46	Synthesis	3	3	21.7±0.72
47	Beregynya	3	3	14.9±0.54
48	P88 128 – 7	3	3	17.7±0.76

Note: 1, 2, 3 is the degree of potato sample infection by the wart agent.

Conclusions

The methods to determine resistance of new varieties and hybrids of potato to wart by UkrSRSPQ enables an operative diagnosis and control of a given agent (wart) and maintenance of phytosanitary safety in a given region.

Acknowledgements

The Proceedings were agreed upon by the scientific Board of UkrSRSPQ on December 7, 2006, Protocol No. 5.

References

Vilkova N.A., Shapiro I.D., Borsheva T.A. Usage of infra-red microscopy for diagnostics of damage and resistance of grains to bugs. From the collection Methods of investigations of plants' pathological changes. - M.:Kolos, 1976. - P. 216 - 219. (Вилкова Н. А., Шапиро И. Д., Борщева Т. А. Использование инфракрасной микроскопии для диагностики повреждения и устойчивости зерновок к клопам. В сб. Методы исследований патологических изменений растений. – М.: Колос, 1976. – С. 216 – 219.).

Dospechov B.A. field experience technique. - M.:Kolos. 1973.- 336 p. (Доспехов Б. А. Методика полевого опыта. – М.: Колос. 1973.- 336 с.).

Zelya A.G., Romaniuk O.P., Kostyshyn S.S., Melnyck P.O. Declarative patent on the discovery 38722A from 15-05-2001 . Method of determination of potato resistance to wart agent Synchytrium endobioticum (Schilb). Perc. In: Industrial property. 2001 No.4. P. 14. (Зеля А. Г., Романюк О.П., Костишин С.С., Мельник П.О. Деклараційний патент на винахід 38722А від 15-05-2001 р. Спосіб визначення стійкості картоплі до збудника раку Synchytrium endobioticum (Schilb.) Perc. In: Промислова власність. 2001 p.No.4. C. 14).

Kiray Z., Kliment Z., Shoimoshi F., Veresh J. Methods of phytopathology. - M.:Kolos, 1974. (Кирай З., Климент З., Шоймоши Ф., Вереш Й. Методы фитопатологии. – М.: Колос, 1974.).

Melnyck P.O. Potato wart ethiology, bioecological background for its profilactics and development limitation measures. (in Ukrainian)- Chernivtsy: Prut, 2003. - 284 p. (Мельник П. О. Етіологія раку картоплі, біоекологічне обгрунтування заходів його профілактики та обмеження розвитку. – Чернівці: Прут, 2003. – 284 с.)/

Saltykova L.P., Tarasova V.P. Methodical instructions (indications) on potato testing for wart resistance. - L., 1982. - 51 p. (Салтыкова Л. П., Тарасова В. П. Методические указания по испытанию картофеля на ракоустойчивость. – Л., 1982. – 51 с.).

Saltykova L.P., Jacovleva V.I. Laboratory diagnostics of potato wart resistance by a method of seedlings' infection from fresh wart outgrowths: the methodical instructions. - M., 1979. - 8 p. (Салтыкова Л. П., Яковлева В. И. Лабораторная диагностика ракоустойчивости картофеля методом заражения ростков от свежих раковых наростов: Методические указания. – М., 1979. – 8 с.).

Jacovleva N.N., Kadyrmatov I.N. Way of potato wart resistance definition copyright certificate No.380264 from 15-05-1973, bulletin No.21. (Яковлева Н. Н., Кадырматов И.Н. Способ определения ракоустойчивости картофеля: Авторское свидетельство No.380264 от 15-05-1973 г., бюллетень No21.)/

Biological system of potato protection from phytopathogens in the republic of Karelia

Z.P. Kotova
State Scientific Establishment Karelian Research Station of Agriculture Russian academy of agricultural science; Centralnaya Street 12, 185506 Karelia, Prionezhskiy district, s. Novaya Vilga, Russia

Abstract

We have examined the influence of different preparations on the quality of potato production. We have defined the most effective plant protection system from diseases.

Keywords: buck eye rot, potato scab, perishability, Russia

Introduction

Researchers from all over the world have a great interest in the production of potato, that takes the 4th place after wheat, corn and rice (Barash and Gorbatenko, 1984; Horton and Sawyer 1985). The level of crop capacity and the quality of potato tubers mostly depend on prevalence and development of harmful pathogen on its planting. Annual harvest shortage caused by diseases and pests according to the scientists of Institute of Potato (Russia) is the limits 23...29% and during the years of epidemics it exceeds 50% (Popkova and Volovik1980). A high intraspecific changeability of pathogens, different resistance of potato sorts to diseases produce a different effect from using plants defence facilities in different soil-climatic conditions.

A stable harvest of healthy tubers is achieved with the help of a modern plant defence system. Lately there has been a great turn to the side of chemical control with hazardous organisms, although it's widely known that a lot of problems are solved thanks to a more ecologically safe and effective methods of defence: managerial, biological, selective-genetic and others that have been still paid little attention. (Banadisev and Ivanuk, 2005). Now preparations, that act not on the principle of biocide, but on the principle of activation of formation in plants biologically active protective agents that lower pathogens viability, gain popularity.

The purpose of our work was to find the most effective potato defence system to the main diseases that show up in the conditions of Karelia.

Materials and methods

The research was conducted in 2002-2003, the method was a two-factor experiment in the early-ripe sort Pushkinec; the square of record plot is 10m^2 (factor B - spraying) and 50 m^2 (factor A - treatment); is fourfold. The previous crop was annual grass. Soils are turf-podzol, easy-loamy. There was a tuber treatment 3 days before planting combining it with a chemical one (the 1st and the 2nd are treatment with the help of systematic preparation, the 3 with the help of contact), biological (thrice-repeated usage of chitozar F 7% VRK (preparation on the basis of chytozan); biologized (1st treatment with systematic preparation, 2 with chytozar f 7% VRK, 3 contact) treatment of vegetative plants. Tubers were sprayed by chemical solutions: Maxi, KS; Tekto; Phytohit - VRP (water-soluble powder on the basis of chytosan); Phenoram - super; Chytozar VRP 44,6% and also Chytozar's mixtures with microelements (M). The control was treatment with water. The development of symptoms of tuber disease was estimated by the 6 grade scale NIIKH (3). Potato was judged by the indexes of the level of development (C) and prevalence (R) in percents.

Results

The analysis of the results showed that the examined preparations didn't influence the passing of the main phonological phases that, to a great extend, depended on the characteristics of the variety and meteorological conditions of the year. Judging by the morphometric indices on average, for the last two years, a reliable increase of plants' height was registered in the treatments with the treatment of Chyzator tubers and Chyzator and Tekto mixture (44,2 and 46,3 cm against 39,6cm in control). In productivity on average for the last 2 years of research the most essential addition, in comparison with control (24,5 t/ga), caused by the pre-planting tuber treatment with the preparation Chytozar in mixture with (M) and Tekto. Excess over this control was 2,7 t/hectare or 9,9% (Table 1).

Weather conditions of field seasons cause a usual display of buck eye rot with a gradual growth by the end of vegetation. The phytopathologic estimate of potato resistance to diseases that was done after tubers collection, shows a high effectiveness in the limitation of development *Actinomyces scabies* Gussow of the preparation Chytozar with (M) in the combination with treatment of vegetative plants: decrease of the level of development of diseases symptoms on the tubers decreased to 19.7% in comparison with control (31.8%). The maximum fungus inhibition *Rhizoctonia solani* Kühn was seen after tuber treatment with Chytozar with (M) with Tekto - decrease of the level of development of disease was 2.2% (15.8% in control). Decrease of the amount of symptoms *Phytophthora infestans* (Mont.) de Bary on the tubers (from 2.3... to 7.2 times) was seen after using fungicides Tekto, Phytohit, Chytozar and also mixtures of Chytozar with (M) with Tekto (Table 2).

Conclusion

Potato tubers' treatment with the preparation Chytozar with microelements in a mixture with the treatment Tekto in combination with a plant defence system from buck eye rot reduces the development of potato scab by 30.4%, Rhizoctonia by 2.5 times, and provides harvest increase of 9.9%.

Table 1. Potato crop yield when using plants defence system, t/h (average for 2002-2003).

Factor A	Factor B	2002	2003	average
Control	Control	19.8	29.2	22.7
	Chemical	16.2	33.5	23.3
	Biological	19.4	31.2	23.6
Maxim	Control	24.3*	28.5	24.8
	Chemical	21.3	27.9	22.6
	Biological	17.9	26.7	20.8
Tekto	Control	19.6	33.6	25.0
	Chemical	16.9	23.5	20.8
	Biological	22.3	33.7*	25.7
Phytohit	Control	18.2	28.1	22.9
	Chemical	23.0	30.9	25.6
	Biological	18.6	29.4	23.7
Phenoram-super	Control	21.7	27.8	21.7
	Chemical	18.4	33.7*	22.2
	Biological	16.5	34.3*	22.0
Chytozar	Control	22.2	27.0	24.6
	Chemical	22.5	32.2	27.4
	Biological	17.9	27.6	22.8
Chytozar (M) + Tekto	Control	28.6*	25.8	27.2
	Chemical	25.7*	29.5	27.6
	Biological	27.1*	27.7	27.4
	Biological-organic	25.1*	37.6*	31.4
	NCP_{05}	4.6	4.6	-

*deviations from control are statistically significant

Table 2. The indexes of potato tubers' receptivity to diseases (on average for 2002-2003).

Factor A	Factor B	*Actinomyces scabies*		*Rhizoctonia solani*		*Phytophthora infestans*	
		R %	C %	R %	C %	R %	C %
Control	Control	98.5	31.8	52.6	15.8	30.5	8.7
	Chemical	98.1	35.1	47.0	18.3	26.3	6.3
	Biological	95.0	28.3	39.7	13.1	14.3	2.9
Maxim	Control	96.2	29.4	38.6	10.4	15.3	3.0
	Chemical	96.9	29.6	50.6	18.5	21.3	5.7
	Biological	91.9	26.8	41.4	10.9	31.5	5.7
Tekto	Control	97.1	24.7	47.5	13.6	25.3	6.0
	Chemical	96.4	32.4	52.6	21.4	39.7	10.9
	Biological	95.4	27.1	50.6	16.9	16.8	3.7
Phytohit	Control	98.0	27.8	50.3	18.2	25.5	5.6
	Chemical	99.1	28.5	37.8	11.6	28.7	6.6
	Biological	98.3	27.6	58.9	16.7	20.3	3.5
Phenoram-super	Control	92.9	25.2	36.2	6.2	26.7	6.3
	Chemical	97.2	24.7	27.4	5.5	27.3	6.1
	Biological	96.1	22.4	45.5	12.3	27.4	5.7
Chytozar	Control	95.7	24.1	56.6	17.7	2.1	1.4
	Chemical	100.0	20.6	34.2	7.9	5.0	0.8
	Biological	97.0	20.5	48.2	11.5	12.0	2.7
Chytozar (M) + Tekto	Control	99.0	22.4	8.4	2.2	25.4	7.5
	Chemical	94.9	22.2	40.7	7.2	2.1	0.9
	Biological	94.0	19.7	33.6	5.3	21.0	5.1

C: the level of disease development %
R: disease prevalence %

Acknowledgements

Authors express gratitude to Doctor of Science in biology, professor Stanislav Leonidovich Tjuterev for giving preparations to perform experiments.

References

Banadisev, S.A., Ivanuk, V.G. (2005). The problems of phytosanitary in Belarus and the ways of their solutions. /Mat. Intern. Scientific- pract. Conference 'Urgent problems of potato defense, fruit and vegetable crops from diseases, pests and weeds'. Minsk, 8-13

Barash, S.I., Gorbatenko, L.E. (1984). World production of potato (Solanum L.) and the climate in XX century/Tr. on botanist teaching, gen. and sel. VIR, 87, 96

Methodic instructions on estimation of selective potato material to check resistence to phytoftoroz, rezyctoniz, diseases caused by bacteria to mechanical failures (1980). Edited by Volovik , A.S.:VASHNIL, 52

M.K. Chohryakov, V.I. Potlajchuk, A.Y. Semenov, M.A.L., (1984), Diseases determinant of agriculture crops (1984) Kolos, 304

Popkova, K.B., Volovik, A.S. (1980). Potato diseases. M.: Kolos, 304

Tuterev, S.L. (2002). Induced plants disease-resistance and the ways of its activation with substances of biogenic and abiogenic nature /Scien. Mater. I All-Russia conference on plants immunity to diseases and pests. St. Pet.,154-155

Horton, D., Sawyer, R.L. (1985). The potato as a world food crop, with special reference to developing areas In: Potato Physiology. New York, 2-34

Curriculum vitae – Z.P. Kotova

Zinaida P. Kotova graduated from Kirghiz Agriculture Institute of Skryabin K.I. in 1986. I had been working in the capacity of a manager in Alakulskiy state sort-tests area in the republic of Kazakhstan for 7 years. There were tests of different crops: grain-crops, soy, sugar-beet, maize, Lucerne and other crops. I have been working in Karelian state agriculture of experimental station of Russian academy of science in the capacity of a branch manager of potato seed-growing for 12 years. Now potato seed-growing department works in the following main directions: the system of seed-growing conduct; schemes, methods of highest-quality potatoes production and methods of seed potato growing; technological methods and means of harvest increase and the quality of potato seed in best specimens seed-growing in the conditions of European North of RF. In 2003 I defeated a Ph. D. thesis in the field of physiology and biochemistry of plants under the direction of doctor of biological science S. N. Drozdova. I'm the author of more than 30 scientific articles including 'The growth of potato in the conditions of European North (by the example of the republic of Karelia).'

Profile – State Scientific Establishment Karelian Research Station of Agriculture Russian academy of agricultural science

State Scientific Establishment Karelian Research Station of Agriculture Russian academy of agriculture science is a scientific establishment that belongs to Russian academy of agriculture science. The main directions of scientific and productive activity of the station include the development, control and application of progressive resource-saving technology in the fields of plant growing and cattle breeding; organization of primary seed-growing and production of best specimens of perspective and zoned potato sorts and perennial grass; methodical supervision for pedigree cattle of aishirskiy breed growing, application of modern methods of breeding on basis of integration of methods of immune, population genetics and large-scale selection; scientific supply of agroindustrial production of republic of Karelia.

Advances in breeding

Potato breeding in Russia: history, general trends and achievements

E.A. Simakov, I.M. Yashina and N.P. Sklyarova
All-Russian Potato Research Institute, 23, Lorch Street, Kraskovo, 140051, Moscow region, Russia

Abstract

The main stages of potato history in Russia as an important food crop and as an object of breeding are considered in the article. The brief analysis of results of potato breeding in the period 1930-1975 is presented. It was minutely analyzed the results of some authors' research, which were devoted to revealing the source of PVY resistance and their further utilization in the programme of breeding. The scheme of backcrossing and new cultivar creation on the base of five revealed sources of PVY resistance and one – PVX resistance is represented. There are some cultivar pedigrees reflecting the differences in the methods of their breeding. Separate parts characterize the results of the investigations devoted to some polygenic traits as field resistance against Late Blight and improved starch properties. The main selection methodological particularities, applied in these directions, are based on screening recombinant genotypes in the progeny and their following utilization in crossings. The results of investigations devoted to selection for Potato Golden nematode resistance and the selection of cultivars suitable for processing into potato products were briefly analyzed.

Keywords: wild species, backcross, field resistance, inheritance, selection, recombination, starch concentration, processing

Introduction

Potato is one of the most important agricultural crops in Russia. Russian people consider potato as one of the basic food products. Potato is cultivated on the whole territory of Russia, in different climatic zones, from the southern borders to the polar circle. Today more than 90% of potato production is concentrated at small individual farms, which were historically formed as early as in the middle of the nineteenth century. Most part of the Russian population consumes fresh potato, but in recent years the volume of potato processing began to grow.

Different natural conditions and a large number of potato producers greatly influence on the purposes and tasks of the potato breeding. The main trend of a potato breeding is a creation of tasty table varieties with different maturing time and non-darkening tuber tissues. Small potato growers are mainly interested in mid-ripening and middle-late cultivars with the good taste, high starch concentration and increased field resistance to the common diseases. A new trend is the creation of cultivars suitable for the potato processing industry. The most important task of the creation of new cultivars is the combination of economically important features with the resistance to different viruses, *P. infestans*, *Alternaria solani*, potato cyst nematode, and stress factors.

Different potato breeding trends use such starting material as varieties from the World Collection and wild and primitive Solanum species, that are used according to the classical selection schemes. Breeders actively use special parental lines, created on the basis of the interspecies hybridization, selection of recombinants among the backcross generations and their progeny. This paper presents the basic results of the breeding work of the All-Russian Potato Research Institute for the whole period of its functioning.

History

Potato was first imported into Russia 120-130 years after it appeared in Western Europe; it immediately began to be cultivated as a food plant. The spontaneous expansion of potato in western Russian provinces began in 1676-1687 from the territory of neighbouring Poland and Latvia. In central and northern provinces potato was imported from England, Holland, and Germ as individual introductions. According to the data of the Free Economical Society, the first introduction was imported from Holland in 1697-1968: Peter the Great sent one bag of potato personally to Field-Marshal B.P. Sheremetyev (Lekhnovich, 1973). In 1735-1738 inhabitants of St.-Petersburg suburbs began to cultivate potato in their kitchen gardens.

An intensive expansion of the new crop was promoted by the policy of Catherine the Great and her government, which was interested in the new food crop, suitable for northern Russia, where the yield of cereals was permanently low. A systematic potato promotion to the peasant's fields and kitchen gardens started with the Decree of the Senate, appropriated funds for the purchase of a potato seed material and its distribution within the country (1765-1767). In addition to potato tubers, certain Siberian regions also received true potato seeds. In this period the guidance on potato production and storage was published in Russia. As a result, by the end of the eighteenth century, potato was cropped not only on kitchen gardens, but also on peasant's fields, though it still occupied small areas (1-2 ha). In this period potato finally dominated over turnip, which earlier was the main crop on the peasant's kitchen gardens.

In the middle of nineteenth century three new decrees, issued by the government of Nicholas I, provided the new stage of potato expansion in Russia. The first of these decrees provided for the production of seed potato on fields of state peasants with the further distribution of this seed material within the country. The second one introduced incentive fees for peasants, succeeding in potato breeding. According to the third decree, new agronomical recommendations in potato production should be published. The first decree caused some disagreement among the state peasants in eleven provinces of the central Russia and also in Ural and North-Western regions, which went down in history as 'potato uprisings'. The reason of these disorders was a misunderstanding of governmental actions, intended to provide the potato production on the state land; peasants thought these actions provided some kind of bondage for them.

The production of seed potato on fields of state peasants continued till 1844. By this time the annual potato production in Russia reached 20 kg per person. Due to governmental efforts, in this period, potato became the main food crop of the rural population and the people inhabiting suburbs. The new crop later was called 'the second bread'. Until now potato significantly influences the Russian economy.

In the first half of the 19th century Russian peasants grew foreign potato cultivars and cultivar mixtures from true potato seeds. Industrial seed breeding began to develop in 1850-1860, when first Russian potato-breeding companies appeared. One of such companies was Grachev's Family Com. The founder, E.A. Grachev (1826-1877), bred potato on his farm near St.-Petersburg. He created more than 20 new commercial potato cultivars with different ripening time and they were very tasty. The best known of these cultivars were Svetlo-rozoviy, Minotavr, Original, and Suvorov. Characteristics of these cultivars were described in the comp's catalogues (1879-1900). There are some data on the successful tests of Grachev's cultivars on the experimental stations (1885-1907), that confirmed their good characteristics.

In the beginning of the 20th century the company changed its field of activity, and its potato collection (250 samples) was bought by Nikitinskiy, who carried out some breeding work in his estate Kostino (Ryazan province). Later his collection grew up to 400 samples and became to be the main source of the seed potato in Russia.

After the October Revolution (1917) the surviving part of the collection was transferred by Lorch, the known Russian scientist, to the Korenevskaya experimental station, established in 1920 and later reorganized into the Potato Research Institute (PRI). The main purpose of the research program of this station was a selection of new potato cultivars with economically sound characteristics. To enlarge the gene pool of initial material, scientists of the station ordered the common potato collection (CPC) in Germ. Using a varietal hybridization in 1921 they created two mid-late cultivars, Lorch and Korenevskiy, which in 1931 were adapted to regional conditions and became widespread. The Lorch cultivar is still producing now.

During 1930-1934 Russian potato breeders began to use in their work some wild species, collected by scientific expeditions of Bukasov and Yuzepchuk in 1925-1927 and delivered to the breeders. Later PRI created the working collection of wild species, which was replenished from time to time by new samples from the All-Union Plant-Production Institute. The species of Demissa series were widely used in the breeding program of PRI. Pushkarev involved *S. demissium* genotypes into the rapid scheme of selection for potato resistance to *P. infestans* (late blight (LB) resistance). In 1936 the new cultivar Fitoftoroustoichiviy 8670 was bred from the large number of backcrosses. This cultivar was not widespread, but served as the starting material for the further selection. Using this cultivar, the known Agronomicheskiy and Kameraz cultivars were bred (1949-1958), which became precursors of m modern Russian potato cultivars.

In 1936-1955 Filippov used different genotypes of *S. semedemissium* in his selection work. Among backcross generations he selected hybrids, relatively resistant to *P. infestans* and promising for the further use in the selection of potato with high starch concentration. Basing on these hybrids, breeders created such potato cultivars as Istok, Gatchinskiy, Moscovskiy ranniy, etc.

During 1950-1975 an intensive selection was carried out concerning the supersensitive type of resistance to *P. infestans*. Using hybrids, originated from *S. demissium*, Russian breeders created a group of high-yielding cultivars with good characteristics, such as Lyubimets (R_1), Druzhniy (R_1), Smena (R_1), Veselovskiy (R_2), etc. Basing on cv. Olev (R_1R_4 + field resistance), breeders created several cultivars, combining R genes with the field resistance (FR): Stoloviy-19 (R_4 + FR), Sotka (R_1R_4 + FR), Loshitskiy (R_4 + FR), Temp (R_1 + FR), etc. The crossing between cvs. Veselovskiy and Kandidat allowed Osipova to bred cv. Nevskiy (R_1R_2 + FR), that is now cultivated on the major part of potato fields in Russia.

In 1965 our institute began a breeding program, intended to obtain new potato cultivars with an elevated resistance to different diseases and pests, high starch concentration, early ripeness and suitability for the use in the potato processing industry. This program included a wide range of tasks, such as a search and use of new genetic sources, analysis of inheritance of the above-mentioned properties, interspecies hybridization, comparative study of backcrossing variants, and the breeding of new cultivars. The purpose of this program was also the creation of special parental lines for main breeding trends and the revealing of optimal variants of the evaluation, testing, and selection of the breeding material. Results of this long-term program are presented in this paper in accordance with the trends of the breeding work.

Breeding of cultivars, resistant to viruses

The most harmful potato virus in Central Russia, where the climate is rather continental, is potato virus Y (PVY), which causes potato degeneration, especially in the case of mixed PVY-PVX infections. Thus, the main objective of our study was to find and isolate sources of resistance to PVX and PVY. The working wild species collection was examined for resistant genotypes using the artificial infection of seedlings, obtained by the standard self-fertilization technique. The resistance to PVY^0 was registered for two genotypes of diploid *S. chacoense* species (2n = 24), which demonstrated the presence of a dominant gene after the splitting (Sklyarova and Yashina, 1973). The analysis of splitting the resistance to viruses was carried out for the F_1 generation and several backcrosses (Yashina and Sklyarova, 2000). Characteristics of the sources, breeding scheme and results of selection are shown in Table 1. To introduce the sources into the crossings, based on 55d and 58d samples, we obtained autotetraploids and amphidiploids, respectively.

We found and isolated three sources of resistance to PVY in plants, obtained by self-fertilization of several backcrosses, carrying R_{ysto} gene (F_2B_n). The seeds were granted by Dr N.W. Simmonds (England) in 1965. Resistant genotypes showed using artificial infection of seedlings (mechanical inoculation, grafting on tomato, and diagnostics with the help of indicators) and the field evaluation of the tuber generation. Three five-species hybrids, immune to PVY, PVX, and PVS and resistant to PLRV, were obtained from Dr Sh. Sharvari (Hungary). The cv. Fanal was used as an additional source of resistance to PVY, containing the R_{ysto} gene.

The cv. Safir, containing R_{xacl} gene, was used as the source of resistance to PVX.

Table 1 shows that the total number of cultivars, bred on the basis of these sources, was 38, and 27 of them have been included into the State Register. The evaluation of backcross generations allowed us to obtain new data on the breeding value of the sources. The progeny of Hungarian hybrids and *S. chacoense* 55d and 58d was resistant to early blight and the hot and droughty weather. Several lines, originating from 55d, were relatively resistant to Colorado beetle (cvs. Peresvet and Nikulinskiy). We also isolated three highly fertile genotypes (15c-11, 128-6, 80-1)

Table 1. Characteristics of the sources of viral resistance and results of their use in the breeding programmes.

Source of resistance and its genotype	Stages of the involvement in the breeding	Scheme of use in the breeding program	Number of backcross generations	Cultivars isolated		Most widespread cultivars
				total	registered in the State Register	
S. chacoense f. garciae 55d (2n=24) R_{Ychc}	Production: - autotetraploids, - F_1 hybrids, - B_1-B_3 backcrosses (F_1 x tbr $^{1-3}$)	55d (2n=24) → treatment with colhicine → 55d (2n=48) 55d (2n=48) x Agra → F_1 (33-29, 33-38, 33-70) 33-29 x Kameraz, Olev → B_1 (174m,178m);33-70 x Kameraz →$B_1$172m; 33-38 x Kameraz → B_1 (173m-142, 328-22, 328-49) and other cvs	3	9	4	Veteran, Bryanskiy nadezhniy, Nikulinskiy, Utenok
S. chacoense f. commersonii 58d (2n=24) R_{Ychc}	Production: - amphidiploids, - three-species hybrids ($3F_1$) - backcrosses - (B_1→B_2)	F_1 S. vernei x S. chacoense 58d (2n=24) → treatment with colhicine → amphidiploids 71dd-74dd (2n=48) crossing 71-74dd x tbr → $3F_1$ (425m, 426m, 445m, 535m) B_1 71-74dd x tbr $^{2-3}$(504m, 512m,591m,596m)	2	6	4	Bryanskiy delikates, Nakra Slava Bryanschiny Golubizna,
Progeny of self-fertilized backcrosses of *S. stoloniferum* (2n=48) Generation F_2B_n R_{Ysto}	Isolation of resistant genotypes, F2Bn x tbr1-3 crossing	Three genotypes: F_2 Bn → 189y-56, 189y-95-1,189y-61 189y-56 x Agronomicheskiy → 15c-11,15c-31 189y-95-1 x Kameraz → 174-4 x Karat → Kumir 189y-61 x Oberartbacherfrue Chernskiy Eros x 15c-11 → Bryanskiy ranniy 15c-31 x Beryozka → 128-6 x (different cultivars) Gatchinskiy x 128-6 →Golubizna	1-3	12	9	Ramzai, Chernskiy, Effekt

Table 1. Conyibued.

Source of resistance and its genotype	Stages of the involvement in the breeding	Scheme of use in the breeding program	Number of backcross generations	Cultivars isolated		the most widespread cultivars
				total	registered in the State Register	
Hungarian five- species hybrids (-br-acl-adg-dms-sto) $R_{Ysto}Rx_{acl}R_{Sadg}$	Hybrid-cultivar crossing. Use in the crossings of new cultivars	69. 5403. 259 x Smena → Resurs KE-23 x 1169-2 → Sokolskiy KZ-1001 x 733-65 (Dezire x Smena) → Kolobok Resurs x Aksenovskiy → Yubilei Zhukova Resurs x Zarevo → Bryanskiy krasniy	1	6	6	Resurs Sokolskiy Kolobok Yubilei Zhukova
Fanal R_{Ysto}	Crossing with cultivars. Use of new cultivars	Fanal x Smena → 80-1 Different cultivars x 80-1 → Lira, Kalinka, Pransa Pransa x Aksenovskiy → Zhigulevskiy	2	5	2	Lira Zhigulevskiy
Saphir R_{Xacl}	Crossing with cultivars	Saphir x Katadin → (350-68,350-77 x 173m-45($B_{1,chc}$) x Anoka	2	2	2	Goryanka
			Total	**38**	**27**	

among the backcrosses, obtained by the self-fertilization of English sources, and the progeny of cv. Fanal; such high fertility is atypical for the carriers of Rysto gene. In all cases we used pollinators with a high pollen fertility (Beryozka and Smena). The backcross 128-6 was the most valuable; we used it to obtain several PVY-resistant cultivars, and it is widely used in the breeding program until now.

The use of the autotetraploid of *S. chacoense* 55d (2n=48) for a breeding of new cultivars required three back crossings; the whole breeding cycle took 28 years (1966-1994). In the case of the amphidiploid *S. vernei* × *S. chacoense* (2n=48) we performed two back crossings, and the whole breeding cycle took 23 years (1977-2000). In the case of the back crossing of different F_2B_n sources, obtained by the self-fertilization, we required 1-3 back crossings. The use of the late LB-resistant cultivar Beryozka increased both the number of backcross generations and the level of LB-resistance of the progeny (Figure 1A). The whole breeding cycle, including 3 backcross generations, took 27 years (1965-1992).

Thoroughly selected Hungarian hybrids produced new cultivars already in the first generation of all crossings. In the case of cv. Fanal we selected valuable genotypes only in the second generation. To date we bred several cultivars, carrying resistance genes from different sources used: Veteran (chc-sto-sto), Russkiy Suvenir (chc-sto), Slava Bryanschiny (chc-vrn-sto), etc. The cultivars and hybrids we obtained from different PVY-resistant sources are widely used now for different breeding purposes, increasing the genetic diversity and heterozygosity of the breeding material.

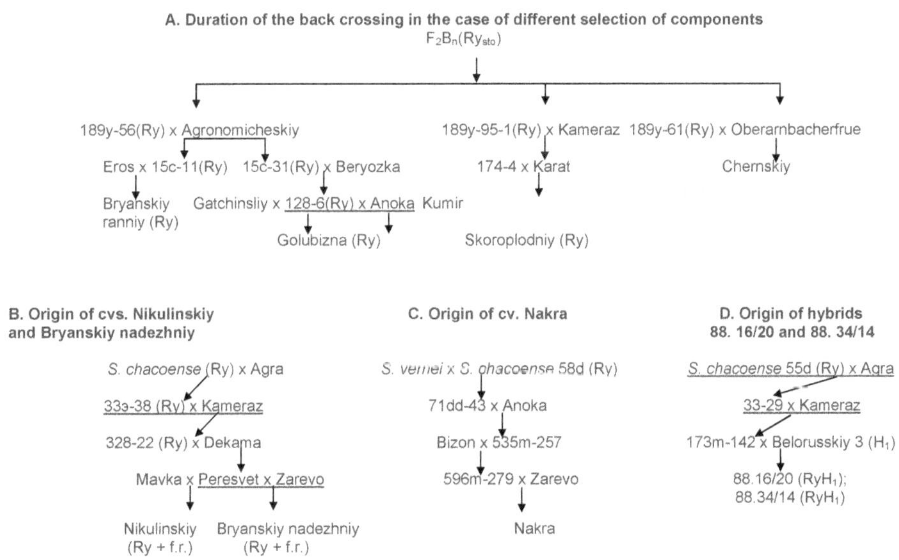

Figure 1. Origin of the cultivars and hybrids, most widely used in the breeding program (A, B, C, D).

Breeding of cultivars with an elevated field resistance to *P. infestans*

Breeding in this direction started from a study of the inheritance of field resistance. In 1965 (during the epidemic) we carried out a field evaluation of the first tuber generation of r-genotypes and showed that the field resistance is controlled by additively acting polygenes, determining the transgressive progeny splitting and the existence of the intermediate stage for the inheritance of this property (Yashina, 1968).

The type of inheritance determined the basic breeding methods we used, i.e. saturating crossings with resistant partners and the selection of transgressive recombinants (TR hybrids), combining a high field resistance with the complex of economically valuable characteristics. To identify TR hybrids, we used an artificial inoculation of detached leaflets with the high-virulent *P. infestans* race (1. 2. 3. 4. 5. 6. 7. 8. 9. 10 XYZ), annually obtained from the All-Russian Research Institute of Phytopathology. The infection load was 25 conidia per microscopic field (120×).

The efficiency of saturating crossings was tested experimentally during back crossing of PVY-resistant hybrids. Based on the non-resistant F_1 hybrid (*S. chacoense* 55 d × Agra), we carried out three back crossings and bred new middle-late cultivars (Nikulinskiy and Bryanskiy nadezhniy) with the high field resistance of tubers and leaves (Figure 1B). Both cultivars were PVY-resistant. In the back crossings we used cultivars with average (Kameras, Dekama) and high (Mavka, Zarevo) field resistance.

Similar results were obtained during the breeding of cv. Nakra (Figure 1C). After the back crossing of the non-resistant three-species hybrid (*S. vernei* × *S. chacoense* 55d) × Anoka with two resistant late-ripening cultivars (Bison, Zarevo), we obtained a mid-ripening cultivar with a high field resistance to *P. infestans*. These data show that after 2-3 saturating back crossings the further back crossing causes an accumulation of polygenes and an improvement of the field resistance.

Now, when we breed new LB-resistant cultivars, we pay special attention to the selection of recombinant genotypes, or TR hybrids. To evaluate them in the first clonal generation, we used the artificial inoculation of detached leaflets with the subsequent long-term field evaluation. Basing on TR-hybrids, we created parental lines, combining a high field resistance to *P. infestans* with a high productivity and other valuable characteristics (in 2006 our collection of TR hybrids contained 92 lines).

The frequency of TR genotypes in the progeny depends on the crossing type and the level of parental resistance, combining ability of parents regarding the productivity and other valuable characteristics. The analysis of 35 hybrid populations, evaluated in different time, showed that the average frequency of TR hybrids is quite stable and makes 2.8-2.2% (Table 2). However, this frequency is greatly varied in different populations. Therefore, to select the best parental genotypes regarding the field resistance, we should individually evaluate the progeny regarding the frequency of TR hybrids.

Table 2. Characterization of two groups of hybrid populations regarding the frequency of TR hybrids.

Evaluation period	Number of studied		Frequency of TR hybrids, %	Range of frequencies of TR hybrids in individual populations, %
	populations	hybrids		
1993-1996	10	1904	2.8	0.7-7.5
2001-2003	24	2450	2.2	0.98-6.4

The highest TR-frequency was registered in the progeny of parents with high and average LB resistance. In the case of the crossing of resistant and susceptible genotypes, the percent of TR hybrids sharply decreases. However, the use of susceptible parents is necessary, since such cultivars and lines represent mainly early-ripe genotypes. The introduction of polygenes, determining the early ripeness, into the starting breeding material provides a possibility of the breeding of early-ripe cultivars with relatively high field resistance to *P. infestans*. Such type of selection is quite possible. For example, cv. Udacha is a natural TR hybrid, obtained by the common breeding scheme and selected from the Vilnya × Anoka population.

The analysis of 52 selected TR hybrids with a high field resistance (7-8 scores) showed that the most part of these hybrids (90.4%) was bred using middle-late cultivars, including those obtained by crossing within this group (44. 2%) and with mid-ripening (25%), middle-early (11.5%), and early (9.6%) genotypes. The share of TR hybrids, obtained by the crossing of mid-ripening parents with the middle-early and middle-late ones, was equal to 5.8%. In the case of the crossing of middle-early and early parental genotypes, the share of TR hybrids was only 3.8%.

Characteristics of selected TR hybrids, obtained by different crossing types, are shown in Table 3. Hybrids, obtained by crossings between the mid-ripening cv. Lugovskoy and several middle-late cultivars (Belousovskiy, Zarevo, Nikulinskiy, Peterburgskiy), demonstrated a high field resistance to *P. infestans*. The R x MR and R x S types of crossings usually caused a loss of 1-2 resistance scores, especially during the severe epidemic in 2005. Nevertheless, the tuber resistance of all hybrids, obtained from the early-ripening parents (Udacha, Utenok, Rossiyanka), was sufficiently high, and many of these hybrids had enough time to produce a high yield.

Since the frequency of recombinants with the high field resistance and early maturity is quite low, we need about 2-3 generations to select a combination of these properties. We use TR hybrids with a high field resistance to *P. infestans* and other valuable characteristics as parental lines and, at the same time, evaluate them according to the breeding scheme.

Breeding of cultivars, resistant to potato cyst nematode (pcn)

The most widespread pathotype of *Globodera rostochiensis* in Russia is R_{0-1}, so the breeding material is evaluated concerning this pathotype. In 1977 the first sources of resistance to this pathogenic organism represent CPC cultivars, especially German ones: Hydra (three new cultivars – Bezhitskiy,

Table 3. Characteristics of TR hybrids, resistant to P. infestans.

Origin and crossing type (concerning the resistance to *P. infestans)**	**Resistance, scores****					**Productivity, g/plant**	**Starch concentration, %**
	to *P. infestans*				**to viruses (visual evaluation in the field)**		
	artificial inoculation		**field evaluation during the epidemic**				
	leaves	**tubers**	**2005**	**2006**			
Nikulinskiy × Belousovskiy (R × R)	76	9.0	8.0	8.0	8.5	980	18.6
	7.4	9.0	8.5	8.0	8.5	930	19.3
	7.2	8.0	7.0	8.0	8.5	1070	16.5
Nikulinskiy × Peterburgskiy (R × R)	7.5	9.0	8.0	8.0	8.5	900	18.5
	7.2	7.6	7.0	8.0	8.5	1000	15.6
	7.0	9.0	7.0	8.0	9.0	1000	14.8
1977-76 × Zarevo (R × R)	8.9	9.0	7.5	8.0	9.0	745	19.3
	8.3	6.0	7.0	8.0	8.0	740	16.2
Nikulinskiy × Kalinka (R × MR)	7.7	9.0	7.0	8.0	7.5	717	18.6
	6.5	9.0	7.5	7.5	8.0	990	18.6
Nikulinskiy × 88. 16/20 (R × MR)	6.7	9.0	7.0	8.0	8.5	810	14.2
Lugovskoy × 88. 16/20 (R × MR)	9.0	9.0	7.0	8.0	8.0	920	14.5
Udacha × 88. 16/20 (R × MR)	5.5	9.0	5	8	8.5	1020	12.1
	6.0	9.0	6.0	7.0	8.0	750	12.6
	5.0	8.5	3.0	7.5	8.5	1030	14.8
Udacha × Utenok (R × S)	6.5	7.5	7.0	8.0	7.5	700	14.0
88. 17/72 × Rossiyanka (R × S)	8.3	7.3	6.5	7.0	8.0	760	18.1
Nikulinskiy × Ausonia (R × S)	5.0	9.0	6.0	8	8.5	1130	18.1
Udacha (early)	5.6	7.3	3.0	5.0	8.5	1030	14.6
Nevskiy (mid-early)	5.8	4.2	2.0	7.0	7.5	933	13.3
Lugovskoy (mid-early)	7.0	8.0	5.0	8.0	8.0	1000	18.4
Nikulinskiy (mid-late)	6.3	8.3	5.0	8.0	9.0	1080	18.8

*R - resistant, MR - moderately resistant, S - susceptible.
**9 – resistant (no affection), 1– susceptible (100% affected)

Zhukovskiy ranniy, and Rossiyanka – originated from this cultivar), Gelda (Lukyanovskiy and Zavorovskiy), and Kardia (Aspiya, Solnechniy). The use of each of cvs. Elvira, Kulpe, and Oktavia allowed breeders to obtain one new cultivar. Recently we bred new nematode-resistant cultivars, basing on our own cvs. Malinovka and Krepysh.

To reveal new sources of resistance to nematode, we analyzed the collection of wild *Solanum* species and evaluate them on the Vilnius base station of the All-Russian K.I. Skrjabin Institute of Helminthology. As a result, we detected only one source, *S. chacoense* 58d, that we have already used during a breeding of PVY-resistant cultivars. Amphidiploids 71dd-74dd (*S. vernei* × *S. chacoense,* 2n=48) were repeatedly crossed with cultivars. In 1978-1980 we analyzed a splitting of 22 populations, obtained by the crossing of three-species hybrids (vrn × chc) × tbr, and 13 backcross populations (vrn × chc) × tbr^2 regarding the PCN resistance. The results confirmed that this property is determined by a single dominant gene. Among the backcrosses we selected several resistant hybrids, such as 591m-29, 591m-54, 596m-30, etc. (Yashina *et al.*, 1987). Two cultivars - Bryanskiy delikates (originated from the backcross 591m-29) and Nakra (originated from the backcross 596m-279) - demonstrated the PCN resistance (Figure 1C). Basing on the *S. chacoense* 58d, we bred cv. Slava Bryanschiny, resistant to R_{0-1}.

Now we use two specially created parental lines, backcrosses 88. 16/20 and 88. 34/14, in our breeding program to obtain PCN-resistant cultivars. These lines have a high fertility, high resistance to PVY, and the moderate resistance to *P. infestans*, Colorado beetle, and stress factors (heat and drought). The origin of these backcrosses is shown on Figure 1D.

Since 2005 we use a special collection of cultivars and hybrids, resistant to R_{0-1} pathotype and including 36 genotypes of different origin; 18 samples from this collection are original hybrids, combining the R_{0-1} resistance with the early and late blight resistance and also the resistance to some viruses.

We use new PCN-resistant cultivars, bred in Russia (Pushkinets, Nayada) and Belarus (Albatros, Atlant, Rosinka, etc.), as the starting material; in our crossings we also use some foreign cultivars, such as Ausonia, Sante, Aroza, Maestro, etc.

To increase the efficiency of the laboratory testing of the resistance to R_{0-1} pathotype, we improved the assessment methods, that, in turn, improved the accuracy of the resistance assessment, carried out on the basic stages of the breeding process (Simakov *et al.*, 2006).

Breeding of cultivars with a high starch concentration

The starch concentration in potato tubers determines consumption and technological characteristics of a cultivar, so this property is important for the breeding work. In Russia a high starch concentration is traditionally associated with the good taste of potato tubers and their mealyness, so the optimal starch concentration for the table cultivars is considered to be equal to 18-20%. However, adverse weather conditions and a short vegetation season, typical for the central Russia, often decrease the value of this parameter by 2-3%.

Our program of breeding of new cultivars with a high starch concentration started from the study of the inheritance of this property. During 1964-1976 we evaluated 43 populations, obtained by the self-fertilization and the crossing of different parental genotypes. We established that the starch concentration in tubers is determined by a small number of additively acting genes. Basing on the analysis of the ratio between the frequencies of hybrids with different starch concentration in the splitting progeny, we determined two independent loci, responsible for the manifestation of this property (Yashina, 1982). We also registered a symmetrical distribution of frequencies, a presence of transgressions, and a high correlation between the average starch concentration in the tubers of parental genotypes and the progeny (the correlation coefficient for the studied populations varied from 0.950 to 0.850).

The way of inheritance of the starch concentration and the small number of loci, responsible for this property, allow us to use several simple methods for the improvement of this characteristic. These methods include the use of parental lines with the heightened starch concentration and the selection of positive transgressions in the progeny. However, a negative correlation between a high starch concentration and a high yield complicated such breeding work. The analysis of 28 hybrid populations (1965-1973) showed that the correlation coefficient between these properties varied from weakly positive (0.394) to weakly negative (-0.248) values. The positive correlation was registered for the populations with the lowered starch concentration. When the average starch concentration for a population was 15-21%, then there was virtually no correlation between these properties (the value of the correlation coefficient varied from 0.083 to -0.109). However, within the groups of hybrids, selected by the starch concentration (equal to the average population value or higher), we observed a negative correlation (the appropriate correlation coefficient varied from -0.485 to -0.718). This fact explains the low frequency of recombinant hybrids, combining a high starch concentration with the high yield, in the most of populations.

The evaluation of 30 hybrid populations, carried out in 1993, showed the average frequency of recombinant genotypes with a high yield (>1000 g/plant) and a high starch concentration (> 18%) was only 0.5% (Table 4). The progeny of the crossing of two genotypes with a high starch concentration was excelled by the higher recombination frequency. If even one of the parents was characterized by a low starch concentration, then the progeny did not include recombinants. We also registered significant differences in the frequency of recombinants in populations of different origin (from 0 to 8.69%). Therefore, the preliminary estimation of hybrid populations concerning the frequency of recombinant genotypes can be very important for successful breeding.

Table 4. Frequency of recombinant genotypes with the high yield and starch concentration in different populations.

Crossing type*	**Number of studied**		**Recombinants selected**		**Recombinant frequency range in individual populations, %**
	populations	**hybrids**	**number**	**%**	
H × H	11	1778	21	1.18	0.41-8.69
H × M	12	3478	12	0.34	0.13-1.8
H × L	6	1252	3	0.24	0.20-0.78
M × L	3	452	0	0	0
Total	30	6979	36	0.5	0.13-8.69

*H – high starch concentration; C – moderate starch concentration; H – low starch concentration

The maximum number of recombinants was selected in the progeny of the following crossings: Effect × 1632-21 (8.69%) and Effect × Zarevo (6.06%). The next period (1998-2000) we widely used cv. Zarevo to work in this direction, since it demonstrated the high starch concentration (22-26%) and field resistance to *P. infestans* and, at the same time, had relatively large tubers. In 2000 we examined 18 hybrid populations, 11 of which originated from this cultivar. We selected 11 hybrids with the high starch concentration (18.4-23.3%) from 9 of these populations; the frequency of recombinant hybrids was 0.43-0.83%.

Most of cultivars with the high starch concentration are late-ripening, so one of the important directions of our work is breeding of early and mid-early cultivars with the heightened starch concentration. Today, basing on the radiation selection and mutagenesis, we created a new starting material, characterizing by the combination of the mentioned properties (Simakov *et al.*, 1991). We also obtained first hybrids with potential valuable characteristics; their testing will allow us to determine the most effective breeding methods.

Breeding of new cultivars for the potato processing industry

Breeding of potato cultivars for the potato processing industry is a relatively new direction of our breeding work. The first studies were carried out in 1980-1983, when we assessed the starting material and analyzed the inheritance of this property. The suitability of samples for processing was estimated by the colour of chips using the nine-point scale (9-7 - highly suitable, 6-5 moderately suitable, 4-1 - unsuitable). Our study was based on the theoretical investigations of Accatino *et al.* (1973), resulted in the revealing of two dominant genes, responsible for the suitability of tubers for the reconditioning.

According to our data, based on the analysis of different hybrid populations, the suitability of tubers for processing is determined by two dominant genes, interacting in an epistatic manner (Yashina and Yurieva, 1992). Later (2003-2004) we estimated 36 hybrid populations and established that the frequency of suitable hybrids in the progeny of two suitable parents is about the same as in the case when one of the parents is unsuitable (Table 5). For the most of populations (8 of 13) we registered the appearance of suitable genotypes in the latter half of the storage period (4,8-42,1%). In 5 populations we revealed hybrids, suitable for processing without reconditioning (5,3-14,2%).

Similar data were obtained in 2003-2005, during the analysis of 15 hybrid populations. The frequency of hybrids, suitable for processing in the first and the latter halves of the storage period varied in different populations from 20.3 to 50.5% and from 1.7 to 52.2%, respectively. Three populations included hybrids, that do not require reconditioning before processing; their frequency varied from 3.4 to 13%.

The analysis of 36 hybrid populations allowed us to specify the period of evaluation of the breeding material for its suitability to processing in different periods of storage. We also specified methods of selection of parental genotypes and established that the optimal breeding should include one or more 'suitable' parental genotypes in every crossing.

Table 5. Evaluation of hybrid populations according to the frequency of hybrids, suitable for processing (2003-2004).

Origin of populations	Crossing type*	% of hybrids with 7-9 scores in different estimation periods		
		with reconditioning		without reconditioning
		28-12-2003	7-03-2004	22-02-2004
Effect × Granola	S × S	45.8	0	0
Effect × Zarevo		28.2	42.1	0
Effect × Divo	S × MS	14.3	15.8	6.8
Effect × 128-6	S × U	6.8	29.4	0
Ladi Rozetta × 88. 16/20		17.5	10.8	5.3
88. 16/20 × Rossiyanka	U × S	5.8	0	11.7
93. 20-149 × Rossiyanka		12.8	14.3	0
Baraka × Ausonia	MS × MS	18.7	0	0
Resurs × 128-6	U × U	7.5	4.8	0
Vestik × Peterburgskiy	S × S	28.5	0	0
Vestik × Zarevo		0	25.0	14.2
Sante × Belousovskiy		0	13.3	0

* S - suitable, MS - moderately suitable, U - unsuitable.

According to the results of estimation of 36 hybrid populations, we selected 20 hybrids, highly suitable for processing over the whole period of storage. These hybrids are excelled by the complex of other valuable characteristics and are used as the starting material in this field of selection.

Improvement of breeding methods

The successful breeding is determined not only by the expansion of the genetic base of a breeding material, but also by the use of reliable methods of evaluation and selection of the valuable genotypes. The testing of the adaptability of breeding material to uncontrolled environmental factors and different pathogens is especially important.

To study the influence of the selection background, we developed the program of parallel evaluation of identical hybrid populations in different soil and climatic zones of Russia. Initially (1972-1973) this program was carried out at VNIIKH and three experimental stations, in 1980 it was performed by 16 scientific organizations, and now this long-term study is carried out by 10 institutions. To evaluate the breeding, our breeding centre formed 3-4 identical sets for each studied population and then sent them to different breeding institutions of European part of Russia, Siberia and Far East. All these institutions tested the populations according to the common breeding scheme.

The efficiency of selection was estimated by the number of cultivars, registered in the State Register. Such trials showed that new cultivars are selected from different populations in different ecological regions. Only one case of the sibling selection was registered. Due to such parallel trials in different regions, the share of cultivars, selected from the same populations, significantly increased (by 20.8% and 40% to 1998 and 2002, respectively; Simakov *et al.*, 2003). During the last decade (1998-2007) our institute bred 21 new cultivar, and other 10 cultivars were selected from the same populations in a collaboration with other breeders; the total number of created original cultivars, suitable for different soil and climatic conditions, is 52. The efficiency of selection from the estimated populations increased more than twice.

Some cultivars, selected in distant regions, were shown to be suitable for the Moscow region, where the identical populations have been tested. This fact testifies a significant influence of random factors, decreasing the value of selection. As a whole, the results of performed studies confirm the use of various selection backgrounds to be more promising, than the work with the single background and in the single ecological region. A multiple use of identical hybrid populations in the selection increases the breeding efficiency and significantly decreases financial expenditures for the breeding of the new cultivar.

In recent years we use some new molecular and genetic methods, including PCR analysis, for the genotyping of different potato cultivars, investigation of the variability of meristematic lines, identification of mutations, detection of viruses and viroids, and investigation of the breeding material using DNA markers of genes, determining the resistance to different pathogens. We also started a program of creation of transgenic potato cultivars.

References

Accatino, P., Peloguin, S.J., Cipar, M.S. (1973) Inheritance of potato chip color at the diploid and tetraploid levels of ploidy. Am. pot. J. 50, 335.

Lechnovich, V.S. (1973) A short history of cultivated potato. In: Genetika kartofelya (Potato genetics), M: Nauka, 3-13.

Simakov, E.A., Yashina, I.M., Loginov, I.Ya. (1991) Results of the radiation selection of potato. Selektsia i semenovodstvo, No. 6, 2-4.

Simakov, E.A., Yashina, I.M., Sklyarova, N.P. (2002) New strategy of potato breeding development and its main results. Abst. of Papers and Posters. In: Potatoes Today and Tomorrow, 2002, July 14-19, Hamburg: p. 292.

Simakov, E.A., Glez, V.M., Manankov, V.A., Zhuravlev, A.A., Ilyicheva, A.A., Pisarenko, E.K. (2006) Laboratory methods of estimation of the potato resistance to potato cyst nematode. M: RASHN, p. 19.

Sklyarov, N.P. and Yashina, I.M. (1973) Inheritance of the resistance to virus Y in potato wild species *Solanum chacoense* Bitt. and *Solanum commersonii* Dun. In: Cytologia i Genetica (Cytology and Genetics), Kiev, 132-135.

Yashina, I.M. (1968) Inheritance of the field resistance to *P. infestans* in different potato cultivars and hybrids. Genetika, 4(6), 5-12.

Yashina, I.M. (1982) Genetic prerequisites for the breeding of potato cultivars with a high starch concentration. Genetika 18(7), 1135-1143.

Yashina, I.M., Kirsanova, E.V., Palamarchuk, T.Yu. (1987) Use of wild potato species in the selection for the resistance to potato cyst nematode. Plodoovoschnoe hozyaistvo, No. 5, 31-35.

Yashina, I.M., Yurieva, N.O. (1992) Genetic principles of the potato selection for the suitability to the processing. Selektsia i semenovodstvo, No. 1, 11-15.

Yashina, I.M. and Sklyarova, N.P. (2000) Mendelian relations revealed in a segregating progeny of the auto- and allotetraploid species and hybrids of potato for the economically valuable features. Poster Abstract. Mendel Centenary Congress, March 7-10, 2000, Brno, Czech Republic, haft 47, 40.

Yashina, I.M., Sklyarova, N.P., Simakov, E.A. (1998) Results and prospects of potato genetic resources application to the breeding in different ecological and geographical zones. In: Proc. Inter. Symposium 'Breeding Research on Potatoes', June 23-26, 1998, Gros-Lusewitz, Jargang, H-2, pp. 85-89.

Curriculum vitae – Evgeniy A. Simakov

Evgeniy A. Simakov studied at the Agricultural University in Penza where he received a Master of science degree in 1972. In 1977, after short period of practical work, he was invited to the All-Russian Potato Research Institute, where in 1980 he received a PhD in Biology, in 1977-1985 Evgeniy Simakov worked in the Potato Research Institute as a researcher. In 1985-1996 he worked as a Head of Breeding Laboratory. In 1997-2004 – Head of Breeding and Seed Potato Centre. Since 2004 until present– Director of All-Russian Research Potato Institute. Simakov is a specialist in breeding and seed potato production.

His current research interests include problems of increasing genetic variability, breeding of new potato varieties and rapid multiplication of virus-free original material and basic seed potatoes. At present he is coordinator of the new program of potato research in Russia for the period 2006-2010. E. Simakov took part in national and international potato conferences and congress: Romania (1977), Finland (1999), Poland (2000), Germany (2002, 2006).

Profile – All-Russian Potato Research Institute (VNIIKH)

The All-Russian Potato Research Institute (VNIIKH) was founded in 1930 in order to improve yield and quality of potatoes in Russia. The main objectives of All-Russian Potato Research Institute are coordination of the national potato research and development program; improvement of genetic variability and breeding of new varieties; production and delivering original seed potato of high biological quality; development and promotion of modern potato technologies (growing, harvesting, storage); national and international cooperation development. The Institute includes the following research departments: Genetics, Breeding, Biotechnology, Seed potato, Agrotechnics and Fertilizers, Plant protection, Technology and Mechanization, Storage and processing, Marketing and management, Scientific and Technical Information.

For 75 years the All-Russian Potato Research Institute has been responsible for improving the quality of potatoes in the Russian Federation. The Institute's core business in the coordination of the national potato research program, the improvement of existing potato varieties and breeding of new varieties. Other key tasks are the production of high quality starch potatoes and the development and promotion of modern technologies in the field of breeding, harvesting and storing of potatoes.

The main results of All-Russian Potato Research Institute are the following: The breeders of the Institute created more than 90 potato varieties, amongst which there are new high yielding disease-resistant and very popular varieties, such as Jukovskiy Ranniy, Udacha, Lukianovskiy and others. The Institute annually produces about 80,000 micro-plants *in vitro,* over 1,000,000 mini-tubers on which basis the Institute's experimental farms produce about 10,000 tonnes of original seed material (pre-basic seed material of the varieties recommended for growing in different regions). Up to 2,000 diagnostic kits for testing potatoes for viruses and bacterial pathogens are produces annually for the seed certification program development. Fifty young specialists and farmers and 120 regional trainers annually attend the practical training course organized by Institute.

Breeding of potato cyst nematode resistant varieties in Japan

M. Mori[1], S. Tsuda[1], N. Mukojima[1], A. Kobayashi[1], C. Matsuura-Endo, A. Ohara-Takada[2] and I.S.M. Zaidul[1]
[1]National Agricultural Research Centre for Hokkaido Region, Shinsei, Memuro, Hokkaido, 082-0071 Japan
[2]National Institute of Crop Science, 2-1-18 Kannondai, Tsukuba, Ibaraki, 305-8518 Japan

Abstract

Potatoes were introduced to Japan at the beginning of the 17th century, but extensive cultivation began in Hokkaido, the northernmost island of Japan, at the end of the 19th century. The cultivation area and production volume in Japan are approximately 87,000 ha and 2,750,000 t, respectively. The average nationwide yield per hectare is 32 t, but the yield in Hokkaido surpasses 40 t per ha, the highest level of productivity in the world. In 1972, the potato cyst nematode (PCN) was confirmed for the first time in the Shiribeshi region of Hokkaido and this pathotype was identified as Ro1. The infested area has been continuing to increase and now exceeds more than 9,000 ha. Beginning with 'Tunica' which was introduced in 1981 from the former East Germany, since then more than 30 PCN resistant varieties have been bred in domestically and also introduced from other countries. The cultivation area of PCN resistant varieties has exceeded 8,000 ha and this is increasing every year.

Keywords: cyst nematode, resistant variety, breeding, cultivation, pathotype

Introduction to Japan

Potato is said to be originated from the Andes, which are about 3000 to 4000 meters above the sea level in South America. There were some civilizations leading to the Inca Civilization, and the diet in those civilizations was based on potatoes and corns whose origins were also in South America. When the Spanish conquered the Inca Empire at the end of 16th century, they brought potato to Europe with them. At first potato was raised for admiration. As potato grows well even in cool weather, it went around the whole of Europe and also spread to different places in the world owing to the overseas expansion of Holland and other European countries (Burton 1989). In Japan, it is said that potatoes were first brought by a Dutchman from Java (Jacanta), Indonesia to Nagasaki in Kyushu (Southern island) at the beginning of the 17th century. Since then potato had not been widely cultivated in Japan, except for the cool mountainous areas of the Honshu (Main Island), until the Meiji Government promoted European style agriculture in Hokkaido at the end of 19th century. Potatoes became an important food for settlers in the early days of Hokkaido's development. Taking advantage of World War I in Europe, starch became an international commodity to export from Japan, and as a result, the planting of potatoes for producing starch were rapidly increased. Hokkaido has become a major place for production since then. In contrast, potatoes are cultivated as a fresh vegetable in the warm area mainly from April to July, which is the off-crop season in Hokkaido (Mori 2001, Mori and Hayashi 2006).

State of domestic production and consumption in Japan

In 2005, the yield per ha in Japan as a whole was 32 t (Table 1). In the case of only Hokkaido, however, it exceeded 40 t and the yield, which is the highest level in the world. The cultivated area of potato is about 87,000 ha, of which about 64% (i. e., 55700 ha.) is in Hokkaido, and the total production is about 2,750,000 t. The productivity has been raised for the following three reasons. Firstly, multiplication systems of healthy successful seed potato, which are not damaged by virus diseases and insects, and a plant protection organization to check the soundness of successful seed potato, have been constructed. Secondly, chemical fertilization and agricultural chemicals have become available at reasonable prices and the technologies of suitable fertilizing and of pest controls in cultivation have advanced (Nakao and Sayama 2001). Finally, these production bases have provided a foundation for breeding high yield varieties, which have enough quality to accommodate the trend variations of demand and supply as well as the pest resistance (Mori 2004).

The per capita annual consumption was 13.4 kg in 1970, since then, consumption has increased gradually and reached 18.0 kg in 2000, but in recent year. This trend is changed to decrease slightly. The increase of consumption has been accompanied by the changes in consumption styles also. In the past, fresh potatoes were purchased mainly for cooking at home, but at present, less fresh potatoes are cooked at home and more processed foods are purchased (Ohara-Takada and Matsuura-Endo 2001). The consumption of potatoes for processed foods has greatly increased since 1970 (Figure 1). The consumption of potato chips in 1980 became ten times compared to that of ten years before and it has been maintaining the same level since then. The consumption of frozen fried potatoes has also increased because of the development of hamburger restaurant chains, but most of the frozen foods are imported from the United States

Table 1. Demand and supply of potato as food and its cultivated area with yield per ha in Japan.

Fiscal year	Cultivated area	Yield per ha	Domestic production	Foreign trade		Amount for domestic consumption				Gross consumption per capita
				Import	Export	Total	Table	Starch	Processing	
	1,000 ha	t	1,000 t	1,000 t	1,000 t	1,000 t	1,000 t	1,000 t	1,000 t	kg/year
1970	159	23	3.611	0	6	3.605	1.390	1.365	0	13,4
1980	123	28	3.421	211	0	3.632	1.159	1.417	581	14,9
1990	116	31	3.552	392	2	3.942	1.183	1.280	944	17,2
2000	95	31	2.899	820	3	3.716	936	1.023	1.356	18,0
2005	87	32	2.749	807	1	3.555	792	1.073	1.304	16,7

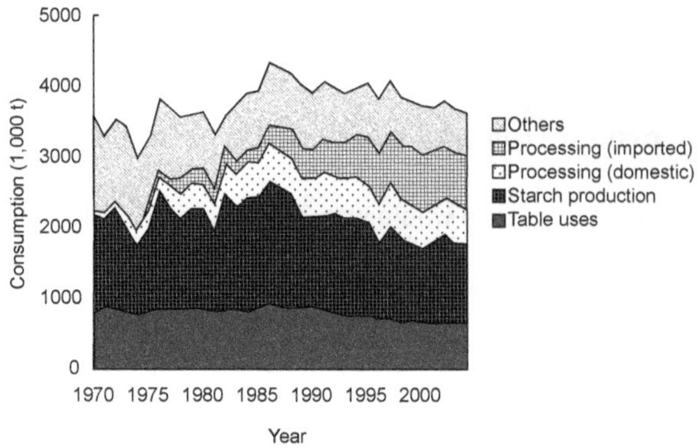

Figure 1. Consumption of potatoes for various uses in Japan.

and other countries because those are cheap as well as supplied stability. Since 1990, more frozen croquettes and packed salad have been consumed and the demand of first processing, including peeling and cutting, has increased in food business sector. These products are mainly consumed at fast food restaurants, at convenient stores and others which feature take-out food (delicatessen). The consumption of processed foods, which was negligible in 1970 but drastically increased to about 1,304,000 t (including imported) in 2005.

Development progress of potatoes in the 20th century

The Hokkaido Development Commission encouraged the cultivation of potato as a suitable crop, and many varieties were introduced from overseas and evaluated from the end of the 19th century to the beginning of the 20th century (Table 2). In 1901, the cultivated area reached about 17,000 ha and potato occupied an important position as a major upland field farming product. In 1905, the Hokkaido national agricultural experiment station designated seven recommended varieties including Early Rose, Snowflake, Rural New Yorker No. 2, Green Mountain, Chicago Market, Bobby, and Early Beauty of Hebron. Around 1915, Early Rose and Snowflake occupied more than half of the cultivated area, but they gradually became less popular and were not cultivated anymore around 1950. On the other hand, Danshakuimo (Irish Cobbler) was introduced around 1908 and May Queen was introduced around 1917. They are the major varieties for fresh consumption till now (Table 3, Mori 2001).

Cross-breeding between varieties and between related species was started in 1918 and 1939 respectively. Benimaru, which is very high-yielding and easy for cultivation, was bred in 1938 by cross-breeding, and was popularized to replace the position of Pepo (introduced in 1922) and Cimbal's Phöenix (introduced in 1905). This remains as a major variety for producing starch even at present. In 1943 Norin No. 1 was bred. This variety has excellent resistance to diseases and is suitable both for fresh consumption and for producing starch. This replaced the position of the introduced varieties, including Green Mountain and Vermont Gold Coin, and cultivated over a larger area. Later Eniwa (bred in 1961) with high starch value and strong resistance to tuber rot, became popularized. Konafubuki (bred in 1981) with resistance to Y virus and high starch yielding capability has considered for cultivation in a larger area. For fresh consumption,

Table 2. Chronology of introduction of potatoes into Japan and its breeding.

Date	Details
1600	Potato introduced into Nagasaki (presumably in 1576, 1598 or 1600).
1700	Cultivated in the cool mountainous areas on the Honshu (Main Island)
1800	Cultivated in a coastal area of Hokkaido (Northern island). Meiji Government (1868-) promoted European style agriculture in Hokkaido
1900	Many varieties were introduced from foreign countries. National breeding program was started in 1902.
	Danshakuimo (Irish Cobbler, 1908) , May Queen (1917?) were introduced. Benimaru (1938), Norin no.1 (1943) were bred by crossing methods
1945	Defeated in World War II, Japan ruled by General Head Quarter of United Nations.
	Foundation seed farm was constructed (1947). Breeding for double cropping in warm areas started (1947). Quarantine of seed potato started under the Plant Quarantine Law (1951).
1972	Potato Cyst Nematode (PCN) was recognized first time.
	Dejima (1971), Nishiyutaka (1978) for double-cropping were bred. Toyoshiro (1976), Konafubuki (1981), Hokkaikogane (1981) were bred.
	Tunika (1981) with resistance to PCN was introduced from the former East Germany. Kita-akari (1987), Touya (1991) and many varieties with resistance to PCN were bred.
2000	Many varieties with resistance to PCN were bred by public and private programs. Some varieties were introduced from Europe and USA by private companies.

early-maturing Waseshiro was bred in 1974 and has been cultivated mainly in northern Japan (Mori, 2001).

The increase in the demand for processed foods after 1970 was remarkable. First the consumption of potato chips increased and later the consumption of frozen fried potatoes followed. To meet the requirement, Toyoshiro was bred in 1976 and occupies about 90% of the raw materials for potato chips. Hokkaikogane, bred in 1981, is used for producing frozen fried potatoes, has had a tough game with imported potato products.

In warmer areas i.e., the South West, double cropping in spring and autumn, excluding hot summer, is possible. A potato harvested in a cultivation season of spring is used as a seed potato of an autumn cultivation season. To this end, dormancy period of double cropping potato varieties has to be short almost as same as 60-80 days. Also in warmer areas, the varieties should have some heat tolerance, such as less vulnerability to bacterial wilt which tends to break out in hot temperatures, even if they are cultivated only in the cool seasons (Nakao and Sayama, 2001). In Nagasaki, potato has been cultivated since the Meiji period, and Nagasakiaka (Triumph) and Nagasakishiro (details of its introduction is unknown) had been selected among the introduced varieties and been cultivated as popular indigenous varieties. In 1947, the breeding program of the varieties suitable for double cropping in the warm area was started, and in 1955, Unzen and Tachibana were bred and replaced the indigenous varieties. Later, Dejima (in 1971) and Nishiyutaka (in 1978) were bred and have become the leading varieties. At present, most of the potatoes produced in the warm area are shipped from winter to summer,

Table 3. Planted area of potato by variety in Japan (2005).

Variety	Planted area (ha)	Ratio (%)	PCN[1]	Uses
Irish Cobbler	22.453	26	s	Table
Konafubuki	16.865	19	s	Starch
May Queen	10.574	12	s	Table
Toyoshiro	9.253	11	s	Processing
Nishiyutaka	5.870	7	s	Table
Kita-akari	2.612	3	R	Table
Dejima	2.415	3	s	Table
Hokkaikogane	1.935	2	s	Processing
Waseshiro	1.802	2	s	Processing
Benimaru	1.520	2	s	Starch
Norin no.1	1.356	2	s	Processing
Touya	1.203	2	R	Table
Kitahime	1.181	1	R	Processing
Sayaka	1.097	1	R	Processing
Others res.[2]	1.737	2	R	
Others suf.[3]	5.042	5	s	
Total	86.915	100		

[1]Potato Cyst Nematode, R: resistance or S: susceptible.
[2]Others resistant to potato cyst nematode.
[3]Others suffer to potato cyst nematode.

which is the off-crop season in Hokkaido, and they are mostly used for fresh consumption (Mori, 2001).

Breeding scheme of potatoes in Japan

Potato, a crop multiplied from a seed potato piece in vegetative propagation and the major breeding technique is crossing, in the same way as seed propagation crops. It has however, the following unique characteristics. Its gene structure is tetraploids heterozygous and the phenotypes are widely segregated in the later generations in crossing. A genetically heterozygous individual, which is selected from the F1 group by phenotype, can be bred directly as a variety. In the case of inbreeding crops, one crossing seed can be bred through segregation and fixation. But in the case of potato, the number of crossing seeds is equal to the number of segregated variations and the most important thing in breeding is to get many crossing seeds in the first stage and to select a desirable phenotype carefully. Furthermore, many combinations cannot be suitable for crossing because of the difficulty in budding and blooming or of male sterility. Only the plant raised from the crossing seeds is reproductive sexual generation, and the later ones are the vegetative generations (Figure 2). Seedlings are selected only for the qualitative characteristics such as pest resistance, but in the later individual clonal selections, the qualitative characteristics, such as the color of fleshy substance and the shape are the key factors for selection because their

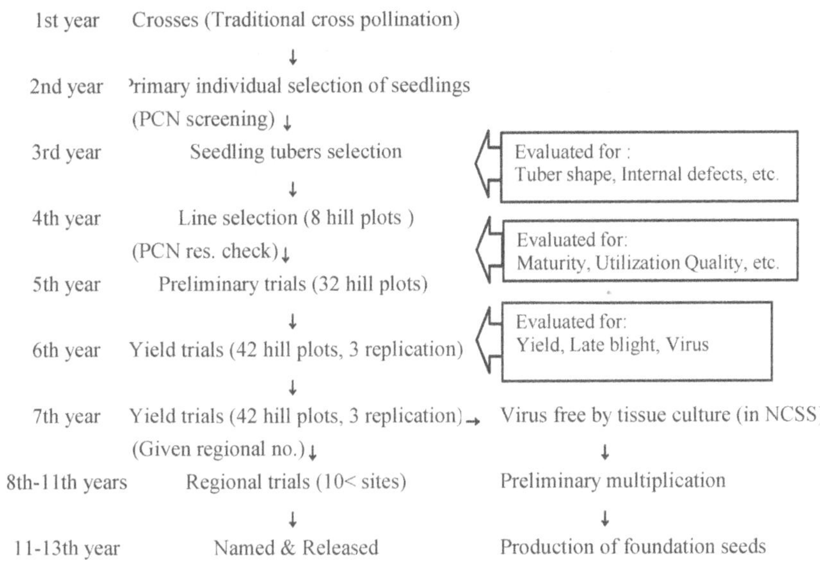

Figure 2. Potato breeding flow chart of NARCH in Japan.

environmental variations are wide. Because of the multiplication rate for one cultivation, is as low as from 10 to 20 times, the multiplication and the selection are conducted at the same time. For selecting the quantitative characteristics like yielding, it takes longer cultivation years. A new variety is bred from about 10 thousand cross seeds in average after more than 10 times of cultivation from first seedlings sowing (Mori, 2001).

Outbreak of potato cyst nematode

In the summer of 1972, the first outbreak of the potato-cyst nematode (PCN, *Globodera rostochiensis* (Woll.) Mulvey and Stone) was identified in Makkari Village, Abuta County, in Hokkaido. Later in 1977, the PCN was also found in the eastern part of Hokkaido. At present, the infested area has been enlarged and it is more than 9,000 ha in the major places of production in Hokkaido, Nagasaki and Aomori prefectures (Figure 3). Since some viable cysts of this PCN were discovered in the guano which was imported from Peru, and had been used in those PCN infested areas that were assumed to be one of the possible means of invasion of this nematode in Japan. The time of invasion was presumed to be in around 1960 for Makkari fields (Inagaki 1984). Furthermore, the pathotype of the PCN population occurring in each place was identified as Ro1, other pathotype was not detected (Nishuzawa *et al.*, 1980).

The cyst (containing eggs) of this nematode has a strong resistance chemically and physically, and maintains its activity for more than 10 years under the natural conditions in the soil. This became a source of infection by moving with a potato and the soil attached to a potato. Therefore, this is considered as one of the most harmful nematode for international plant quarantine, and seed potato production is prohibited in the infested fields (Spears, 1968).

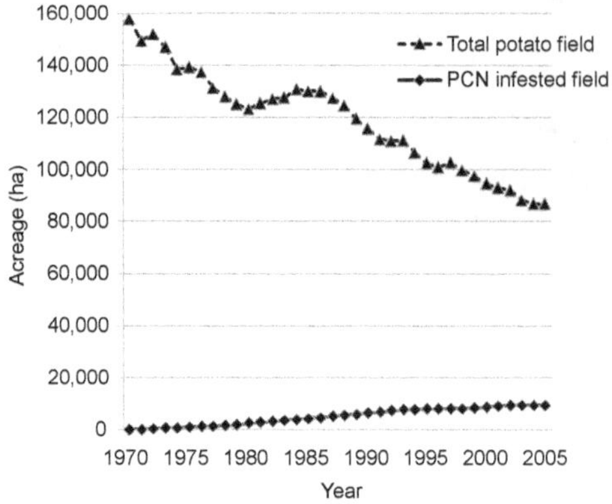

Figure 3. The acreage of total potato field and PCN infested field in Japan.

Breeding of PCN resistant varieties

When the varieties with resistance to PCN are cultivated in infested fields, the density of the nematode in the soil was decreased without chemicals. In addition, no cyst is attached to a produced potato tuber itself. If the movement of the soil attached to the potato can be prevented, area expansion of infested fields through the potato shipped for fresh market can be small. The use of a resistant varieties are effective technique with synergy same as crop rotation and the use of chemicals.

Preliminary examination revealed that all of the Japanese potato varieties were susceptible to the PCN occurring in Hokkaido. Many potato varieties and wild species of *Solanum* had already been introduced from Germany, England, Holland, and other countries, for general breeding purposes. Some of them were used for parental materials in crossing this time. More than 10 varieties with resistance to the PCN, including Tunika introduced in 1981 from former East Germany, were evaluated. Tunika was released as the first PCN resistant variety in Japan by examination of its good result for several years (Inagaki, 1984).

Common method for screening nematode resistance is to grow individual seedlings in nematode infested soil directly. This seedling method needed some unexpected time and labour to make the tubers produced on Fl plants free from the nematodes which were introduced into the pots as inoculums. To overcome this point a new method named 'paired tuber method' was invented (Figure 4). The F1 seedlings, after being screened for the maturity and some tuber traits, were grown in pots in the greenhouse. The tubers produced were harvested individually. Two tubers produced on one Fl plant, were given the same code number and divided into two groups. One group (A) was used for resistance screening test and the other group (B) was stored. The tubers of group (B) which had the same code numbers to those of group (A) which were judged to be nematode resistant in the screening test, and were selected and used for further agronomical examinations. This paired tuber method perfectly got rid of the danger of nematode contamination of the breeding field. The inoculation efficiency was raised up

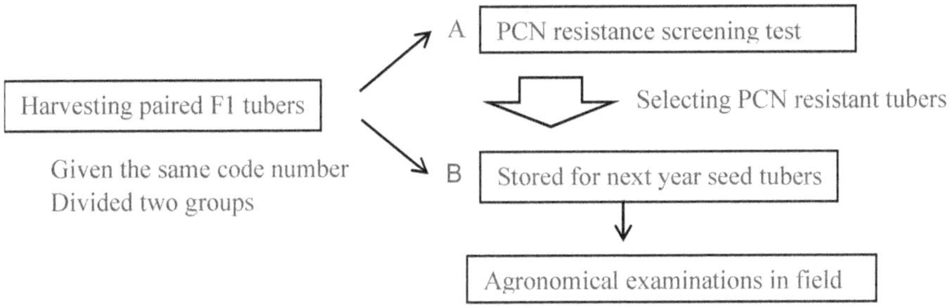

Figure 4. Paired tuber screening method for potato cyst nematode resistance.

because transplanting injury that was avoided although there are some inconveniences may occur (Inagaki, 1984).

True seeds 'R392' derived from a cross of Hudson x Wauseon were provided by Cornell University. R392 found to be resistant more than 90%. This high percentage of resistant individuals was explained by the genotype combination of duplex (Hudson: HHhh) and simplex (Wauseon: Hhhh). From this combination genotypes in F1 were thought to be HHHh (3), HHhh (15), Hhhh (15), and hhhh (3). Then the ratio of resistant individuals was calculated as 33/36, that is, 91.7%. The value obtained in the screening test is 91.4%, which is fitted very closely to this theoretical value. This result in turn indicated the accuracy and high reliability of the screening tests involved in the present study. The tubers obtained from these resistant individuals have been used as parental materials for further breeding programs under the kind permission by Cornell University. From these resistant individuals, triplex line R392-50 was selected. Afterwards many PCN resistant varieties (Table 4) were bred by using R392-50 and its progeny lines (Mori and Umemura, 1992).

Breeding of new varieties with high qualities

Although the enhancement of pest and disease resistance and the yield properties of potatoes are the common goal but at the same time improvement of characteristics and quality according to end use are the challenges to be addressed. Potatoes for fresh consumption, which centre on market shipment, are required to contain

characteristically suitable for different types of preparation, e.g. boiled and mashed potatoes, depending on texture, degree of disintegration, etc., and their appearances must be attractive and appeal to the sensitivity of consumers. On the other hand, potatoes to be used as processed food materials are inextricably linked with the profitability of corporate management, and efficiency is important. It is desirable that potato chips do not brown during deep frying and that product yield is high, and for frozen fried potatoes, that dry matter distribution inside potatoes tubers are even and potatoes are large in size in addition to being suitable for deep frying. Potatoes to be used as starch production require unique starch quality characteristics in addition to high yield (Ohara-Takada and Matsuura-Endo, 2001, Mori, 2004).

Table 4. Potato cyst nematode resistant varieties in Japan.

Year	Table	Processing	Starch
1980	Kita-akari		Tunika*
	Ezo-akari		Toyo-akari
1990	Touya	Musamaru	Astarte*
	Beni-akari	Atlantic*	Sakurafubuki
	Sayaka	Yankee Chipper*	Early Starch
	Hanashibetsu		Prevalent*
	Fugenmaru		
	Star Queen		
2000	Tokachikogane	Kitahime	Natsufubuki
	Star Ruby	Hikaru	
	Haruakari	Yukistubura	
	Aiyutaka	Ohotukuchip	
	Saya-akane	Ranranchip	
	Kitamurasaki	Koganemaru	
	Northern Ruby		

*introduced variety.

The varieties with resistance to potato-cyst nematode include Kita-akari (bred in 1987), whose yellow and powdery fleshy substance is delicious, Touya (bred in 1991) which is early-maturing enabling early shipment, and Sayaka (bred in 1995) which is suitable for salad (Table 5). The cultivated area of these new varieties, which meet the requirements of the consumers' demand and tendency, exceeds 8,000 ha in total and is still increasing (Figure 5).

Toward the future

Firstly an extensive program should be taken to increase the ratio of PCN resistant varieties.

The intensity of PCN infested fields enlarging will continue to increase. Promotion of resistant varieties to overcome this problem requires satisfying the demands for potatoes for commercial and food processing uses with high quality as well as supporting consumption by meeting the demands of consumers and manufacturers through the enhancement of stability from both the production and supply perspectives.

Luckily, the pathotype of PCN population occurring in Japan is only Ro1 (Aihara *et al.*, 1998, Kushida and Momota, 2005), although different pathotype and different species (potato white cyst nematode) occur worldwide. Therefore, the risk management to prepare for these new PCN invasions, breeding of parental materials in crossing must be continued.

Table 5. Characteristics of major potato varieties for cyst nematode resistance in Japan.

Variety	Released year	Breeding station	Utilization	Characteristics in appearance	Note
Kita-akari	1988	Hokkaido	fresh market	flat round, pink eyes	mealy texture, early maturity
Touya	1995	Hokkaido	fresh market	round, shallow eyes	smooth texture, early maturity
Beniakari	1997	Hokkaido	croquette	flat oval, red skin	high yield, suitable for mashing
Sayaka	1998	Hokkaido	salad, fresh	oval, white skin	low discoloration, smooth texture
Early Starch	2000	Hokkaido	starch	short oval, large	high yield
Kitahime	2005	Hokuren	chip (crisps)	round	long storability
Aiyutaka	2006	Nagasaki	fresh market	oval, smooth skin	suitable for double cropping
Nastufubuki	2006	Kitami	starch	round	20% < starch value
Ranranchip	2007	Hokkaido	chip (crisps)	short oval	good flavour
Koganemaru	(2008)	Hokkaido	fry, fresh	long oval, large	high yield, mealy texture

Hokkaido: National Agricultural Research Centre for Hokkaido Region.
Nagasaki: Nagasaki Agriculture and Forestry Experiment Station.
Hokuren: Hokuren Federation of Agricultural Cooperatives.
Kitami: Hokkaido Prefectural Kitami Agricultural Experiment Station.

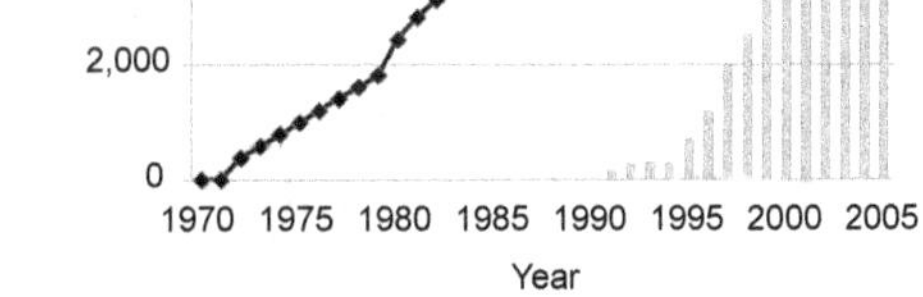

Figure 5. The acreage of PCN infested field and PCN resistant potato varieties in Japan.

Acknowledgments

We thank potato breeders and potato cyst nematode researchers in Japan for the all achievements.

References

Aihara, T., T. Sumiya and K. Suzuki (1998). Studies on the Pathotype of the Potato Cyst Nematode (*Globodera rostochiensis*) in Nagasaki Prefecture. Research Bulletin of the Plant Protection Service, Japan 34: 71-79. (In Japanese).

Burton, W.G. (1989). The Potato 3rd ed. Longman Scientific & Technical: 1-37.

Inagaki, H. (1984). Studies on the ecology and control of the potato cyst nematode, *Globodera rostochiensis*. Research Bulletin of the Hokkaido National Agricultural experiment Station 139: 73-144.

Kushida, A. and Y. Momota (2005). Multiplication of Japanese populations of the potato cyst nematode, Globodera rostochiensis, on H1 resistant potato cultivar. Japanese Journal of Nematology 35(2):87-90. (In Japanese).

Mori, M. and Y. Umemura (1992). Potato breeding for cyst-nematode resistance. Progress in Nematology, A commemoration of the twentieth anniversary of the Japanese Nematological Society: 272-276. (In Japanese).

Mori, M. (2001). Breeding of potato varieties to meet changing demand. Farming Japan 35(5): 10-15.

Mori, M. (2004). Establishing a stable potato production system in Japan and developing new varieties with specific end-use characteristics. Food & Fertilizer Technology Center Extension Bulletin 556: 1-9.

Mori, M. and K. Hayashi (2006). A lecture, Potato (Part 1). Journal of the Japan Association of Food Preservation Scientists 32: 291-296. (In Japanese).

Nakao, T. and M. Sayama (2001). Cultivation and disease and pest control of potato. Farming Japan 35(5): 16-23.

Nishizawa, T., K. Okamoto and M. Ichinohe (1980). An identification of the potato cyst nematode pathotype, Studies on controlling the potato cyst nematode. Agriculture, Forestry and Fisheries Research Council Secretariat. Results of Research 127: 21-34. (In Japanese).

Ohara-Takada, A. and C. Matsuura-Endo (2001). Consumption, use and quality of Potato. Farming Japan 35(5): 24-29.

Spears, J.F. (1968). The golden nematode handbook. Survey, Laboratory, Control and Quarantine Procedures. USDA, ARS, Agric. Handbook No. 353: 81.

Curriculum vitae – M. Mori

Motoyuki Mori – Born in Gifu prefecture, Japan – studied Plant Genetics and Breeding at Gifu University from which he received a Bachelor in 1982. He then joined National Agricultural Research Center for Hokkaido Region (NARCH) as a researcher (1982-1992) and a senior researcher (1993) specializing in potato breeding. He was transferred to Nagasaki Agriculture and Forestry Experiment Station as a chief (1994-1996) of potato breeding and cultivation section and led a research program on the potato double cropping in spring and autumn. After he returned to NARCH as a chief (1997-2005) of Potato Breeding Laboratory. By the organization reorganization, he was promoted to team leader (2006-) of Potato Production and Protection Team. He is leading research programs on the potato breeding of pest and disease resistance and yield properties as well as high quality traits. He bred more than 20 varieties i. e., Touya, Sayaka, Tokachikogane, Ranranchip etc, and that cultivated area as a whole is exceeded 3,000 ha and increasing in every year.

Profile – National Agricultural Research Center for Hokkaido Region

In Hokkaido, the northernmost island of Japan, the agriculture is managed under climatic conditions of cool summers and cold winters. In summer, Hokkaido has favourable agricultural conditions, with long day-length, plenty of sunshine and large day-night temperature differences. Large-scale dairy, paddy rice and upland crop farming has developed utilizing vast land resources, and Hokkaido has become the most important food-supply base in Japan. The National Agricultural Research Center for Hokkaido Region (NARCH), a part of the National Agriculture and Food Research Organization (NARO), carries out research for the development of agricultural techniques suited to Hokkaido such as: breeding of highly adaptable cultivars in Hokkaido; management of agro-environments; control of plant diseases and weeds, physiological research of crop quality; establishment of the utilization and storage of upland crops; investigation of regulatory mechanisms for chilling and cold tolerance. Memuro Upland Farming Research station, one of the departments of the NARCH, has the Potato Breeding Laboratory. We have been developing new potato varieties using hundreds of varieties and lines as the genetic resources. Recently we have bred unique varieties including potatoes with colourful skin/fresh, potatoes that create little glycoalkaloid, and potatoes that are early maturation and can be harvested before late blight damages.

Potato breeding to improve storage and quality

N.W. Kerby[1], M.F.B. Dale[2], J.E. Bradshaw[2] and M.A.Taylor[2]
[1]Mylnefield Research Services Ltd, United Kingdom
[2]Scottish Crop Research Institute, Invergowrie, Dundee DD2 5DA, Scotland, United Kingdom

Abstract

This paper describes recent research at the Scottish Crop Research Institute (SCRI) to combine the development of breeding strategies to produce new improved potato cultivars with enhanced storage and quality traits, specifically reduced low temperature sweetening and reduced potential for the formation of acrylamide in processed potatoes.

Keywords: diagnostics, disease, dormancy, breeding, low temperature sweetening

Introduction

Compared with many other crops, progress in potato breeding appeared to be relatively slow and less successful during the 20th century. This has been partly attributed to the narrow genetic base of the crop, but may also have been due its naturally high harvest index, complex inheritance patterns resulting from the possession of four sets of chromosomes, and the rather conservative approaches taken by 20th century potato breeders. Potato fresh weight yields in the United Kingdom did, nevertheless, double from 22 to 42.5 t/ha over the forty-five year period from 1960 to 2005 (British Potato Council figures), but these increases are largely attributable to disease control, irrigation and other improvements in agronomy, rather than new cultivars. However, progress in improving the efficiency of potato breeding is evident in a number of areas. The application of molecular techniques as well as a fuller genetical knowledge of the important economic traits promises significant advances. Molecular markers are becoming an indispensable tool in plant breeding, genetic and biodiversity studies. Increasingly, the application of molecular marker techniques will facilitate the characterisation of the potato genome, marker assisted breeding, and the characterisation of important economic traits.

In the United Kingdom over half of the potato crop is stored for a number of months to supply retail outlets and processing factories, so safe and effective storage is a priority for the potato industries. Both the quality and quantity of marketable tubers need to be maintained throughout the storage periods to maximise profits and reduce wastage. Potatoes may be stored for periods of up to 10 months. Tuber quality is defined relative to the intended end-use of the raw product. Potatoes intended for processing into French fries and crisps (chips) must meet specific quality standards in terms of low reducing sugar concentrations, minimum dry matter levels and finished product colour. Final Standards may change geographically depending on national or regional consumer preference. Potatoes destined for the fresh market or seed potatoes do not have the same quality standards. Storage should control sprouting and limit the development of diseases that are important in stored seed tubers.

There is a trade off between low temperature storage, reducing sugar accumulation, sprouting and the quality of the end product. Lower storage temperatures delay sprouting, but reducing sugars, mainly glucose and fructose, accumulate more rapidly. This paper describes some of the research at SCRI in the important areas of storage and breeding to improve the nutritional and processing quality of tubers.

Potato tuber dormancy

The processes that lead to stop-start cycles in the growth of the potato tuber apical meristem are important in the control tuberisation and dormancy. Although tuber apical buds exhibit the phenomenon of endodormancy (that is, meristem growth is repressed under apparently favourable conditions for growth), direct control of the length of the dormancy period is difficult. Premature dormancy break can lead to deterioration in quality during potato tuber storage. Sprouting is accompanied by changes that are detrimental to processing, including increases in reducing sugar content, respiration and water loss (Burton, 1989). The length of the post-harvest dormancy period depends on both the genetic background of the cultivar and the prevailing environmental conditions during tuber development (Kotch *et al.*, 1992). As the potato industry diversifies its product range to satisfy consumer demand for salad potatoes or yellow-fleshed *Solanum phureja* cultivars, tuber dormancy has become even more of a problem. Some *S. phureja* cultivars, were originally selected for continual cropping in South American environments and consequently have extremely low levels of tuber dormancy creating problems for all year round consumer supply. There is a growing awareness that CIPC treatment is expensive and leaves chemical residues in the food product, stimulating interest in the mechanisms underpinning dormancy bud break.

As with many plant developmental processes, roles for the phytohormones in the control of dormancy have been investigated (reviewed in Claassens and Vreugdenhil, 2000; Fernie and Willmitzer, 2001; Suttle, 2004). Recent results have demonstrated that continuous ethylene treatment can be used as an effective sprout suppressant in commercial stores (Prange *et al.*, 2005), whereas short-term ethylene treatment can prematurely terminate tuber dormancy (Rylski *et al.*, 1974). Thus although some effects of the phytohormones have been described, limited information exists on the mechanisms underpinning bud dormancy in potato and particularly on the genetic and molecular processes involved. In particular only limited forward and reverse genetic approaches have been applied to the investigation of the roles of the plant phytohormones, although such approaches are now technically feasible.

Several research avenues are currently being explored to enhance our understanding of the control of tuber dormancy. A multidisciplinary approach recently developed at SCRI led to the concept of the potato tuber life cycle being controlled by cycles of meristem activation and deactivation, mediated via symplastic association and disassociation of the tuber apical bud (Viola *et al.*, 2001). Thus on dormancy release, the apical bud regains symplastic connection with the tuber and growth resumes. Subsequent work examining dormancy release in buds of the mature tuber identified molecular markers in potato tuber buds that were induced or repressed specifically on release from endodormancy, in some cases prior to any visible sign of growth (Faivre-Rampant *et al.*, 2004).

Other studies have demonstrated that potato tuber sprouting can be controlled by manipulation of carbohydrate metabolism (Sonnewald, 2001). Tuber sprout growth is initially supported by energy captured from sucrose breakdown. As inorganic pyrophosphate is a necessary co-factor for sucrose breakdown, removal of pyrophosphate by expression of a bacterial pyrophosphatase in transgenic tubers increases sucrose content and prevents its use as an energy supplier. Consequently, sprout growth is significantly inhibited when sucrose is limited but accelerated when sucrose supply is increased. Transgenic approaches have also started to address the role of the plastid-derived isoprenoid hormones (particularly, cytokinins, gibberellins and abscisic acid) in the control of potato tuber dormancy. For example, over-expression of the gene encoding the first step of the isoprenoid biosynthetic pathway (catalysed by 1-deoxy-D-xylulose 5-phosphate synthase) leads to accelerated sprouting, probably due to an increase in the ratio of cytokinins to gibberellins (Morris *et al.*, 2006a). This provides an understanding of the steps of the pathway that should be targeted for molecular breeding approaches, such as allele mining of populations that exhibit variation in the trait of interest.

Tuber nutritional quality

SCRI has focused on a number of important quality traits, including the isoprenoid metabolism in potato. Isoprenoids are a large group of metabolites that impact directly on nutritional quality (for example carotenoid content) and also on plant growth and development (for example gibberellins, cytokinins). Understanding the regulation of this metabolic network remains a major challenge. Using a transgenic approach SCRI has altered isoprenoid metabolism in potato tubers and obtained several phenotypes of interest such as enhanced tuber beta-carotene content (Ducreux *et al.*, 2005, Morris *et al.*, 2006b). Using profiling methodologies in combination with the transgenics, SCRI is identifying how the isoprenoid metabolic pathways are regulated. In other transgenics we have over-expressed a bacterial gene encoding 1-deoxy-D-xylulose 5 phosphate synthase. Although there was only a small impact on tuber carotenoid profile, major changes occurred in the tuber life-cycle, providing us with clues as to how dormancy and sprouting is controlled. The development of genetic, genomic and metabolomic resources for the potato will be discussed which will allow detailed investigation of important biochemical pathways and will lead to notable improvements in a number of nutritional qualities of the potato.

Acrylamide formation during processing

In 2002 the Swedish National Food Administration announced that acrylamide could be found in foods containing starch that were cooked at high temperatures. Acrylamide is formed in the Maillard reaction as part of the cooking process and can never be completely eliminated. Acrylamides are a family of chemicals known to cause damage to the nervous system of humans at very high levels of exposure and are also potential human carcinogens, though there is no direct evidence that they cause cancer in humans. Acrylamide in foods is largely derived from heat-induced reactions between the R-amino group of the free amino acid asparagine and carbonyl group(s) of reducing sugars (fructose and glucose) at a rate that increases markedly with temperatures (120 to 170 °C). The levels in the food vary depending on the food type, cooking temperature and time. Foods containing acrylamide have been found to be French fries, crisps, coffee, pastries, biscuits, breads, rolls and toasts.

Amrein *et al.* (2003) demonstrated that the acrylamide levels in potato products are strongly correlated with the tuber content of reducing sugars and asparagine. Since asparagine levels are high and exhibit limited variation, the far broader variation of the reducing sugars essentially determines the final levels of acrylamide formation. Reducing sugars are also important substances determining the quality of potato chips and French fries. The storage conditions are even more important than the cultivar. When stored below 8–10 °C, reducing sugar levels in tubers increase. Biedermann *et al.* (2002a, b) found that for the cultivar Erntestolz, 15 days of storage at 4 °C increased the potential acrylamide levels by a factor of 28, i.e. the total sum of the reducing sugars found in the fresh weight increased from 80 to 2250 mg/kg (Biedermann *et al.*, 2002a, b).

The potato variety has a strong influence on the content of acrylamide in potato crisps and French fries. There is a good correlation between the content of acrylamide and glucose levels in potatoes, but little correlation with the content of asparagine is found. The strong relationship between reducing sugar levels and the formation of acrylamide in the variety Agria (r^2=0.9856) has been demonstrated by Amrein *et al.* (2003) (Figure 1). The amount of acrylamide also depends on the nature of the processed product e.g. coarse cut French fries tend to have lower amounts of acrylamide than fine cut French fries.

Breeding efforts in SCRI programmes to produce processing varieties include selection for storage at low temperatures (e.g. 4 °C) for prolonged periods without accumulating reducing sugars. Given the significant relationship observed in Figure 1 it is evident that the reduction in reducing sugar levels in new potato varieties approaching the market may reduce potential acrylamide levels in processed products.

Breeding strategies for lower reducing sugar levels

With an increasing proportion of the European potato crop being processed (25-30%) into value added products with the UK at approximately 50% (British Potato Council, 2004). It is increasingly evident that new products have stringent quality targets and require higher standards of certain attributes of the original raw product/germplasm entering the factories. Potatoes for processing in Great Britain are harvested in autumn and stored throughout the

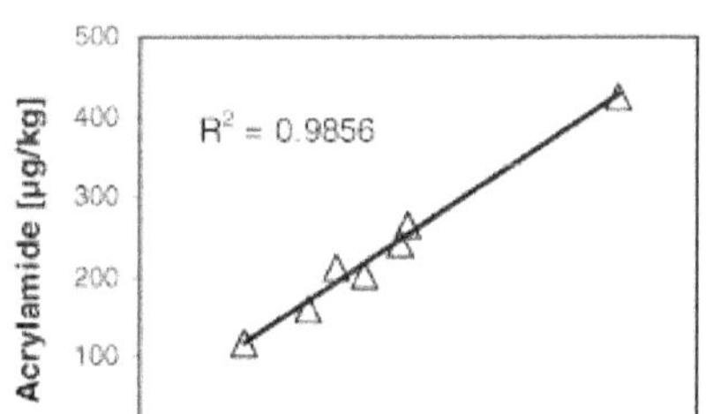

Figure 1. Concentrations of reducing sugars and potentials of acrylamide formation for seven tubers from the same lot of Agria potato (Amrein et al.*, 2003).*

winter and spring to maintain a supply to the factories until fresh produce becomes available again in summer. Tubers have to be stored between 2-4 °C to maintain them in good condition, to prevent sprouting and minimise disease problems. However, at these temperatures starch reserves are normally mobilised and there is a rapid accumulation of reducing sugars. When the tubers are fried, a complex Maillard reaction occurs between the reducing sugars and some amino acids and results in a dark bitter-tasting product. Consequently, manufacturers have to store tubers at higher temperatures (>6.5 °C), thus adding to their costs and exacerbating disease problems.

Within a project to investigate material for processing ability across a range of diverse environments, the reducing sugar contents of tubers from 33 *Solanum tuberosum* genotypes, grown at five sites in Europe in each of two years, were analysed both at harvest and after storage at 4 °C and 10 °C. There were notable differences between genotypes with regard to their processing ability at harvest and after storage at 4 °C and at 10 °C. Significant differences were observed between the genotypes and also between the harvest and post-storage values with regard to levels of reducing sugars and the response of the genotypes to the different storage regimes. A sample of the data illustrating the differences found within the material are presented in Table 1, with some genotypes e.g. line 1 *12601* exhibiting lower reducing sugar levels and superior processing colours when compared to industry standard control varieties following storage at 4 °C.

Previous research at SCRI identified sources of low temperature, sugar stability (i.e. resistance to cold-induced sweetening) amongst its diverse gene pool of the cultivated potato. Crosses

Table 1. Fry colour and reducing sugar contents of a range of potato genotypes following 1.harvest, 2. storage at 10 °C for 10 weeks, and 3. storage at 4 °C for 10 weeks.

Genotype	**Harvest**		**12 weeks - 10 °C**		**12 weeks - 4 °C**	
	Fry Col.	**Sugars**	**Fry Col.**	**Sugars**	**Fry Col.**	**Sugars**
Line 1 *12601*	6.5	0.22	7.0	0.05	5.9	0.65
Line 2	6.5	0.19	7.0	0.09	5.7	0.82
Line 3	6.3	0.23	7.0	0.09	5.1	1.05
Saturna	6.7	0.14	6.7	0.09	4.4	1.53
Hermes	6.9	0.19	7.1	0.08	4.2	1.66
Line 5	6.6	0.26	6.7	0.18	5.1	1.08
Line 6	6.5	0.29	6.7	0.42	5.0	1.41
Line 7	6.5	0.26	7.0	0.14	5.5	0.88
Line 8	6.2	0.34	6.7	0.17	5.5	0.75
Line 9	6.2	0.34	5.8	0.30	3.2	3.13
s.e.d.(sites)	0.24	0.058	0.25	0.078	0.02	0.273

Note: Fry colour:1 = black / unacceptable, 9 = light / preferred processing colour. Sugars: reducing sugar levels as % Freeze dry matter – lower levels desirable.

between these clones and other 'normal' breeding clones and cultivars showed that this trait was heritable and culminated in the submission of two sugar stable clones to official trials. Cultivar Brodick was National Listed in 1990 and cultivar Eden in 1991. Subsequent research showed that progenies (families) with superior low temperature storage characteristics could be identified very early in a breeding programme. By determining the sugar stability (assessed as acceptable fry colour) of tuber progenies during storage at 4 °C after their first field grown year, it is possible to select those progenies, which have the greatest likelihood of containing superior clones within two years of sowing true seedlings in the glasshouse. By applying selection for superior clones within these progenies, SCRI research demonstrated that potential cultivars for processing could be identified more efficiently than by selection of clones with some processing potential in a more general breeding programme. Industry funding allowed this new breeding procedure to be put into practice and resulted in five new cultivars being National Listed from 1999 to 2001 (Montrose, Harborough Harvest, Golden Millennium, Scarborough and Tay) (Mackay *et al.*, 1997). These new cultivars all had Brodick as one parent and three had Eden as the other parent. Recent research has shown how the use of mid-parent values and progeny tests can increase the efficiency of potato breeding for combining processing quality with the disease and pest resistance now required for economical and environmentally sustainable potato production (Bradshaw *et al.*, 2003). The five new cultivars, and other germplasm, are therefore being used as parents in industry funded breeding programmes to achieve these new objectives.

More than one cycle of crossing and selection is required to combine many desirable traits into a new cultivar and this can only be achieved more rapidly through new molecular breeding methods made possible by recent advances in genetics and genomics. A big impact on the rate of progress would be the identification of superior clones genotypically as seedlings in the glasshouse and the use of modern methods of rapid multiplication to progress them to commercialisation, whilst also using them as new parents in further hybridisations. This would require molecular-marker assisted selection, or preferably direct recognition of the desired allele at a genetic locus. Prospects for doing this for cold temperature sweetening have increased now that a potato molecular-function map for carbohydrate metabolism and transport has been published (Chen *et al.*, 2001; Menendez *et al.*, 2002).

Further improvement of successful cultivars using *Agrobacterium*-mediated transformation systems is an attractive proposition, despite consumer concerns about genetically modified potatoes. Transgenic approaches to cold temperature sweetening are possible based on an understanding of primary carbohydrate metabolism. First, as stored tubers maintain the potential for starch synthesis, a possible target to reduce cold temperature sweetening would be to stimulate starch synthesis. This should limit the availability of hexose phosphates for sucrose biosynthesis, and hence reduce sugar formation. Thus, Stark *et al.* (1992) increased tuber starch content and lowered the levels of reducing sugars by expressing an *E. coli* glgC16 mutant gene which encodes for the enzyme ADPglucose pyrophosphorylase and increases the production of ADPglosose, which in turn becomes incorporated into the growing starch granule. Second, and perhaps the most obvious target, Greiner *et al.* (1999) were able to minimise the conversion of sucrose to glucose and fructose by expressing a putative vacuolar invertase inhibitor protein from tobacco, called Nt-inhh, in potato plants under the control of the CaMV35S promoter. However, for commercialisation, demonstration of the economic, environmental and health benefits of genetically modified potatoes will be crucial in convincing sceptical consumers of their value.

Conclusions

To improve the storage of potatoes SCRI is conducting research on tuber developmental processes that impact on storage, analyzing how the isoprenoid metabolic pathways are regulated and is developing superior cultivars that can be stored at low temperatures without the deleterious effects of accumulating reducing sugars.

Acknowledgements

Research was funded by Department Environment, Food and Rural Affairs, Scottish Executive Environment and Rural Affairs Department.

References

Amrein, T.M., S. Bachmann, A. Noti, M. Biedermann, M.F. Barbosa, S. Biedermann-Brem, K. Grob, A. Keiser, P. Realini, F. Escher and R. Amadó (2003). Agricultural Food Chemistry 51 (18): 5556 -5560.

Biedermann, M., S. Biedermann-Brem, A. Noti and K. Grob (2002a). Methods for determining the potential of acrylamide formation and its elimination in raw materials for food preparation, such as potatoes. Mitt. Lebensm. Hyg. 93: 653–667.

Biedermann, M., A. Noti, S. Biedermann-Brem, V. Mozzetti and K. Grob (2002b). Experiments on acrylamide formation and possibilities to decrease the potential of acrylamide formation in potatoes. Mitt. Lebensm. Hyg. 93: 668–687.

Bradshaw, J.E., M.F.D. Dale and G.R. Mackay (2003). Use of mid-parent values and progeny tests to increase the efficiency of potato breeding for combined processing quality and disease and pest resistance. Theoretical Applied Genetics 107: 36-42.

Burton, W.G., (1989) The Potato, 3rd Edition, Logman, London.

Chen, X., F. Salamini and C. Gebhardt, (2001). A potato molecular-function map for carbohydrate metabolism and transport. Theoretical Applied Genetetics 102: 284–295.

Claassens, M.M.J., and D.Vreugdenhil, 2000. Is dormancy breaking of potato tubers the reverse of tuber initiation? Potato Research 43: 347-369.

Faivre-Rampant, O, G.J. Bryan, A.G. Roberts, D. Milbourne, R. Viola, M.A. Taylor (2004). Regulated expression of a novel TCP domain transcription factor indicates an involvement in the control of meristem activation processes in *Solanum tuberosum*. Journal of Experimental Botany 55: 951-953.

Fernie, A.R. and L. Wilmitzer, 2001. Molecular and biochemical triggers of potato tuber development. Plant Physiology 127: 1459-1465.

Greiner, S., T. Rausch, U. Sonnewald and K. Herbbers, (1999). Ectopic expression of a tobacco invertase inhibitor homolog prevents cold-induced sweetening of potato tubers. Nature Biotechnology 17: 708–711.

Kotch, G.P., R. Ortiz, and S.J. Peloquin, 1992. Genetic analysis by use of potato haploid populations. Genome 35: 103-108.

Mackay, G.R., D. Todd, J.E. Bradshaw and M.F.B. Dale (1997). The targeted and accelerated breeding of potatoes. SCRI Annual Report 1996/97, pp. 40–45.

Menendez, C.M. E. Ritter, R. Schafer-Pregl, B. Walkemeier, A. Kalde, F. Salamini and C. Gebhardt, (2002). Cold sweetening in diploid potato: mapping quantitative trait loci and candidate genes. Genetics 162: 1423–1434.

Prange, R.K., B.J. Daniels-Lake, J.C. Jeong, M. Binns (2005). Effects of ethylene and 1-methylcyclopropene on potato tuber sprout control and fry color. American Journal of Potato Research 82: 123-128.

Rylski, I., L.Rappapor, and H.K. Pratt (1974). Dual effects of ethylene on potato dormancy and sprout growth. Plant Physiology 53: 658-662.

Sonnewald, U., (2001). Control of potato tuber sprouting. Trends in Plant Science 6: 333-335.

Stark, D.M., K.P. Timmerman, G.F. Barry, J. Preiss, G.M. Kishore (1992). Regulation of the amount of starch in plant tissues by ADP glucose pyrophosphorlase. Science 258: 287-292.

Suttle, J.C., (2004). Physiological regulation of potato tuber dormancy. American Journal of Potato Research 81: 253-262.

Viola, R., A.G. Roberts, S. Haupt, S. Gazzani, R.D. Hancock, N.Marmiroli, G.C. Machray, and K.J. Oparka, 2001. Tuberization in potato involves a switch from apoplastic to symplastic phloem unloading. Plant Cell 13: 385-398.

Cisgenesis: next step in advanced traditional potato breeding

E. Jacobsen
Laboratory of Plant Breeding, Wageningen University, P.O. Box 16, Wageningen, the Netherlands

Abstract

For world food production, potato is crucial and it is grown at an increasing number of places worldwide. Under these circumstances, breeding of high producing new varieties with sufficient quality traits and (a)biotic stress resistances earns high priority. The classical way of variety breeding is not efficient enough to meet all requirements. Many important traits, an increasing number of which originate from wild species, have to be combined or specific traits have to be added to existing varieties. Modern genomics, including marker assisted selection and genetic modification (GM), is a new potential way to improve the effectiveness of variety breeding. In this contribution, serious attention has been given to a special way of breeding GM-potatoes by using cisgenes instead of transgenes. Transgenes are (partly) originating from other organisms or non-crossable species. They are a new alien gene pool for the plant breeder and it is, therefore, reasonable that new regulations are needed for the introduction and release of them in GM-varieties into the environment in order to avoid potential hazards. Recent developments in plant sciences enabled the production of GM-plants with only cisgenes, which are natural genes from the plant itself or from crossable species, therefore they belong to the existing gene pool of the plant breeder. In this paper the potential of cisgenes for effective potato breeding is discussed and it is strongly argued that cisgenic GM-potatoes should be exempted from the EU Directive 2001/18EC. Cisgenic GM-potatoes can be considered to contain micro-translocations or micro-introgressions without linkage drag which are a great improvement to classical introgression and induced translocation breeding. While on the other hand, classical introgressions and induced translocations are always accompanied with linkage drag in a multiple step approach.

Keywords: genomics, marker assisted selection, genetic modification (GM), existing gene pool, EU directive 2001/18EC

Introduction

Potato breeding is important for safeguarding and stimulating potato production all over the world. The new varieties have to fulfil more and more requirements in order to be a real improvement for the existing varieties. It means that many traits have to be combined and a lot of them have to be found in wild species (Ross, 1986). Introgression breeding is needed for transfer of traits from wild species into cultivated potato but is always connected with linkage drag (Jacobsen and Schouten, 2007). It means that not only variety breeding is important but also the procedure of pre-breeding to remove undesired traits and to combine more introgressions in individual plants (Ross, 1986). For many years, the number of seedlings needed for new varieties has been increasing and it definitely should be gradually minimized.

For more effective breeding, new advanced breeding approaches have to be developed which can be found in modern genomics.

The approach of modern genomics

Recent developments in genomics have led to mapping of many genes and traits in all important crops. In potato, an ultra-dense map with more than 10.000 markers was created (van Os, 2005). These mapping activities have the potential to make potato breeding more effective by marker assisted selection (MAS) but further developments in application for different traits is intensively needed.

At the moment, genomics is not only concentrated on genetic mapping but also on unravelling of the whole genome sequence of model plants like *Arabidopsis thaliana*, main crop plants like rice and potato and of pathogens like *Phytophthora infestans*. The above mentioned high density map (van Os, 2005) plays an important role in the global potato genome sequencing consortium(PGSC) in which at the moment 16 countries are involved (Bachem and Visser, 2005). It means that potentially, all genes and their different alleles will be available for plant breeding (Jacobsen and Hutten, 2006). In that case, next step in breeding research is functional analysis of all these genes and alleles and for biotic stress resistance, molecular unravelling of the interaction between pathogens and crop plant. Selected alleles of genes with known function from crop plants or crossable species can directly be used via transformation for improving already existing varieties.

The bottle necks of GM-varieties with transgenes

In theory, GM-potato breeding has a great potential for improving existing varieties with one or a few traits. However, the biggest bottle necks in GM-breeding are the strict GM- regulations and public acceptance (Schouten *et al.*, 2006 a, b and c). These strict regulations have been developed in a period of time when isolated plant genes were hardly available and always bacterial selection markers were needed. Most of the improvements were expected to come from transgenes, natural or heterologous genes of other organisms, coding, for example, for Bt insect resistance (*Bacillus thuringiensis*), virus resistance (coat proteins), herbicide resistance and antibiotic resistance from micro-organisms to meet the 'requirements' of the grower and the consumer.

Antibiotic resistance genes are needed as selection gene for efficient transformation. These selection genes together with the earlier mentioned agricultural genes created at the beginning a 'new gene pool' of plant breeding, called transgenes. It is reasonable to develop separate rules, like the EU Directive 2001/18EC (Anonymous, 2001), for applying these new gene transfer techniques (process) and for domestication of genes from this new alien gene pool into cultivated plants (product). However, because of public acceptance problems, these new GM-regulations are implemented very strict in many countries. As a result, normal SME breeding companies can not easily go for transgenic crops due to the GM-regulation-related obstacles. It is now an exclusive playing field of multinationals which is a less acceptable situation. It is neither reasonable nor logical to implement the same strict rules in GM-plants with genes belonging to the 'existing gene pool' of breeders in comparison with transgenes belonging to the 'new alien gene pool'.

Prospects of GM-varieties with cisgenes

New developments in plant sciences, like genome sequencing of crop plants, functional analysis of isolated genes, efficient isolation of important functional genes by map based cloning or allele mining and marker-free transformation, did change the situation in GM-plant research dramatically. There are more and more genes both from the plant itself and from crossable species available as well nowadays. These genes belong to the 'existing gene pool' of breeders which have already been used for a long time in traditional breeding. These natural genes from the traditional gene pool of breeders are nowadays called 'cisgenes' in order to distinguish them from transgenes. In contrast to introgression and induced translocation breeding of traits, cisgenes can be inserted into the plant without linkage drag, but they are inserted randomly in the genome, as is also the case with insertions by induced translocation breeding (Friebe *et al.*, 1996). In the present regulations for GM-plants, there is no clear distinction between transgenes and cisgenes. It is less efficient and effective for traditional breeders, using the 'existing gene pool', only to apply the traditional multiple step and linkage drag connected introgression or induced translocation approaches for interspecific gene transfer, without trying the new findings in modern cisgenics which is potentially an efficient one step approach. In addition, cisgenics is very attractive for stacking of multiple traits, originating from different species, because of the absence of the earlier mentioned linkage drag problems (Jacobsen and Hutten, 2006). Therefore, cisgenic plants deserve to be exempted from the GM-regulations for their highly comparability with those of traditional introgression and induced translocation breeding and the absence of the annoying linkage drag. A cisgenic insertion has to be considered as a micro-introgression or a micro-translocation without linkage drag. Therefore, cisgenesis should be added in the annex 1B of the Directive 2001/18EC (Anonymous, 2001), where mutations and protoplast fusions between crossable species have been exempted.

The potential of this new way of advanced traditional potato breeding with stacking of cisgenes will be illustrated with results of our long-term research on sustainable resistance to *Phytophthora infestans* in which *R*-genes from different sources and *Avr*-genes from different isolates of the pathogen are playing a key role.

References

Anonymous, (2001). Directive 2001/18EC of the European Parliament and of the Council of 12 March 2001on the deliberate release into the environment of genetically modified organisms and repealing Council Directive 90/220/EEC. Off. J. Eur. Comm. 106, 1-38 (http://europa.eu.int/eur-lex/pri/en/oj/dat/2001/1_ 106/1_ 10620010417en00010038.pdf).

Bachem, C. and Visser,R. (2005). Update on the status of the potato genome-IC sequencing consortium (PGSC). 2. Solanaceae Genome Workshop, Ischia, Italy. P. 109.

Friebe, B. *et al.*, (1996). Characterization of wheat-alien translocations conferring resistance to diseases and pests: current status. Euphytica 91:59-87.

Jacobsen, E. and Hutten, R. (2006). Stacking resistance genes in potato by cisgenesis instead of introgression breeding. In: Potato Delevopments in a changing Europe (Haase, N.U. and Haverkort, A.J., eds) pp 46-57. Wageningen Academic Publishers.

Jacobsen, E. and Schouten, H.J. (2007). Cisgenesis strongly improves introgression breeding and induced translocation breeding of plants. Trends in Biotechnology 219: 219-233.

Os van, J.G. (2005). The construction of an ultra-dense genetic linkage map of potato. PhD thesis, Wageningen University, The Netherlands.

Ross, H. (1986). Potato breeding- Problems and perspectives, Berlin, Hamburg: Verlag Paul Parey.

Schouten, H.J. *et al.*, (2006a). Cisgenic plants should be excluded from GMO regulations. Nat. Biotech. 24: 753.

Schouten, H.J. *et al.*, (2006b). Reply to: 'Cisgenic' as a product designation. Nat. Biotech. 24:1331-1333.

Schouten, H.J. *et al.*, (2006c). Cisgenic plants are similar to traditionally bred plants. EMBO Rep. 7:1-3.

Curriculum vitae – Prof. Dr. Ir. E. Jacobsen

Evert Jacobsen was born on May 20th ,1947 in Arnhem, the Netherlands.
He finished his BSc and MSc study on Plant Breeding at Wageningen University, the Netherlands and his PhD study on Modern Potato Breeding Techniques using 2n Gametes and Tissue Culture in 1978 at the Friedrich Wilhelm University in Bonn, Germany.

His first job (1974-1978) was on modern potato breeding research at the Max-Planck-Institute for Plant Breeding in Cologne, Germany. In the period 1978-1988, he was a staff member at the Department of Genetics of the Groningen University, the Netherlands. His main activities included genetic dissection of nodulation of nitrogen fixing root nodules of pea by selection of induced mutants and development of biotechnological techniques for unravelling starch biosynthesis in potato genetics. The isolation of the amylose-free potato mutant was a remarkable success. Recently the first amylose-free variety Eliane, based on this starch composition mutant, was released in the world. In 1988, he was appointed at Wageningen University (WU) as full professor in Plant Breeding where GMO research on starch was one of the key issues. In the nineties last century, two amylose-free GM-potato varieties were produced out of that research. At a certain moment over 1500 ha of these GM-potatoes were grown in the Netherlands before the NGO's success with their protest in Europe. In 1991, he started the famous interuniversity graduate school of Experimental Plant Sciences and was appointed as the first director of it until the end of 1998. In the period of 1997-2001, he was the director of the Department of Plant Sciences of WU and from 2001-2004, he was the science director of the Plant Sciences Group of Wageningen University and Research Centre (Wageningen UR). At the moment, he is a professor of Plant Breeding, CSO of the Plant Sciences Group and the science director of the scientific subprogram on Inventions of Transforum Agro&Groen . Within the coming years, he will work worldwide on the acceptance and introduction of cisgenesis and of cisgenic crops, including potato. It is wise to exempt cisgenic crops, in parallel with mutants and protoplast fusions between crossable species, from the Directive 2001/18/EC on GMO's because they are representing a next step in traditional potato breeding.

Profile – Wageningen University and Research Centre: Plant Sciences Group

The Plant Sciences Group is part of Wageningen University and Research Centre (Wageningen UR), a cooperation between Wageningen University (WU), research institutes, a training and advisory centre and van Hall Larenstein Polytechnic. Within Wageningen UR, the university and research institutes are collaborating in five areas of expertise: Plant Sciences Group, Animal Sciences Group, Social Sciences Group, Environmental Sciences Group and the Agrotechnology & Food Sciences Group.

The Plant Sciences Group is a collaboration between the Department of Plant Sciences (WU), Plant Research International BV (PRI) and Applied Plant Research BV (PPO).

Within Wageningen UR, the Plant Sciences Group brings together science education and scientific, strategic and applied research in the fields of biology of plants in relation to their environment, of plant-related organisms, and of plant production. The Plant Sciences Group provides expertise and products in the field of healthy foods, sustainable green production chains, biodiversity and plant and environment to students, governmental institutions and industry.

Conservation, evaluation and use in breeding of potato genetic diversity at the N.I. Vavilov Institute of Plant Industry (VIR)

S.D. Kiru, T.A. Gavrilenko, L.I. Kostina, E.V. Rogozina, O.Y. Antonova, E.V. Truskinov, N.A. Shvachko, E.A. Krylova and A.B. Smirnova
N.I. Vavilov Institute of Plant Industry (VIR) 42-44, B. Morskaya Str., 190000 St. Petersburg, Russian Federation, step_kiru@imail. ru

Abstract

The Potato Collection of the N.I. Vavilov Institute of Plant Industry is one of the biggest in the world. It totals about 9000 accessions including 3000 accessions of wild and 3500 of cultivated species, 2100 of bred varieties, 400 of interspecific hybrids and dihaploids. Accessions of cultivated species, hybrid clones and bred varieties are preserved *in vitro*. One hundred potato cultivars stored under *in vitro* conditions have been genotyped with 10 SSR primers. The initial material with a complex of valuable traits including resistance to fungal, viral, bacterial diseases and pests has a special importance for modern breeding programs. The main source of genes that control resistance to diseases and pests are the wild and cultivated species. Complex evaluation of hundreds of samples is carried out annually at VIR and its experimental stations. As a result, valuable material for breeding has been selected. Crossability of wild potato species belonging to the series *Acaulia* Juz., *Glabrescentia* Buk., *Commersoniana* Buk., *Demissa* Buk., *Longipedicellata* Buk., *Simpliciora* Buk., *Megistacroloba* Card. et Hawk., *Transaequatorialia* Buk. (=*Tuberosa* wild Hawkes, =*Bukasoviana* Gorbat.) with species of other series and bred cultivars has been investigated recently. The first generation of hybrids and their subsequent backcrosses and synthetic crossings have been obtained. Germplasm from the VIR collection is permanently requested and widely used by breeders as initial material for creating new varieties.

Keywords: collection, accession, trait, resistance, breeding

Introduction

The collection of the N.I. Vavilov Institute is widely represented by genofound of *Solanum* genus, section Petota: about 170 wild and cultivated potato species, both Russian and foreign breeding cultivars. The field potato collections maintained using traditional methods suffer significant losses, caused by the influence of extreme environmental factors and accumulation of viral and bacterial diseases in plants. Therefore, *in vitro* collection of the plants cleaned and identified by modern methods is an obligatory and integral part of modern banks of genetic resources.

One doubtless advantage of *in vitro* collections is that many hundreds of collection samples can be stored in controllable conditions and exchanged between various genebanks and breeding companies.

One of the major directions of VIR activity –research in pre-breeding area, searching and creation of sources and donors of various valuable traits, including resistance to most dangerous pathogens. Our institute is focusing on research in pre-breeding area, as well as selection and creation of sources and donors of various valuable traits, including resistance to most dangerous pathogens.

Now, in Russia about 90% of potato is grown at little farms and kitchen gardens, where the improvement of agricultural technology with elements of plants protection appears very difficult. In this situation, now the most important question is the creation of universal cultivars that would combine high efficiency, earliness, high table and processing qualities, with high and stable resistance to the most harmful diseases and pests. The most harmful pathogens for potato grown in Russia are late blight (*Phytophthora infestans* (Mont) de Bary), rhizoctonia *Rhizoctonia solani* Kuhn., common scab *Streptomyces scabies*, potato viruses X, Y, S, M, golden potato cyst nematode *Globodera rostochiensis* Woll and Colorado beetle. Breeders require new genetic sources of resistance to these pathogens, and also sources of high yield, earliness and tuber quality. Therefore, the primary goal of the research which have been carried out over the last 6 years (2001-2006) was to comprehensively study the VIR potato collection, and to determine which genotypes have valuable agronomic features (characteristics, properties, traits) and are resistant to the above-mentioned pathogens.

Material and methods

The comprehensive evaluation of the samples from the potato collection was carried out in collaboration with researchers from the All-Russian Plant Protection Institute (RPPI) and was aimed at finding new genetic sources for breeding on resistance to the main potato pathogens. The samples for screening were selected after a preliminary three-year visual evaluation. Laboratory and field screening on resistance to fungal, viral diseases and golden potato cyst nematode was carried out using methods developed by the VIR (Metodicheskie ukazaniya, 1986), RPPI and All-Russian Potato Research Institute. In total, more than 1900 accessions of the potato collection were screened. The evaluation of commercial traits of breeding cultivars and cultivated species was carried out simultaneously in five VIR's experimental stations in various geographical regions: St. Petersburg, Murmansk, Moscow, Tambov Region, Krasnodar. The breeding varieties selected after the preliminary studying evaluation were evaluated examined (studied) using a new method, based on evaluation of the progeny from the self-pollination. According to this method of pre-breeding is recommended to use an initial material with the donor properties, selected on the basis of multistage screening (four stages). The methods of *in vitro* collection maintenance are described by E. V. Truskinov [1987]. PCR method, microsatellite (SSR), RAPD and organelle-specific primers were described earlier (Antonova *et al.*, 2004; Gavrilenko *et al.*, 2007).

Results and discussion

Maintaining, conservation and evaluation of genetic diversity of the VIR potato collection

The world potato collection of the VIR today account totals 9000 samples, including 3000 samples of wild species, 3500 cultivated species, 2100 – breeding varieties, 400 – dihaploids and interspecific hybrids. About 300 old varieties created 60 and more years ago are unique, because they are maintained only in the VIR Potato Genebank. Currently, studies are being carried out to investigate taxonomically important morphological characteristics of the samples and establish the number of chromosomes in them. In 2006, such studies were carried out for 250 samples of cultivated species and 90 samples of closely related wild species. We have almost finished working on passport and evaluation databases that include information on geographical, taxonomic and biological characteristics of the samples. The collection grows each year through the exchange with other genebanks, breeding companies and research institutes.

The VIR *in vitro* potato collection is not large (compared to field collections), consisting of about 300 samples, and has been created as a doublet collection of the virus-free plants representing unique and most valuable material, and also samples which cannot be reproduced by seeds. The majority of samples maintained in *in vitro* conditions are represented by breeding varieties (Russian and foreign). The *in vitro* collection also includes tri- and pentaploid cultivated species (*S. juzepczukii, S. chaucha, S. curtilobum*) which can not be reproduced by seeds and samples that are of particular interest for geneticists and breeders, including samples of wild species and hybrid clones. The *in vitro* collection is constantly growing as new material cleaned using standard methods of meristem culture and chemotherapy comes in each year. Every 2-3 years, indexation of the *in vitro* collection on presence of the main viral diseases (Y, X, M, S, L and Andean viruses) is performed using method IFA analysis. The *in vitro* collection is maintained under minimal growth conditions protected from light at +20°C as microtubers. The samples of plants are stored for 1–3 years (depending on the genotype) without carrying on fresh nutrient media.

Maintenance of large collections requires control and identification of the material maintained. Therefore, since 2000 the VIR has increasingly applied molecular methods based on the use of DNA markers that make it possible to genotype the samples and evaluate their genetic diversity. The genotyping of breeding varieties and South American aborigine potato cultivars on basis SSR of the analysis is carrying out. Ten nuclear microsatellite markers were successfully used for genotyping 100 breeding and 20 Chilean aborigine cultivars from the VIR collection.

The genetic diversity of 100 grades of Russian and foreign breeding potato varieties was analyzed by PCR with organelle-specific primers. 8 various haplotypes were identified in the analysis set. Potato organelle DNA's were shown to be insignificantly polymorphic – the overwhelming number of cultivars (92,9%) related only to two haplotypes, 62 of them had the same 'cultivated' type of cytoplasm (Gavrilenko *et al.*, 2007).

The genetic stability of 20 *in vitro* samples was confirmed by comparing SSR-and RAPD-spectra of plants maintained *in vitro for long periods of time* and their field analogues. The results of this research confirm the reliability of an *in vitro* sample maintenance method, which does not include tissue and cell culture and consists of consecutive stages of propagation by axilliary buds and maintenance of minitubers protected from light.

In vitro collections make it possible to maintain the improved field samples in controllable conditions and can be a source of material for future cryocollections.

Evaluation of the VIR collection for valuable breeding traits

The comprehensive evaluation of the collection for such commercial traits as yield, earliness, starch content, made it possible to identify valuable initial material for breeding.

Yield. The yield evaluation of breeding cultivars identified both Russian and foreign varieties which within 3-5 years exceeded the standard varieties by 60-80%. The following varieties showed the highest yield (compared to standard, Nevsky and Petersburgsky, Aurora, Akrosia, Bouquet, Elizaveta, Elyseyevsky, Lazar', Malinovka, Russkii souvenir, Sparta, Kholmogorsky, Effect, Youbiley Zhukova (Russia); Blakit, Zhouravinka, Zarnitsa, Zdabytak, Lazourit, Odissey, Talisman, Yavor (Byelorussia); Baszta, Bzura, Koga, Triada, Tristar, PS-17036 (Poland); Alwara, Ivetta, Velox (Germany); Koreta, Korneta (Czech Republic). Varieties Udacha and, Lazurite were equally high and stable in terms of years in different geographical conditions and were considered plastic on yield, earliness and table qualities.

Number of tubers - is one of the components of potato yield. The richest sources of this trait are South American cultivated species *S. andigenum, S. rybinii, S. goniocalyx* and *S. stenotomum*. The wide polymorphism characteristic of these species has made it possible to identify the new sources of high number of tubers, namely *S. andigenum* k-3141, k-8194, k-4629, S. *rybinii* k-1713, k-6406, k-7160, k-8592, k-5935, k-5974, k-8596, *S. goniocalyx* k-6507, k-7048, k-8069, k-3154, k-8162, k-4620, k-17961, *S. stenotomum* k-12782, k-10915, k-13257, k-13250, k-10433. The average number of tuber per one plant in these samples is between 25 and 40.

Earliness. Creation of early varieties is an important breeding objective, especially for North and Northwest regions of Russia. Over the last years, the potato collection has been augmented with new early varieties, many of which combine precocity with other valuable traits. After three years of studying the new varieties, the following have been considered to be the best early varieties: Alyona, Bezhitskii, Bryanskii delicatess, Daryonka, Debryansk, Zhavoronok, Zhukovskii rannii, Lakomka, Lina, Ljubava, Pamyaty Osipovoy, Pogarsky, Russkii souvenir, Snegir', Udacha, Kholmogorsky, Effect (Russia), Yavor (Byelorussia), Aster, Bekas, Irga, Harpun, Lena (Poland), Andra, Bonus, Velox (Germany); Kobra, Korela, Korneta, Krasa, Tegal (Czech Republic) etc. Most of them exceeded the standard for this parameter by 20–40%.

High starch content. Byelorussian varieties are a valuable source for breeding on high starch content. It was confirmed by the results of an evaluation on progeny from self-pollination of the varieties that had shown a high degree of inheritance of this trait (60-80% of seedlings). The following cultivars can also be recommended as high starch varieties: Bryanskii nadjozhnyi, Bryanskaya novinka, Golubizna, Nakra (Russia), Alpinist, Atlant, Garant, Zdabytak, Zoubryonok, Lazurit, Miolavitsa, Synthez (Byelorussia), Zarevo (Ukraine), Asaja, Assia (Germany), Agria, Kardal, Karida, Karnico, Vebeca (Netherlands), Ceza (Poland).

Resistance to late blight. This disease is still one of the most harmful and common disease for potato grown in Russia. Therefore, the problem of creating varieties resistant to late blight is one of the most important tasks in potato breeding. Wild species of the *Solanum*_L genus are the main source for breeding late blight-resistant cultivars. The studies of their wide diversity in potato collections make it possible to find new sources of high resistance, both supersensitive type, and horizontal. Therefore, apart from the species that have already been discovered, namely *S. berthaulti, S. bulbocastanum, S. demissum, S. fendlerii, S. medians, S. microdontum, S. michoacanum, S. pinnatisectum, S. polyadenium* and *S. polytrichon,* which have a lot of resistant genotypes, we found genotypes, which had not been identified before as possessing resistant forms: *S. albicans* k-9813, *S. cardiophyllum* k-16828, k-17380, *S. dodsii* k-20705, k-20709, S. *hougasii* k-10515, k-12165, k-18886, *S. okadae* k-20175. It needs to be mentioned that the following samples can be characterized as being highly late blight-resistant: *S. cardiophylum* k-16828, k-10456 and *S. ocadae* k-20177.

Five years of studies have shown that the following species are highly resistant to late blight: *S. albornzii* k-17621, *S. avilesii* k-20410, *S. berthaultiik* k-7637, k-8787, k-23047, *S. brachycarpum* k-21258, *S. capsicibaccatum* k-15155, *S. immite* k-12168, k-16564, k-18109, *S. microdontum* k-19136, *S. ruiz-ceballosii* k-7370, *S. simplicifolium* k-5399, *S. vernei* k-18158, *k-18161,* k-20125 and North-American species: *S. demissum k-23306, k-2332, S. bulbocastanum* k-19048, k-21275, k-21278, k-22677, k-22998, *S. hougasii* k-18886, *S. iopetalum* k-1052, *S. papita* k-16888, k-9145, k-17454, k-21547, k-22592, *S. pinnatisectum* k-17463, k-19758, k-15252, k-15253, *S. polyadenium* k-15256, k-18139, k-19330, *S. polytrichon k-18925, S. stoloniferum* k-19200, k-19201, *S. verrucosum* k-22609. Accessions *S. berthaultii* k-23047, *S. cardiophyllum* k-16828, *S. neoantipovichii* k-8505, *S. ruiz-ceballosii* k-7370 combine resistance of tubers and leaves to late blight (Zoteyeva *et al.*, 2004).

It is important to emphasize the high value of accessions of wild species that are resistant to several pathogens. Accession of *S. pinnatisectum* k-15253 combines resistance to late blight and to three pathotypes of PVY, Colorado beetle and potato tuber moth (Zoteyeva *et al.*, 2004).

An evaluation of breeding varieties identified accessions with high field resistance to late blight (7–8 points). These are Aspiya, Vestik, Loukyanovskii, Nikulinskii, Udacha, Charodey (Russia); Zdabytak, Lasunak, Souzor'e (Byelorussia); Zarevo, Lugovskoy, Lybid' (Ukraine); Ania, Baszta, Dunajec, Grot, Jantar, Klepa, Koga, Lawina, Meduza, Omulev, Triada, Vistula (Poland), Clarissa (Germany); hybrid clones from USA LBR-1, LBR-7, LBR-18, LBR-46, LBR-47.

Resistance to other fungal diseases. Studies on more than 600 accessions of wild and cultivated species from the VIR potato collection that were carried out in cooperation with researchers from Petrozavodsk State University and were aimed at assessing their resistance to common scab and rhizoctonia showed that the following accessions are highly resistant to these diseases: *S. chacoense* k-21321, *S. kurtzianum* k-20038, *S. fendlerii* k-20011, *S. oplocense* k-19145, S. polytrichon k-20087, *S. andigenum* k-1752, k-764, k-3191, k-3895, k-4709, k-4713, k-4716, *S. rybinii* k-9087, k-16534, k-1815, k-9276, k-3375. On resistance to common scab and silver scurf a sources of these traits are revealed among species *S. boliviense (*k-18766), *S. cardiophylum,* (k-21835), *S. hondelmanii* k-20773, *S. jamesii* k-9155, *S. andigenum*: k-11856 (Zoteyeva *et al.*, 2004).

Resistance to potato viral diseases (PV). Studies on accessions from the VIR potato collection conducted in cooperation with the All-Russia Plant Protection Institute and Institute of Plant Acclimatization and Breeding (IHAR, Mlohov, Poland) over the past few years made it possible to single out new genetic sources of resistance to PVY and PVX.

High resistance to PVX was shown by accessions of samples *S. acaule* k-10678, k-10679, k-18002, k-18007, k-18021, *S. bulbocastanum* k-23167, k-24200, *S. microdontum* k-23434, *S. chacoense* k-23232, *S. demissum* k-23321, *S. fendlerii* k-12158, k-23841, *S. cardiophylum* k-21954, k-23278, k-24203, k-24375, *S. jamesii* k-21456. The accessions of *S. acaule* k-9794, *S. berthaultii* k-23047, *S. demissum* k-3345, k-3362, k-3540, *S. polytrichon*, k-5347, k-5682, *S. pinnatisectum* k-4459 and *S. guerreroense* k-18407 combine resistance to PVX and late blight. The latter showed resistance to all three stamms of PVY).

Resistance to PVY. The screening on resistance to PVY made it possible to identify highly resistant accessions of *S. chacoense* k-2731, k-2926, k-3678, k-4236, *S. demissum* k-3362, k-3540, *S. dolychostigma* k-7610, k-7613, *S. guerreroense* k-18407, *S. michoacanum* k-5763, S. *neoantipovitczii* k-8505, *S. pinnatisectum* k-4455, k-4459, k-19157, *S. stoloniferum*, k-3533, k-3554, k-4226, *S. polytrichon* k-5682. Resistance to three stamms of PV was revealed in the samples of wild species *S. stoloniferum* k-5682, *S. neoantipovichii* k-8505, *S. chacoense* k-4236, *S. dolichostigma* k-7610, *S. pinnatisectum* k-4459, *S. polytrichon* k-5347, k-5682 (Zoteyeva *et al.*, 2004).

Many of the accessions identified as being resistant to viruses combine this trait with resistance to other pathogens. Accession k-7610 of *S. dolichostigma* Buk. was shown to be resistant to PVY and highly resistant to Colorado beetle. Resistance to PVY and late blight was shown by accessions of *S. stoloniferum* k-2534, k-3326, k-3527, *S. neoantipovichii* k-8505, *S. pinnatisectum* k-4459 and *S. polytrichon* k-5347. Accession *S. stoloniferum* k-3326 combines high resistance to late blight with resistance to PVY and partial resistance to PVX.

Resistance to golden potato cyst nematode (*G. rostochiensis* Woll.) remains a priority objective in potato breeding in the nearest years. According to L. Kostina *et al.* (2007), now the total number of breeding cultivars resistant to *G. rostochiensis* in the world is more than 600. For example, 240 resistant cultivars have been created in Germany, more than 180 in the Netherlands, 40 in Poland, 24 in England, 40 in Byelorussia, 19 in Russia, 5 in Lithuania, 4 in Ukraine, etc. In 2006, the State Register List of the Russian Federation included only 12 Russian varieties (Aspiya, Desnitza, Zhukovskii rannii, Zavorovskii, Krepysh, Kristall, Loukjanovskii, Pushkinetz etc.). Given a significant decrease in the level of quarantine control over the distribution of potato cyst nematode (PCN), the problem of breeding varieties resistant to this pathogen in Russia remains crucial. In Russia, the pathotype of *G. rosthochiensis* is common. However, since other pathotypes are common in Europe, it is quite probable that eventually they will spread to Russia (Ro_2, Ro_3, Ro_4, Ro_5). Therefore, Russian breeders should work not only on resistance to Ro1 patotype, but start to use in breeding initial material possessing resistance to various pathotypes of *G. rostochiensis* and of other species - *G. pallida* (Pa_1, Pa_2, Pa_3). As an initial material it is recommended to use varieties which transfer in sexual progeny not only genes of resistance to few pathotypes, but also such commercial traits as high yield, earliness, high content of dry matter, table qualities, etc. The pathotypes se are: (1) resistant to pathotypes $Ro_{1,2,3}$ - Allure, Amalfy, Belita, Cordia, Liseta, Mara, Platina, Producent, Red Scarlett, Vebeca,

Veenster; (2) – resistant to $Ro_{1, 2, 3, 4}$ - Amera, Elkana; (3) - resistant to $Ro_{1, 3, 5}$ - Roeslau; (4) to $_{Ro1, 4, 5}$ - Esta; (5) resistant to $Ro_{1,5}$ - Lyra, Wega; (6) resistant to $_{Ro1,2,3,5}$ - Fox, Hilta, Ute; (7) resistant to Ro_{1-5} - Aiko, Arnika, Franzi, Miranda, Ponto; (8) resistant to $Ro_{1, 2, 4, 5}$ - Turbo etc. Varieties resistant to two species of nematode *G. rostochiensis* and *G. pallida* have particular significance in terms of breeding: Ro_1, Pa_2 - Maritima, Ramos; Ro_1, Pa_3 - Drop; Ro_1, $Pa_{1, 3}$ - Vantage; $Ro_{1,5}$, Pa_2 - Heidrun; $Ro_{1, 2, 5}$, Pa_2 - Benol; $Ro_{1, 3, 4}$, Pa_2 - Danva; Ro_{1-3}, Pa_2 - Karida, Karnico, Pansta; Ro_{1-3}, $Pa_{2, 3}$ - Kantara; Ro_{1-4}, Pa_2 - Atrela, Elles, Producent, Promesse, Sante; $Ro_{1, 2, 3, 5,}$ Pa_2 - Tanja; Ro_{1-5}, Pa_2 - Darwina, Proton. These varieties are also valuable because they combine resistance to PCN with others valuable traits (high yield, resistance to viral diseases or late blight). Evaluation of their progeny from self-pollination made it possible to identify a high share of resistant seedlings in a segregating population. PCN-resistant seedlings of cultivars Alcmaria, Granola, Provita, Quarta, Sagitta and hybrid clone SVP (VT) 2 62-33-3 account to more than 60%. Varieties Agria, Amigo, Berber, Elcana, Panda, Ukama, Van Gogh (Netherlands), Drop, Grot, Koga, Lawina (Poland), Juliver (Germany) also showed a high degree of inheritance of resistant cultivars.

Studies of crossability of wild and cultivated species

One of the basic characteristics directly determining the suitability of wild species for breeding is their crossability with a cultivated potato. Currently, the Department of Potato Genetic Resources is focusing on studying the crossability of wild potato species and interspecific hybrids obtained with their use. The VIR Potato Germplasm was studied and used for the purposes of breeding immediately after the first potato accessions had been obtained by expeditions organized in 1925–1927. The first results of these investigations were described by Kameraz in 1936. At the VIR, studies in interspecific hybridization using species with various levels of ploidity and aimed at improving crossability of species were started before the Second World War. It was then that the polyploid forms (2n=96) of tetraploid species *S. acaule* and (2n=48) and diploid *S. rybinii* were created. Later polyploid plants of 30 potato species of 18 series (Bukasov, Kameraz, 1972) were obtained. The polyploid forms of a wild potato were for the first time used for the purposes of practical breeding by Lebedeva. Hybrid combinations with polyploidy forms of species from series *Transaequatorialia* Buk., *Cuneolata* Hawk., *Megistacroloba* Card. et Hawk. *Longipedicellata* Buk., *Verrucosa* Buk. were synthesized by Zhitlova. Some of them were subsequently used as an initial material for creating new potato varieties in Russia, Byelorussia and Kazakhstan. Vavilova created hybrids by crossing S. tuberosum with species *S. commersonii* (series *Commersoniana* Buk.) and *S. chomatophilum* (series *Chomatophila* Gorbat.) and evaluated them for valuable breeding properties. In order to make diploid wild species suitable for breeding, Budin and Soboleva (Budin *et al.*, 1984; 1988; Soboleva, 1984) hybridized them with dihaploids of breeding varieties and *S. andigenum* Juz. et Buk. clones. Wild species of the far related series *Pinnatisecta* Rydb., *Bulbocastana* Rydb., *Etuberosa* were hybridized by cell engineering. In some combinations, somatic hybrids were successfully crossed with S. tuberosum (Gavrilenko, 2005).

This long and comprehensive research work aimed at using various *Solanum* species in breeding resulted in the creation of several interspecific hybrids that were subsequently used in breeding. Clones of hybrids, in which valuable traits of wild species are combined with valuable commercial traits of cultivated potato, were widely requested by breeding centres. Table 1 shows the wild potato species that were used as sources of resistance to pathogens and the hybrid clones that were created on their basis.

Table 1. Use of wild potato species from the VIR Collection in creation of initial material for breeding on resistance to pathogens.

Series	Species	Interspecific hybrids	Authors
Resistance to *Phytophthora infestans*			
Demissa	*S. demissum*	Y9-95	Kameraz, Patrikeyeva, 1980*
Longipedicellata	*S. stoloniferum*	458-1	Budin, 1989
	S. polytrichon	P-33-1, P-32-13, P-34-6, P-38-3	Zhitlova, Kotova, 1984
	S. vallis-mexici	X9-30, X9-32	Kameraz, Patrikeyeva, 1980
Verrucosa	*S. verrucosum*	B_1 (*S. verrucosum* (2n=48) × *S. tuberosum*) × *S. tuberosum*	Zhitlova, 1989
Transaequatorialia	*S. vernei*	c1-102, c1-103	Bukasov, Kameraz, 1972
Simpliciora	*S. simplicifolium*	B_1 (*S. simplicifolium* 2n=48) × *S. tuberosum*) × *S. tuberosum*	Kameraz, Vavilova, Zhitlova, Ivanova, 1974
Resistance to PVY			
Longipedicellata	*S. stoloniferum*	76-534...542 1980	Naidanova, 1974,
Resistance to PVX			
Acaulia	*S. acaule*	B_1 (*S. acaule* (2n=96) × *S. tuberosum*) × *S. tuberosum*	Kameraz, Vavilova, Zhitlova, Ivanova, 1978
Resistance to PVM			
Transaequatorialia	*S. gourlayi*	B_1 (*S. gourlayi* × *S. tuberosum*) × *S. tuberosum*	Zhitlova, Truskinov, 1984
Megistacroloba	*S. megistacrolob.*	B_1 (*S. megistacrolobum* (2n=48) × *S. tuberosum*) × *S. tuberosum*	Zhitlova 1989
Resistance to *Globodera rostochiensis*			
Transaequatorialia	*S. famatinae*	Л 65/26	Osipova, Evdokimova, 1980
	S. kurtzianum	B_1 (*S. kurtzianum* × *S. tuberosum*) × *S. tuberosum*	Zhitlova 1989
	S. leptophyes	Л 48/67	Osipova, Evdokimova, 1980
	S. multidissectum	B_1 *S. multidissectum* (2n=48) × *S. tuberosum*) × *S. tuberosum*	Zhitlova 1989
	S. vernei	23/71, 25/71-2,11,12	Ponin, 1984

Increased harmfulness of common potato pathogens and, threats from previously insignificant potato pathogens dictate the need to constantly expand the germplasm and search new sources of resistance to extreme environmental factors. In 2004 – 2006. *Solanum* species from South America quite recently described by K. Ochoa that previously had not been sufficiently studied or used in breeding were for the first time used in hybridization by researchers from the N. Vavilov Institute. These are species belonging to series *Tuberosa* Rydb. (Hawkes) or *Bukasoviana* Gorbat. The first generation of hybrids from crossing species *S. alandiae, S. doddsii, S. gandarillasii* with a cultivated potato has been obtained. A number of potentially important hybrid clones (Rogozina 2005) have been selected for breeding purposes. Studies of inheritance of resistance to late blight and PCN in generative progeny of separate clones are now being carried out.

Conclusions

The world potato collection of the N. Vavilov Institute is the basic source of initial material for all breeding purposes. Thanks to the rich diversity of wild and cultivated species made achievable by modern technologies, it is now possible to involve useful traits in relatively short periods of time. Comprehensive evaluation of collection samples makes it possible to identify and create a highly effective initial material for the creation of new cultivars having a complex of valuable properties, including high and stable resistance to the main pathogens. By using the new *Solanum* species in hybridization, initial breeding material can be created.

References

Antonova O.Y., Shvachko N.A., Kostina L.I., Malyshev L.L., T.A. Gavrilenko. Гавриленко. (2004). Geneticheskaya differentziatzia sortov kartofelya s ispolzovaniem SSR markjorov (Genetic differentiation of potato cultivars using SSR markers). Agrarnaya Rossia. N 6, 2004, s. 19-24.

Budin K.Z., Bavyko N.P., Turuleva L.M. Znachenie diploidnykh vidov kartofelya I puti ikh ispolzovania v selektzii (Importance of diploid potato spesies and the ways of their use). In: Diploidnye vidy kartofelya i ikh ispolzovanie v selektzii. Vyp. 145. L. 1984. S. 3 – 8.

Budin K.Z., Kolobayev V.A., Zhitlova N.A. Poisk istochnikov ustoichivosti k fitoftorozu i virusu Y sredi vidov roda Solanum L. In: (Searching sources of resistance to late blight and PVY among SolanumL. species). Selskokhozyajstvennaya biologia. 1988, N 5, S. 40-42.

Bukasov S.M., Kameraz A.Y. Selektzia I semenovodstvo kartofelya (Potato breeding and seed production). Izd. 'Kolos'. L. 1972. 359 s.

Gavrilenko T.A., Antonova O.Y. Kostina L.I. Izuchenie geneticheskogo raznoobraziya kartofelya s ispolzovaniem PCR analiza DNK organell (Studying potato genetic diversity using PCR analizes of organelle DNK). Genetika 2007. (In press).

Gavrilenko T.A. Sozdanie novykh form kulturnykh rastenii na osnove somaticheskoi gibridizatzii (Creating a new forms of cultivated crops on basis of somatic hybridization). In: 'Identifitzirovannyi genofond rastenii i selektziya'. Ed. by Rigin B. V., 2005. St. Petersburg. VIR. S. 628–644.

Zhitlova N.A., Kotova K. 1984, A.S. polytrichon Rydb. – donor Phytophtoroustoichivosti (Solanum polytrichon Rydb. As donor of resistance to late blight)// Diploidnye vidy kartofelya I ikh ispolzovanie v selektzii. Vyp. 145. S. 29-31.

Zhitlova N.A., Trouskinov E.V., 1984, Ob ustoichivosti k virusu M dikogo vida kartofelya *Solanum gourlai* Hawk. i kharaktere eyo nasledovania (About resistance ot PVM of wild potato species *Solanum gourlai* Hawk. and its inheritance. In: Genetika. T. 20. N. 3, S. 463–468.

Zoteyeva N.M., Evstratova L.P. Reaktzia dikikh vidov kartofelya na zarazhenie serebristoi parshoi (*Helmintosporium solani*) (A response of wild potato species to artificial infection by silver scurf (*Helmintosporium solani*). In 'Vestik zaschity rastenii'. St. Peterburg. VIZR. 2004, Vyp. 1, S. 76-80.

Zoteyeva N.M., Khrzhanovska M., Evstratova L.P. Fasulati S.R., Yusoupov T.M. Ustoichivosti obraztzov dikikh vidov kartofelya k boleznyam I vreditelyam (Resistance of accessions of wild species to diseases and pests). In: Katalog mirovoi kollektzii VIR, Vyp. 761. St. Petersburg. VIR 2004. 88 s.

Kameraz A.Y. Vavilova M.A., Zhitlova N.A., V.N. Ivanova. Mezhvidovaya gibridizatzia kartofelya s uchastiem vidov serii Transaequatorialia Buk. i Simpliciora Buk. (Interspecific hybridization of potato with use of species from series Transaequatorialia Buk. and Simpliciora Buk.) In: Trudy po prikl. Bot., gen. I selektz. T. 53, Vyp. 1. 1974. S. 170-193.

Kameraz A.Y. Vavilova M.A., Zhitlova, Ivanova V.N. Mezhvidovaya gibridizatzia kartofelya s uchastiem dikikh vidov serii Acaulia Juz. (Interspecific hybridization of potato with use of species from series Acaulia Juz) In: Trudy po prikl. Bot., gen. I selektz. T. 62, Vyp 1. 1978. S. 92-109.

Kameraz A.Y. Patrikeyeva M.V. Phytophtoroustoichivost' slozhnykh mezhvidovykh gibridov kartofelya (Resistance of complex interspecific hybrids to late blight) In: Bull. VIR. 1980. Vyp. 105. S. 38–41.

Kostina L.I., Fomina V.E., Korolyova L.V., Kosareva O.S. Mnogostupenchatyi screening pri vydelenii iskhodnogo materiala dlya selektzii kartofelya na khozyaistvenno-tzennykh priznaki (Using multistage screening at searching basic material for breeding on commercial traits). in: Sbornik trudov Trudy po prikl. Bot., gen. i selektzii. St. Peterburg. VIR. T. 163. 2007.

Mezhvidovye gibridy kartofelya - geneticheskie istochniki i donory ustoichivosti k patogenam (Interspecific potato hybrids as sources and donors of resistance to pathogens). Katalog mirovoi kollektzii VIR. Vyp. 477. L. 1989. 18s.

Metodicheskie ukazaniya po izucheniyu I podderzhaniyu obraztzov mirovoi kollektzii kartofelya (Guidelines on methodics on evaluation and maintaining of accessions ov potato collection). In: Sost: Budin K.Z., *et all*. L. VIR. 1986.

Najdanova G.M. Ustoichivost' k shtammam virusa Y dikikh vidov i mezhvidovykh gibridov kartofelya (Resistance of interspecific hybrids to any stamms of PVY. In: Selektzia I semenovodstvo kartofelya. Vip. 1 (13). Novosibirsk. 1980. S. 12–19.

Osipova E.A., Evdokimova Z.Z. Iskhodnyi material dlya selektzii kartofelya na ustoichivost' k kartofel'noi nematode (An initial material for breeding on resistance to potato cyst nematode) In: Bull. VIR. 1980. Vyp. 105. S. 20–23.

Rogozina E.V. Yuzhnoamericanskie dikorastuschie vidy kartofelya: Osobennosti ontogeneza i perspectivy ispol'zovania v selektzii (The South-American wild potato species and features of their use in breeding). In: Selskokhozyaistvennaya biologia. 2005. N 5.

Soboleva T.I. Gibridizatzia digaploidov drug s drugom i s diploidnymi vidami (A hybridization between dihaploids and diploid species). In: Diploidnye vidy kartofelya I ikh ispolzovanie v selekytzii. Vyp. 145. L. 1984. S. 14-18.

Truskinov E.V. Podderzhanie I khranenie kollektzionnykh obraztsov kartofelya v usloviyakh *in vitro* (Maintaining and conservation of potato collection in *in vitro* conditions) Metodicheskie ukazaniya. L. 1987.

Curriculum vitae – Stepan D. Kiru

Stepan D. Kiru - studied agronomy at St. Petersburg (former Leningrad) Agrarian University from which he received a MSc in 1977. He then worked for four years for the cooperative agrarian enterprise and after it in N. Vavilov Institute of Plant Industry up to day. He received his PhD degree from the N. Vavilov Institute in 1989 based upon his work on using of diversity of VIR Potato Collection in breeding. The possibility of creating varieties for TPS production with homogeneous progeny, based on effect meiotic polyploidity was shown. He led a research program on evaluation of wide genetic diversity of South American cultivated potato species *Solanum andigenum* Juz. et Buk. Since 2001 he heads the Department of Potato genetic resources of the N. Vavilov Institute of Crop Production, St. Petersburg. He defended a degree of State doctor on biological sciences in 2004 on his work during the last ten years on complex studies of the value of *Solanum andigenum* species for breeding. He is author of over 80 scientific papers. He is a member of the Association of Inergenebank Collaboration (APIC)

Profile – N.I. Vavilov Institute of Plant Industry (VIR) (St. Petersburg, Russia)

The N.I. Vavilov Institute of Plant Industry is the only research institution in Russia whose activities include PGR collection, conservation and study. This Institute, its accomplishments, and role in maintaining the global ex situ collection are well known world-wide. Its global PGR collection represents crop plant diversity encompassing 320 000 accessions of 155 botanical families, 2 532 species of 425 genera. According to the FAO (1998), in terms of collection size VIR occupies the 4th place in the world. This voluminous genetic material serves as a source not only for the national breeding, but for a wider international community as well. VIR also maintains a herbarium of 260 000 specimens. The scientific network of VIR includes the institute's Headquarters with 9 plant resources departments, 13 fundamental research laboratories and 12 experiment stations in different geographic zones of Russia. VIR has established efficient cooperation with more than 40 countries of the world supplying them with various accessions. In results of long-term activity of institute on studying a genetic diversity of cultivated crops and their wild relatives numerous sources of valuable traits for breeding were used in creation of thousands Russian varieties of agricultural crops

Ecological (zonal) models of table potato varieties and gene resources for breeding in Western Siberia

B.N. Dorozhkin[1], N.V. Dergachova[1], L.S. Anoshkina[2], A.D. Saphonova[3] and S.N. Krasnikov[4]
[1]Siberian Research Institute of Agriculture, 644012, Omsk, av. Korolev, 26, Russian Federation
[2]Kemerovski Agricultural Research Institute, 650510, Kemerovskaya oblast, v. Novostroika, Russian Federation
[3]Siberian Plant Breeding and Crop Production Research Institute, 630500 Novosybyrskaya oblast, v. Krasnoobsk, Russian Federation
[4]Siberian Research Institute of Agriculture and Peat, 636464, Tomskaya oblast, v. Kolpashevo, Russian Federation

Abstract

Western Siberia is one of the largest regions of Russia, its population exceeds 15 million residents. The climate is distinguished by a short vegetative period (100-120 frostless days), shortage of rainfalls (190-260 mm in May-August) and extreme high day temperatures in July. Potato is grown here, mainly for food purposes, on 0.3 mln ha, predominantly in the forest-steppe zone, where black soils dominate. Potato breeding is carried out by four regional institutes in three ecological zones. Coming from breeding purposes, ecological (zonal) empirical models of early and second early table potato varieties are developed for forest-steppe and taiga. To construct models, results of regional potato collections studies; breeding material and commercial variety assessment were used. The authors start from the assumption that it is unachievable for present breeding to significantly surpass cultivating variety's potential simultaneously by a number of traits. In fact, substantial progress is reached for some traits, while the rest stay at the former level or undergo small progress. Following this concept, the models exceed standard varieties significantly (at 1.5 score) only for 1-2 traits. For realization of the models by means of breeding, above 100 varieties are selected from regional potato collections as gene sources. Of special interest are varieties of wide adaptivity which are selected in every zone: Alyona, Zarevo, Lazar, Lyubava, Nevski, Pushkinets, Svitanok kievski, Romano, Rosara, Sante and others.

Keywords: table potato, empirical variety model, ecological zones

Introduction

Western Siberia, one of the largest Russian regions (Figure 1), is inhabited by more than 15 mln residents. Potato is grown here on 0.3 mln hectares and is predominantly used for food purposes, its average per capita consumption totals 150 kg. Principal acreages under the crop are located in forest-steppe zone (55° NL) known for its rich black soils.

Because of the short vegetative period, early and second early potato varieties are prevailing. Early tuber production is of priority in potato breeding work carried out by four research centres

Figure 1. Geographical location of West-Siberian research centres engaged in potato breeding

located in various environmental conditions. As soon as about 95% of Siberian potato is grown by private producers, mainly for own consumption, cooking qualities are also very important.

Nature of Western Siberia and potato

Widely spread, various by its climatic and soil conditions, the agricultural territory of the region includes zones from taiga to steppe. The climate is of the continental type. Its character is unstable, because frequently cold air from the North and dry, infrequently hot, from the South, result in sharp fluctuations of the weather at any time interval. The temperature regime is characterized by a severe cold winter, warm (and hot in the South) summer, short-term spring and autumn. The frostless period in the taiga is on the average as short as 105 days, in the forest-steppe zone it amounts to 115-118 days. Accordingly, the mean precipitation rate is equal to 254 and 194-235 mm (Ivanov, 1970). Precipitation and temperature dynamics in the decades of May-August in four locations of the region are shown in Figure 2.

Soil types are diverse in the region, black soils (5-9% of humus content) dominate in the forest-steppe and steppe, whereas podzolic (1,5-2,5% of humus, low content of hydrolysable nitrogen) are prevalent in taiga. Rather common in the region are diseases of fungal (late blight, black scurf, common scab, early blight, fusarium dry rots) and viral origin (PVX, PVS, PVM, PVY, PLRV). Potato cyst nematode (PCN) is extending last decades.

In spite of the severe climate of Western Siberia, its resources in forest-steppe zone (latitude of Omsk, Novosybirsk, Kemerovo) are favourable enough and, in accordance with van der Zaag's calculations (personal correspondence to B. Dorozhkin), permit to grow up to 80 t/ha of potato tubers. In practice, tuber yield reaches 30-40 t/ha in the advanced farms and it is 2-3 times fewer as regional average, mainly because of improper technological providing.

Of more than 50 potato varieties cultivated in the region, the most popular ones are Alyona, Yermak, Zhukovski ranni, Lyubava, Preobski, Pushkinets, Rosara (early), Lyna, Nevski, Svytanok

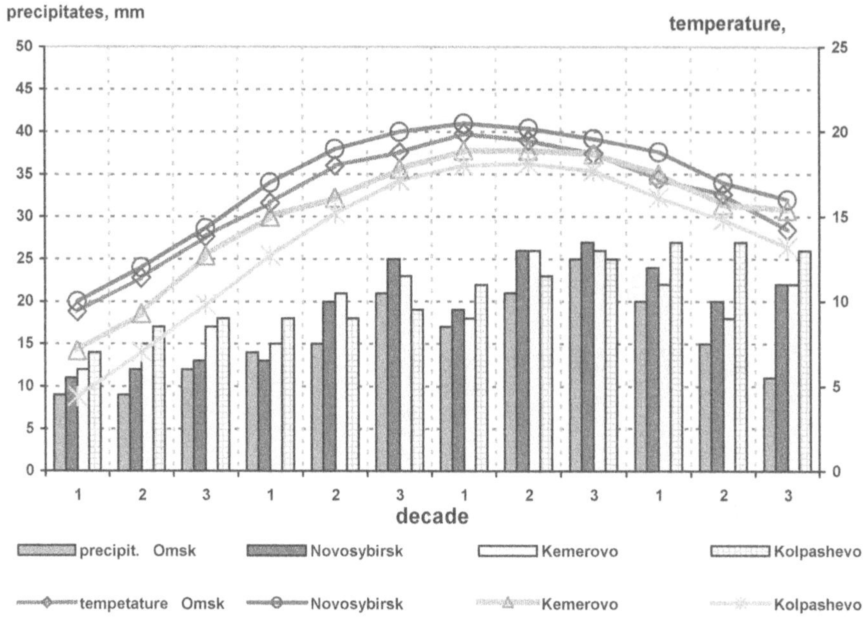

Figure 2. Average data for precipitates and air temperature of vegetative period in four locations of Western Siberia.

kievski, Romano, Secura (second early), Lazar, Lugovskoi, Nakra (intermediate). Despite of wide choice of varieties grown, breeding work is needed to reach higher standards in yield potential and its stability, tuber quality, resistance to most harmful diseases and tolerance to negative abiotic environmental factors.

Variety model

Modelling as a method is widely used in breeding work. Different formulations of the term are proposed, diverse principles and methods are pronounced, various approaches and ways of model realization are used.

Gilbert (1922) noted that plant breeding is worthy of the name only as it sets definite ideals and is able to attain them. Ideals for the perfect potato will change in the different localities. Certain attributes, however, are universally desired: high yield, good quality, disease-resisting capabilities, good keeping qualities etc.

Altergot *et al.* (1974) using the term 'variety model' implied a technical plan for breeding of certain variety. The plan must include detailed description of agronomical, morphological and physiological properties of the future variety and the ways leading to the expected parameters (combinations, methods of selection etc.). It was noted that mathematical models were ineffective for practical breeding purposes and that resulted in the development of descriptive models. The last name the list and quantitative parameters of characters for selection in course of breeding process.

Foltyn (1977) considered that ideotype (which equated to him with a model) must include all important quantitative and qualitative characteristics. The first are to be expressed in absolute values (number, area, weight, ratio), whereas the second – in points on a 9-point scale. The ideotype should not be the result of speculation; its determination should ensue from the confrontation of economic objectives of breeding, ecological conditions of the region, and characteristics of varieties in the world collection. Empirical method, use of science and industry data were permitted at model parameters development.

In compliance with Kumakov (1982), ideotype (ideal type) is ideal plant, variety of future in which high yield, good quality, resistance to unfavourable growing conditions are combined. Variety model is determined as scientific prediction ascribing balance of plant traits needed for ensuring targeted model parameters. The model must include characteristic of growing conditions; confirmation the reality of planned yield; description of the traits bred, their genetic analysis; indication of known gene sources or donors for principal traits etc. Among the feasible methods, empirical one was considered as principal. Following it, experience of practice, breeding and variety trials are generalized, the results are extrapolated.

In the monograph 'What is a variety model?' (Koval *et al.*,2005) it is affirmed that 'breeding remains an art, based on subjectivism, not strictly on set rules'; it is slightly touched by science progress. The authors mark out mathematical, descriptive and speculative model types. It is admitted that constructing a model using precise methods is unreal in near future. In absence of scientific base for variety model constructing we are to resort to more simple designs giving an answer to the principal question: what genotypes are to be selected? Most of the breeders produce their own models heeding agro-ecological conditions and including in them possibly great number of agronomical and biological traits. As soon as it is unreal at present to breed an organism considerably exceeding the existing ones by all the parameters, a model is based upon the principle of gradual variety improvement. The authors of the monograph exceed from the postulate of redundant productivity potential of practically all crops. Each model is to be zonal; each zone needs proper models.

Borodye (2006) considers that a variety must conform to common ecological situation of a concrete region. For example, in Western Siberia draughts of June are typical and that is to be considered in a variety model.

Some authors declared requirements to potato variety model and named parameters of principal traits in accordance to appropriate zones of Russian Federation - Volga river basin (Keeper, 1972), Far East (Brodyanoi, 1977), Non-black soil zone (Recommendation...1983), Mid-Ural (Koksharov, 1985) and of Belarus Republic (Kolyadko, 2003). At that Keeper (1972) noted: 'the requirements mentioned is a single whole from which no chain may be pulled out with no impunity'.

A great many research works are devoted to scientific foundation of models from positions of plant morphology, genetics, physiology etc. (Hruška, 1977; Konyayev, 1980; Dambroth, 1984; Kumakov, 1985; MacKerron, 1985; Haverkort and Kooman, 1977; Yin *et al.*, 2003 a.o.). There is a general agreement among plant breeders and physiologists that physiological knowledge can be applied to improve breeding efficiency in the future (Jackson *et al.*, 1996). At the same time, the present day breeder stands in need of well-grounded zonal variety models of agricultural crops for practical use and those models are to include a whole set of actual agronomical characters.

Materials and methods

Elaborating potato models of early and second early varieties for forest-steppe and taiga zones of Western Siberia, the empirical method of approach was used.

To substantiate the models, the principal results of potato research in the region were used. Potato collection and breeding material study in the course of the last 30 years (above 3 thousand collected cultivars from more than 40 countries, 1,5 thousand hybrid populations) permitted to study the variability of basic traits and to formulate general principles and regional priorities of potato breeding (Dorozhkin, 2004).

The model parameters were set based upon expert approach. There were considered regional trial results of potato collections and breeding material (four research institutes); advanced potato hybrids (12 State Trial Stations). The experience of potato farms and small (family) potato producers of Western Siberia also was taken into account.

It was used 9-score scale of characters evaluation; besides facilitating manipulating with a good lot of a figures, the score-to-score interval of it equals 11,1% and that, to some extent, is heeding high level of variability of quantitative characters.

Modelling is done for three zones where potato breeding is carrying out: forest-steppe of Ob and Irtysh rivers basin (Omsk, Novosybirsk); forest-steppe of Kuznetsk Hollow (Kemerovo) and taiga (Kolpashevo,Tomsk region). As a model standards, early varieties of Siberian origin: Alyona, Lyubava, Antonina and second early Nevski (which is bred in Leningrad region) are taken.

Results and discussion

Before proposing concrete models, the authors have formulated general requirements to early and second early table potato varieties for Western Siberia. For that, there were analyzed results of regional potato breeding, faults of the commercial varieties, and availability of varieties (gene source) for models breeding realization. Permissible parameters of principal 26 traits are presented in Table 1.

Parameters of the characters presented in the table can be attained for regional breeding. The best of the commercial varieties cultivated in the region possess or exceed most of the parameters presented in the table, but none for whole set of 26. E.g. early variety Alyona is noted as early tuber yield producer, it has large tubers of good market and cooking qualities. At the same time, the variety has some faults: low tuber number, liability to outer cracks formation, susceptibility to late blight. Second early Nevski corresponds the requirements to table potato variety at yield, tuber appearance, resistance to late blight and viruses. But, its cooking qualities and starch content are not satisfactory.

Table 1. General requirements to table potato varieties in Western Siberia (early and second early groups).

No.	Traits	Maturity type	
		Early	**Second early**
A. Common			
1	Planting-harvesting time interval, days	70-80	81-90
2	Early tuber productivity (70th day after planting), t/ha	30	25
3	Early tuber productivity, % from the total yield	80	60
4	Standard tubers (> 50g), number	5-8	7-12
5	Standard tuber %	100-140	90-130
6	Tubers uniformity, score	6-8	6-7
7	Tuber shape (length/width index)	1,0-2,0	
8	Eyes depth	shallow to medium	shallow
9	Tuber skin colour	not regulated	
10	Tuber flesh colour	homogeneous	
11	Plant height	middle-low	middle-tall
12	Leafiness	moderate to dense	
13	Growth habit	erect-semi erect	
B. Utilization characteristics			
14	Starch, %	11-17	14-20
15	Ascorbic acid, mg/%	> 15	
16	Row protein, %	> 1,3	
17	Tuber taste	satisfactory-good	satisfactory -excellent
18	Cooking type*	A/B	A/B/C
19	Boiled flesh darkening	none to trace	
C. Resistance to diseases, score			
20	Late blight (foliage)	5-6	6-7
21	Late blight (tubers)	6	6-7
22	Early blight	5-7	
23	Common scab	5-7	
24	Black scurf	5-7	
25	Wart (Dahlem pathotype)	9	
D. Resistance to pests, score			
26	Potato golden cyst nematode (PCN, race 1)**	9	

Note: *is for use as: A – salad; B – frying; C – cooking (Batsanov, 1970)
**desired character

Significant improvement of to-day potato variety simultaneously by a number of traits is unachievable at present because of complex and diversified heredity nature, multiple and discrepant links between the characters. High variability level makes it difficult to identify genotypes in need. In practice, substantial progress is reached for some traits only, while the rest stay at the former level or undergo small progress.

Following this concept six models were worked out, taking into account the peculiarities of steppe and taiga of Western Siberia (Figure 3). The models presented, early and second early for each zone, make no provision for sharp rise of most parameters of standard varieties, they stay unchanged or are progressing at 0,5-1,0 score. Significant (1,5 score) increasing is provided for late blight resistance of early variety foliage (standard variety Alyona) and for starch content and tuber taste of second early variety (standard variety Nevski). As gene source of late blight resistance of early varieties may be used: Pushkinets, Svitanok kievski, Lena, Sentyabr, Odysei, Alvara a.o. Starch content and cooking qualities of second early variety may be improved using Svitanok kievski, Yermak, Alaya zarya, Sentyabr, Adretta and some other varieties as parental forms.

The only characteristic (it is mentioned as desirable in the table), whose rate is increased from 1 (susceptible) to point 9 of the scale, is PCN resistance. It is to be explained, firstly, due to susceptibility of all the standard varieties used and, secondly, with a qualitative status of the character and (in accordance to Yashina, 2000), the dominant mode of its hereditability.

Presented variety models may be used by breeders in forest-steppe and taiga of Western Siberia. The models are to be considered as temporary: after their breeding realization, new improved models are to be worked out.

For realization of the models by means of breeding, above 100 varieties are selected from regional potato collections as gene sources. Of special interest are varieties of wide adaptivity which were distinguished in four breeding centres: Alyona, Zhukovski ranni, Lyubava, Pushkinets (early yield); Lyubava, Nevski, Sante (total yield); Nevski, Romano, Rosara (tuber appearance); Lazar, Nakra, Svitanok kievski, Adretta (starch content); Svitanok kievski, Sentyabr, Adretta (cooking qualities). As a parental forms for breeding to resistance, are recommended Zhukovski ranni, Pushkinets, Agria, Latona, Rosara, Sante (PCN); Zarevo, Lazar, Lugovskoi (late blight); Alaya zarya, Lazar, Svitanok kievski, Amsel, Escort, Katahdin, Romano (viruses); Zarevo, Lazar, Lugovskoi, Svitanok kievski, Agria (early blight); Lazar, Nakra, Svitanok kievski, Rosara, Simphonia (common scab); Alyona, Lazar, Romano, Sante (black scurf). Some varieties from above mentioned are good by a number of agronomical traits: Alyona, Zarevo, Lazar, Lyubava, Nevski, Svitanok kievski, Romano, Rosara, Sante a.o.

Acknowledgments

The research described in this paper was founded by the Russian Academy of Agricultural Sciences. The authors are indebted to Mr. V. S. Doronenko for his kind guidance in virtual reality.

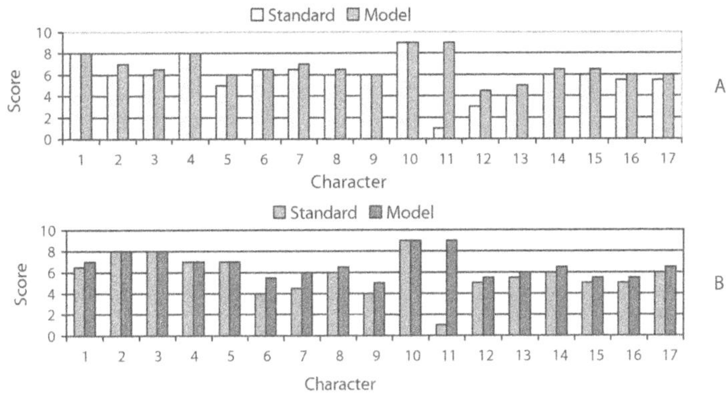

I) Forest-steppe zone - Omsk, Novosybirsk (standards: A - var. Alyona; B - Nevski)

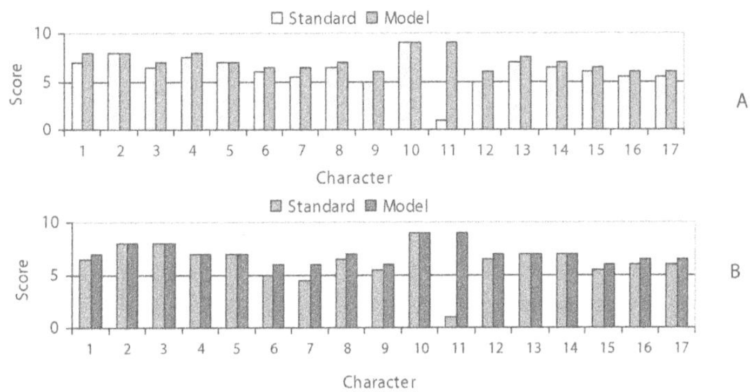

II) Forest-steppe zone of Kuznetsk basin - Kemerovo (standards: A - var. Lyubava; B - Nevski)

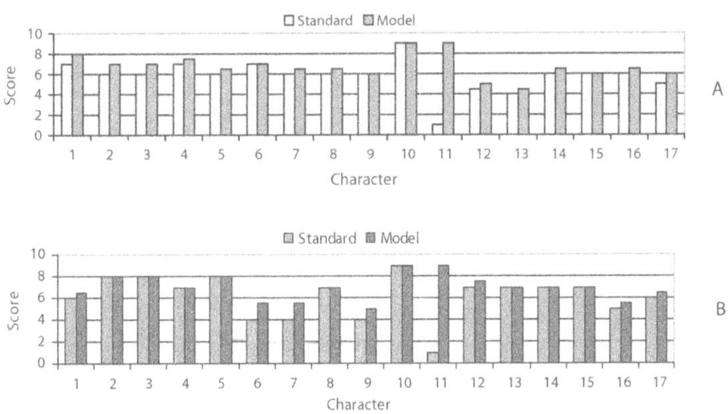

III) taiga - Kolpashevo (standards: A - var. Antonina; B - Nevski)

Figure 3. Ecological models of early and second early table potato varieties (scores).
Note: A – early, B – second early;
The characters – 1 – maturity; 2 – productivity; 3 – tuber appearance; 4 – average tuber weight; 5 – tuber number/plant; 6 – starch content; 7 – tuber taste; 8 – cooked flesh darkening; 9 –mealyness; 10 – wart resistance; 11 – PCN resistance; 12 – late blight resistance (foliage); 13 – late blight resistance (tubers); 14 – virus resistance; 15 – early blight resistance; 16 – common scab resistance; 17 – black scurf resistance

References

Altergot V.F., S.F. Koval, S.S. Mordcovich (1974). Ecological and physiological model of intensive wheat variety. In: Proceedings: Variety and fertilizer. Irkutsk. SiBFIBR, the USSR, 72-82.

Batsanov N.S. (1970). Potato. Ed. by N. S. Batsanov. Moscow. Kolos, the USSR, 376 p.

Borodye Y.G. (2006). A model of intensive variety of spring bread wheat and barley for droughtly forest-steppe and steppe zones of Southern Western Siberia and Northern Kazachstan (physiol..-agron.-breeding foundation). Barnaul, Russian Federation, 390 p.

Brodyanoi V.M. (1977). Potato breeding for earliness. In: Proceed. of DalNIISH, Khabarovsk, the USSR 23: 73-75.

Chiru S., G. Olteanu, T. Bianu, G. Pamfil (1988). Morphological ideotypes with high productive capacity in potato breeding. Biul. Acad. Sc. Agr. Forest. 17: 83-88.

Dambroth M. (1984). Modell einer Kartoffelpflanze mit effektiver Aufband-Ertrag-Relation. Kartoffelbau 35(6): 260-264.

Dorozhkin B.N. (2004). Potato breeding in Western Siberia. Monograph. Omsk: RASHN. Sib. Branch. SibNIISH, Russian Federation, 272 p.

Fishman S., H. Talpaz, R. Winograd e. a. (1985). A Model for simulation of potato growth on the plant community level. Agr. Syst. 18(2): 115-128.

Foltyn J. (1977). Determination of the Quantitative Characteristics of Wheat and Barley Ideotype for Central Europe. Sci. Agric. Bohemoslovaca 9(1): 13-19.

Gilbert A.W. (1922). The Potato (Assisted by M.F. Barrus and D. Dean). New York, The MacMillan Company, 318 p.

Haverkort A.J and Kooman P.L. (1997) The use of systems analysis and modeling of growth and development in potato ideotyping under conditions affecting yields. Euphytica 94: 191-200.

Hruška L. (1977). Přispévek k ideotypu bramborů. Acta Univ. Agr. Fac. Agron. Brno 25(3): 45-52.

Ivanov V.K., S.V. Ivanova, T.M. Korzuchina (1970). Climate of Omsk: Results of 40-years observations. Meteorol. Sta. OmSHI. Omsk. OmSHI, the USSR, 54 p.

Jackson, P., M. Robertson, M. Cooper, and G. Hammer. 1966. The role of physiological understanding in plant breeding, from a breeding perspective. Field Crops Res. 49: 11-37

Keeper I.M. (1972). Breeding and seed production of early potato. Moscow, the USSR, 116 p.

Koksharov V.P. (1985). Breeding of early potato varieties and seed potato production technology. Thesis Dis... Dr. Agric. Sci. 'Breeding and seed production'. Novosybirsk, the USSR, 40 p.

Kolyadko I.I., V.L. Machanko, L.N. Vologdina (2003). Potato breeding to earliness. In: Proceed. of Internat. Conf. devot. to 75-year Jubilee of Belorussian Potato Research Inst. Minsk. Merlit, the Belarus Republic, 1: 34-40.

Konyayev N.F., N.P. Vachrameev (1980). Leaf area formulas and productivity of some potato varieties. In: Proceed.. Novosib. SHI, the USSR, 131: 21-25.

Koval S.F., V.S. Koval, V.M. Chernakov, R.A. Zielke, S.D. Bogdanova (2005). What is variety model ? Monograph. Omsk, Publ. House FGOU VPO OmGAU, Russian Federation, 277 p.

Kumakov V. A. (1982). Models of agricultural crops development. In: Proceed. of USSR science-techn. confer. October 19-21, Odessa. Moscow, the USSR, 35-39.

Kumakov V.A. (1985). Physiological foundation of wheat varieties models. Agropromizdat, Moscow, the USSR, 268 p.

MacKerron D.K.L. (1985). A simple model of potato growth and yield. P. 2. Validation and external sensibility. Agr. Forest Meteorol. 34(4): 285-300.

Yashina I.M. a.o. (1983). Recommendations to potato variety model for Non-black-soil zone of USSR and physiological-biochemical methods of breeding material assessment in practical breeding process. In: Use of physiological methods at breeding material assessment and modeling new crop varieties. Moscow, the USSR, 125-129.

Yashina I.M. (2000). Development and genetical evaluation of new initial material and its effective use in breeding work. Diss...Dr of Agric. Sci. VNIIK, Moscow, Russian Federation, 68 p.

Yin X., P. Stam, Kropff M.J. and A.H.C.M. Schapendonk (2003). Crop modeling, QTL mapping, and their complementary role in plant breeding. Agron. J. 95: 90-98.

Curriculum vitae – Boris N. Dorozhkin

Boris N. Dorozhkin – born in 1938 – studied agronomy at Zhitomir Agricultural Training Institute (Ukraine). His PhD research work was done in Lviv Agricultural Training Institute. It was devoted to table potato breeding in Western Ukraine. In 1967 the young specialist was invited to the Presidium of All-Union Academy of Agricultural Sciences (VASHNIL), Moscow. Here he was engaged as Science Secretary of the Council for Mutual Economic Aid, the Agriculture and Forest Industry Commission.

Since 1972 up to present day Dr. Dorozhkin is working in a potato department of the Siberian Research Institute of Agriculture (SRIA), Omsk. In 1981 the first potato breeding laboratory was developed in the Asian part of the USSR, and B. Dorozhkin was proposed to be the head. From 1994-1998 he was leading the Potato Department of the Institute. After 1998 he combined research work in the Potato Department with administrative activities, being the science director of SRIA. At present he is deputy director of the Institute being responsible for the regional science policy in agriculture and innovative activity.

B. Dorozhkin is the author of over than 150 science publications, monograph 'Potato breeding in Western Siberia', methodical recommendations, textbooks for agricultural universities, public editions. B. Dorozhkin among other reported results of his research work at international conferences, symposiums, work meetings in Peru (CIP), USA (Maryland, Fairbanks Universities), England, Scotland, Denmark, the Netherlands,. Potato varieties bred by the researcher and his SRIA colleagues are grown in Russia and Kazakh Republic. Under his guidance eight PhD potato breeders were educated and now work in the Ural, Western and Eastern Siberia.

Profile – Siberian Research Institute of Agriculture

The Siberian Research Institute of Agriculture (SRIA) was founded in 1828, first as a small experimental station. Now it is a State Research Institution functioning within the framework of the Siberian Branch of the Russian Academy of Agricultural Sciences (RASHN). Approximately hundred researchers of the Institute are engaged in research activities, including plant breeding, zonal farming technologies, animal husbandry, engineering and economy.

The Plant Breeding Centre of SRIA carries out breeding and seed production with more than a dozen agricultural crop such as common bread wheat, durum wheat, winter rye and winter wheat, barley, oats, pea, perennial grasses, potato. Specialists of genetics, physiology, biochemistry, immunology, grain quality are participating in the breeding work.

The dominating crop in the Southern Siberia and Northern Kazakhstan is common spring wheat; over 8 million hectares in Western Siberia and the Kazakh Republic are sown yearly by the varieties bred by the Institute. Over 12 thousand tons of elite seeds of common spring wheat are produced by experimental farms of SRIA situated in a forest-steppe zone.

The main task of the Potato Department of the Institute is breeding of table potato varieties adapted to West-Siberian soil and climate conditions and seed tuber production. The early table potato 'Alyona' is widely grown, from the Perm region to the Far East of Russia and Northern Kazakhstan.

Conditions and prospects of the development of variety selection and seed production in Kuzbass

N.A. Lapshinov, V.I. Kulikova and L.S. Anoshkina
Kemerovo Scientific-Research Institute of Agriculture, the city of Kemerovo, Russia

Abstract

The climate conditions of the Kemerovskaya Region are sharply continental, a frostless period makes it possible to grow the varieties of early and middle-early groups of ripeness successfully. For many years the yield capacity is kept at the level of 9-11 t/h. The main part of the potato crops is in a private sector and 700 ha from 36,000 ha are only in farms and agricultural enterprises. Aiming the rise of the effectiveness of potato producing, the technical re-equipment of potatoes-growing and vegetables-growing farms was done in the Region since 1991 up to 1995, practically all the farms adopted a European technology of potato cultivation and storage. But in spite of the food, forage and technical value and the introduction of new technologies, it hasn't been able to stop the reduction of planting area in the short term and sharply increased the yields. The important role in the increasing of yields and consumer's qualities of potatoes is in creating the varieties adapted to local conditions. The potato selection goals in Kuzbass include the breeding of varieties of early and middle-early groups of ripeness that could have the complex of economically useful qualities for the sharply changeable climate. To keep the selection at a comparatively high and stable level is only possible on the base of new technologies. At present the selection is run with the use of the bio-technology: the hybridization of the improved material; improvement and speeding reproduction of perspective hybrids in the selection nurseries. The most complete realization of the genetic potential of potato varieties is only possible with the production of the seed material of high quality. As a result of a long standing research of the Institute a special technology was elaborated. It is the technology of the improvement and reproduction of the improved given material of high economic quality with the use of the complex of field and laboratory methods, the technology that include several stages. And as a result, the Bank of the Improved Varieties of Potatoes (BISP) was established. In BISP in the crop '*in vitro*' there are 27 adapted to the region and perspective varieties of our own selection. They entered the state list of selection achievements among 10 regions: Lubava, Nevskiy, Tuleevskiy, Udalets, Nakra, Kemerovskiy.

Keywords: biotechnology, improving, pathogens, processing, reproduction

Area and yields

Potato growing has begun to develop in Kuzbass as long ago as the beginning of the last century. Areas planted with potatoes constantly increased, thus it was 5 thousand ha in 1913, 43 thousand ha in 1945, and 85 thousand ha in 1984 including 55 thousand ha belonging to state farms and collective farms (35%). Record potato harvests were also achieved in Kemerovskaya region. In 1942 the team of Anna Kondratevna Yutkina obtained 133,3 t/ha.

During the period of change of Russian economy from command into market the total area planted with potatoes were noticeably reduced. When the average potato productivity in the world was 15 t/ha, it was 9-11t/ha[1] in Russia. That was considerably lower than the potato productivity in Kemerovskaya region, though the main areas belonged to individuals, 5.4% falls to the share of agricultural enterprises (Table 1).

The highest reduction of the areas planted with potatoes falls in the period of 1988-1996. In this period worked through years the classic seed-growing scheme was practically destroyed. The absence of this valuable product in the market in sufficient quantities leads to both economic and social tension.

Variety selection

The transfer of potato production to the private sector leads to changes in the variety composition of the crop. The quality of seed material and phytosanitary conditions of plantations deteriorated. That is why the work of creating potato varieties with certain quality indices is carried out. More and more attention is paid to the varieties having disease resistance, with smooth tubers of good shape and with high gustatory qualities.

The natural conditions of the region allow obtaining potato yields of 30-40t/ha. For that enough quantity of precipitation falls. Soils are rich with humus and retain water well. One of the reasons of low productivity is a slow change and renewal of varieties that leads to the accumulation of pests, viruses, bacterial and fungal infections.

The creating of the varieties of local selection is dictated by the requirements of producers. It is necessary to take into consideration that at present potatoes are basically grown in the

Table 1. The potato production in Kemerovskaya Region.

Year	Area 1000 ha	Agricultural enterprises		Gross yield 1000 t	Agricultural enterprises	
		1000 ha	%		1000 t	%
1984	85,7	30,5	35,6	859	274	32
1988	73,3	27,1	36,9	689	223	32
1992	63,9	17,2	26,9	719	164	23
1996	65,1	4,6	7,0	741	50,2	6,7
2000	64,4	3,9	6,1	862	58,1	6,7
2003	70,8	3,0	4,2	847	44,8	5,3
2005	68,5	3,7	5,4	718	44,6	6,2

private sector. That is why potato varieties must have complex disease and pests resistances, accumulate less pathogens in monoculture, combine high productivity, fast-ripening and quality indices.

It is generally known that the potato varieties created in certain soil-climate conditions and having a complex of positive signs are able to increase potato production. The selection value of local varieties is in their high adoptive potential relatively to a certain region, in an appropriate complex of consumers' qualities and also in a high prominence of some of them.

For the successful selection of certain signs it is necessary to sort out original ancestral genotypes correctly. Taking into consideration all the time growing requirements to potato varieties, it becomes necessary to involve new sources of valuable traits into the selection process. For this annually the complex of useful traits of the ancestral material received from the world collection of All-Russian Institute of Plant-Growing and other scientific institutions is studied. Genetic sources are singled out considering the traits of productivity, early ripening, the content of starch, the disease resistance, the adaptive qualities, the suitability to processing. On the basis of the obtained data a number of perspective hybrids containing useful signs was created. Hybrid 128-99 (Nevskiy×Zarevo) was passed on the State Variety-Testing under the name of Kuznechanka in 2005.

To reduce the breeding grounds of golden nematode of potato (*Globodera rostochiensis*, Woll) it is offered to include only resistant to this disease varieties in the State List of Selection Achievements. In connection with this beginning with 2 000 the ancestral forms having the gene of the resistance to this pest are included into the hybridization. There were obtained hybrid numbers 24-02 (Udacha×Prolisok), 29-02 (Udacha×Prolisok) and 450-03 (Nevskiy×89-1-12) that combine a number of useful traits with the resistance to golden nematode of potato.

A perspective trend in the potato selection is the creation of new varieties suitable for industrial processing and making food semi-finished products. Special requirements are made to the potatoes that go to the processing, that is why it is necessary to select and grow special varieties, the most suitable for making food semi-finished products that can also be used for technical aims.

It is only possible to keep up the selection of a relatively high and stable level using new technologies. At present the selection is carried out with the application of a biotechnology-hybridization of the improved material; improving and rapid reproduction of perspective hybrids in selection nurseries.

According to the agreement with All-Russian Scientific-Research Institute of Potato Growing of scientific and technical co-operation hybrid potato seeds were obtained. As a result of the selection process the hybrids with a complex of positive signs were chosen. Such additional study of hybrid populations at the first stages of the selection allows using the genetic variations of these populations fuller, because in different conditions the genotypes adapted to different conditions of selection are singled out. It gives the possibility to create the varieties with the zone sensibility to ecological conditions. As a result of selection study of hybrid populations the potato varieties of different groups of ripening (Lubava, Nakra, Tuleevskiy, Udalets) have been created and put into the State List of Selection Achievements since 1990 up to 2006. These varieties differ in their signs by high productivity, plasticity, disease resistance. Their short description is below.

Lubava

An early ripening variety of a food purpose. The potential productivity is 55 t/ha. The starch content is 13-17%. It is relatively resistant to scab and Phytophthora infestans. It is weakly infected by p. Alfernaria. The tubers are large ovally rounded, the eyes are of middle depth, the rind is red, from middle to rough the pulp is white. It has a good taste. It is kept well, it has an extended period of storage.

Nakra

A mid-ripening variety of a universal purpose. The potential productivity is 40 t/ha. The starch content is 20-24%. It is highly resistant to scab, relatively resistant to *Phytophtora infestans* and p. Alternaria. It is weakly infected by Colorado beetles. The taste qualities are good. It is kept well. It is suitable for cooking crisp potatoes.

Tuleevskiy

A mid-early ripening variety of food purpose. The potential productivity is 50 t/ha. The starch content is 12-15%. It is relatively resistant to *Phytophthora infestans,* scab and p. Alternaria. The taste qualities are good and excellent. It is kept well.

Udalets

A mid-early ripening variety of food purpose. The potential productivity is 60 t/ha. The starch content is 12-15%. It is relatively resistant to *Phytophthora infestans,* scab and Alternaria. The taste qualities are good. It stores well.

Seed production

The production of high-quality seed material is a central branch of potato production where the genetic potential of varieties is fully realized. The objective of seed potato -production is a variety selected by the complex of economically valuable traits and characteristics. As a rule, a rather large quantity of pathogens (viral, bacterial and fungal, especially in latent form) is accumulated by the moment of transferring a new variety to the State Variety-Tasting. That is why freeing of seed potatoes from pathogens and also the preservation of reproductive characteristics of varieties are ensured by a system of measures of improving the seed material.

The potato improvement is a difficult complex process combining the use of both field and laboratory improving methods including the use of a crop tissue and covering all the cycle of obtaining the original material.

A system of virus free potato seed growing was worked out in the institute. It is based on the use of a plant of an apical meristem in combination with other bio-technological methods. The farms of our region have managed to increase the productivity 1,5-2 times, using the improved seed material for their inner potato seed growing. The reproduction of the improved varieties and the perspective hybrids is realized in the limits of (original) initial potato seed-growing.

The basis of the original seed growing is the liberation of perspective varieties and hybrids from pathogens and other factors of degeneration. The primary stage of improving is obtaining of ancestral improved plants. When creating a perspective hybrid there is a search of a basis clone on a selection plot.

Each plant is subjected to a careful individual visual estimation. And each plant is additionally checked in the period of budding and at the beginning of flowering by the method of immunoferment analysis and by leave tests. Especially prominent clones (with high productivity and big quantity of tubers, free of pathogens by visual estimation and in latent form) are a source to form a laboratory collection *in vitro*. The other part of clones is a source of making initial seed growing till a reliable collection of improved plants is formed (Figure 1).

It is very important for an institute where the selection is led to have a seed potato stock meeting the requirements of the State Standard before making the variety local. At this stage of growing super-super best specimens the method of improving named Clone Selection in combination with the method of immunoferment analysis are used. So to the moment of putting in the State List in 2006 the varieties Tuleevskiy, Udalets, 12th and accordingly 8th of super-super best specimens of seed material meeting the requirements of the State Standard 7001-91 were obtained.

The infecting of potato plants of the selection material of newly created varieties with viral and bacterial diseases in a latent form has reached 5.4% (the variety Tuleevskiy), 48.4% (the variety Udalets) during the first year of improving by the method 'clone selection + immunoferment analysis'. During the following years of growing, clone selection in combination with a visual estimation and immunoferment analysis as an improving method was effective for the variety Tuleevskiy and ineffective for the variety Udalets because of a quick accumulation of infections. This improving method is used for newly created varieties of high value and is used till a reliable stock of improved variety specimens is made (SIPV).

As a result of a many years' research we offer the technology of improving and reproduction of demanded by the market improved ancestral material of high quality. The technology includes a complex of field and laboratory methods and consists of several stages:

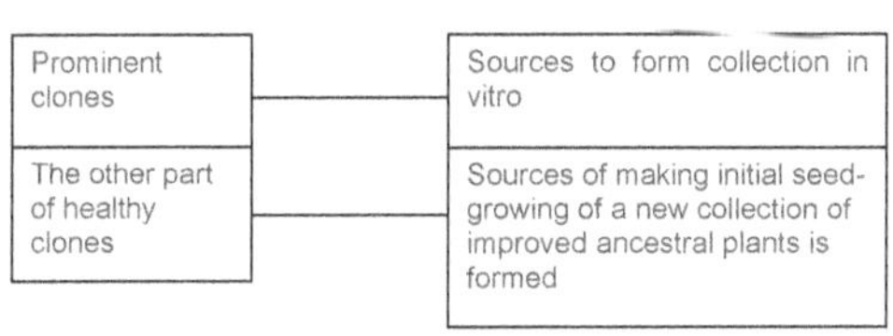

Figure 1. A diagram of the use of selected plants (clones).

- bringing of healthy productive material *in vitro* using the method of apical meristem, thermotherapy after the field evaluation of clone seed material that includes variety typicalness, productivity, fungal infections, viral and bacterial diseases;
- multiple control of latent bacterial and viral infections of plant generations *in vitro* using the method of immunoferment analysis;
- microclone reproduction of variety specimens *in vitro* in nutrient mediums with a periodical renewal of the material (in 2 years);
- obtaining of improved ancestral potato tubers with the use seedling plants grown in non-sterile conditions in boxes and in peat-compost pots with the use of growth regulating preparation chlorcholine;
- revealing of viroids (fungoid viruses) in newly improved seed specimens both in a crop *in vitro* and in a seedling crop in a specially equipped room with a heightened temperature;
- revealing of latent pathogens in improved seedling plants in hothouse and field conditions using the method of immunoferment analysis during harvest time as the first estimation; and the second estimation of the obtained ancestral tubers (clones) using the methods of immunoferment analysis + indexation;
- field reproduction of improved ancestral tuber materials, the production of super-super best specimens in quantities defined by programmes.

In our research the rotation of the improved meristem lines promotes the delay of the process of degeneration of lines and prolongs their vitality, because the most prominent clones are selected for improving. It is necessary to notice that the lines of the same variety kept in a collection can be considerably different in economical terms. Thus the varieties Nakra and Lubava have differences in the quantity of tubers, the mass of a plant and starch content.

In the stock of improved potato varieties (SIPV) in crops *in vitro* there are 27 local and perspective varieties of our own selection put in the State List of Selection Achievements of the region: Lubava, Nevskiy, tuleevskiy, Udalets, Nakra, Kemerovskiy, Kuznechanka.

Clone reproduction of improved potatoes solves the second important task-obtaining of improved material.

A rather big amount of methods of rapid reproduction is known. Their use in protected grounds provides the obtaining of high coefficients of reproduction when producing ancestral tubers. There were obtained in a crop *in vitro* the following amounts of plants: 7466 in 2001, 8000 in 2002, 7500 in 2003, 8500 in 2004, 8300 in 2005, 7600 in 2006, 8500 in 2007.

The next stage is to grow seedlings in specially equipped rooms in conditions *in vivo* in boxes and in peat-compost pots with the use of growth regulating preparation chlorcholine. Spraying of seedlings in the phase of 3-4 leaves raises the amount of tubers 1.2-18.8 times.

The use of different methods of rapid reproduction including the method of reproduction by cuttings, that increases a reproduction coefficient like 1:208-1:262, growing of seedlings from indices allow to reproduce newly created varieties quickly.

Table 2. The volume (t) of production and selling of original seed potatoes.

Year	Area, ha	The volume of production and sales of seed, t		
		total	reproduction	sold
2001	17.8	457.59	92.7	364.80
2002	17.0	240.10	73.55	166.60
2003	18.6	247.60	69.60	178.0
2004	21.4	491.80	139.40	352.40
2005	23.6	489.40	113.50	375.90
2006	23.0	409.0	87.0	322.0

In 2004 our diagnostic laboratory was accredited as an experimental laboratory in the system of potato seeds certification. It is the only laboratory in West Siberia.

In the research of aphids populations in potato crops it was ascertained that the activity of aphids infected with viruses depends on climatic conditions of a year. Melilot plays an active role of 'aphids trap' in a crop rotation. When planting melilot on the land the biggest part of aphids annihilates, and a part migrates to potato plants. The species composition of aphids changes from year to year. So at the period of 1999-2002 potato aphids and ordinary aphids prevailed (60-70%) and beginning from 2003 ordinary aphids and peach aphids prevailed (60-90%).

The periods of the treatment of potato reproduction nursery gardens with a desiccant and mowing of leaf tops are fixed taking into account peak periods of aphids migration, the resistance of varieties to different pathogens and the dynamics of productivity.

The producing of the original potato material will allow to provide with improved seeds not only potato growing farms of Kemerovskaya region but also of other regions: Tomskaya, Novosibirskaya, Chelyabinskaya regions, Krasnoyarskiy, Altaiskiy regions.

So, providing the market both agricultural producers and the population with improved potato material taking into account raised requirements of potato quality can be fulfilled completely. The research leads to the creation of productive and qualitative varieties resistant to various pathogens.

References

Russian Potato Varieties (a catalogue) / E.A. Simakov, B.V. Anisimov, N.P. Sklyarova, I.M. Yashina / All-Russian Scientific-Research Institute of Potato Farming, Russian Agricultural Academy. – M., 2005. – 127p.

Podgaetskiy, A.A. Genetic Potato Resources: The Materials of International Anniversary Scientific Practical Conference Devoted to 75-Years of the Institute of Potato Growing of National Academy of Sciences of Byelorussia. – Minsk: Merlit, 2003. – p. 180-189.

Bukasov, S.M. Potato Selection and Seed Growing / S. M. Bukasov, A. Ya. Kameraz. Potato Selection and Seed Growing. – L.: Kolos, 1972. – 352 p.

Simakov E.A./ The Effectiveness of Selected Choice in Estimation of Hybrid Potato Populations in Different Ecological and Geographical Conditions / E. A. Simakov, N. P. Sklyarova, I. M. Yashina: The Materials of International Anniversary Scientific Practical Conference Devoted to 75-Years of the Institute of Potato Planting of National Academy of Sciences of Byelorussia. – Minsk: Merlit, 2003. – p. 1 – p. 92-100.

Simakov, E.A. The Results of the Work of All-Russian Station of Potato Testing of Canker and Nematode / E.A. Simakov, A.A. Ilcheva, V.V. Manakov and others In: Questions of Potato Seed-Growing. Actual problems of Science and Practice: Scientific Works / All-Russian Scientific-Research Institute of Potato Planting. – M., 2006. – p. 114-118.

Sklyarova, N.P. New Potato Varieties of the Selection of All-Russian Scientific-Research Institute of Potato Planting / N.P. Sklyarova, E.A. Simakov, I.M. Yashina, H.H. Apshev: The Materials of International Anniversary Scientific Practical Conference Devoted to 75-Years of the Institute of Potato Planting of National Academy of Sciences of Byelorussia. – Minsk: Merlit, 2003. – p. 1 – p. 144-149.

Breeding of potato for earliness in the North-East of European Russia

N.F. Sintsova and Z.F. Sergeeva
Falenki breeding station of N. V. Rudnitsky North-East Agricultural Research Institute

Abstract

The study of dynamics of gain in weight of plants of potato of different groups of maturity was carried out to get to know the basic principles of crop formation. The primary influence of humidity on the degree of gain in tuber weight (r = 0.923-0.913) down to stop of normal tuber growth in adverse humidity periods was established. The influence of temperature was expressed slightly, as high day temperatures about 32°C (at which there was discontinuity of physiological processes) was compensated by lower night temperatures, when there was a basic outflow of nutritious substances into tubers. The close relations between earliness and phenological phases was not found, probably because of a strong infection of seedlings of early varieties by rhizoctonia. The dependence of efficiency on its elements during growth period undergoes changes. In the initial period of tuber formation of the early varieties the relation of efficiency with total tuber number is more expressed, then this relation weakens, but simultaneously relation of efficiency with amount of commodity tubers increases. At the end of the growth period the weight of one commodity tuber took the first place. The early varieties have simultaneous initiation of the maximal tuber number, the efficiency accumulates at the expense of increase of weight of separate tubers. Under conditions of drought there is a reduction of tuber beginnings. At late varieties the gain in tuber number continuous during the whole growth period. Therefore at selection of hybrids at early stages of breeding it is necessary to pay attention on tuber size. The analysis of dynamics of gain in weight under fluctuations of weather factors has shown, that under conditions of the Volga-Vjatka region in addition to technique of the definition of ripeness group of hybrids by trial digging up, accepted on the Falenki breeding station, we can begin digging potato on the 50th day after planting.

Keywords: breeding, potato, early-ripening, tubers

Introduction

Successful introduction of potato into northern areas of the European part of Russia, including Volga-Vjatka region, occurs due to cultivation of varieties with intensive formation of tubers during a short period. The early varieties allow to bypass peaks of adverse weather conditions, such as long cold weather in May, drought in June - July, which abound in the climate of the Volga-Vjatka region.

Therefore one must choose the samples most adopted to certain ecological conditions that produce commercial production in short time within all cultural varieties and newly created selection material of potato with different type of crop formation. In avoidance of culling of early varieties at the initial stages of breeding and because of insufficient amount of planting material it is better to spend selection by indirect characters.

Kol'adko *et al.* (2003) have chosen two indirect characters: the size of the largest tuber and uniformity of tuber nest. Morphological characters of potato plants can serve this aim too. Loginov S. P. (1999) has established correlation between identification of hybrids earliness by herbage type and by results of trial diggings up to 66.6-71.6%.

The morphological type of potato plants meets its ecological type. As a basis of plant grouping Kustarev A. I. (2002) has put conformity of variety to one of main ecological factors - moisture. The groups of hygromorphous, xeromorphous, and mesomorphous ecological morphotypes and polar ecologic morphotype were designated by him.

However the most significant way of definition of earliness is conduction of trial liftings. 'The Technique of research in culture of a potato' (1967) defines optimum term of first trial digging at the 60-th day after planting, when early varieties should generate commodity tubers yield. The disadvantage of this method is that it is impossible to spend the careful analysis of rates of tuber formation at early stages of selection because of small amount of tubers.

The purposes of our research were studies of dynamics of gain in plants weight for potato varieties of known ripening groups and hybrids with prospective earliness on a background of varying weather conditions, definition of morphological features of herbage and tubers of investigated samples, that will allow to define the basic features of growth and development of varieties of potato of different ripening groups under exact conditions, to choose indirect characters for selection of the early varieties at the beginning of the breeding process.

Materials and methods

To carry out the research 18-23 samples were selected. Early varieties – Vesna, Udacha; mid-early varieties - Nevsky and Elisaveta, mid-season variety – Chaika were used as checks. The hybrids showing the highest results in the previous year by the data of trial diggings were taken from preliminary nurseries up to competitive varietals test nurseries. However diggings at 40-th day has shown, that five hybrids can be recognize as late varieties, therefore they were excluded from the further study.

The varieties were planted in 4 multiple replications with 10 tubers in each. During the growth season phenological investigations were carried out. Study of dynamics of gain in weight of tubers and herbage was carried out by trial diggings everyone 10 days after planting with two plants in test. Morphological changes were fixed by calculation of amount of tuber, amount of main stems in a bush, measurement of plant height, calculation of leaves, stolons and lateral stem number.

Soil was turf-podzolic, middle clay. The humus contents was 2.6-2.7%, pH - 5.4, phosphorus – 34.4 mg/100 g of soil, potassium – 21.2 mg/100 g of soil. The planting pattern was 70 x 35.

The experiment was carried out in 2 years (2004 - 2005), differed because of weather conditions. Only May was similar, moderately warm and dry with daily average temperature 2.0-4.2 °C above the norm and precipitation in May dropped almost 2 times below the norm. Therefore planting began in optimum terms into warm enough soil.

The temperature development of the growth period of 2004 was close to average long-term meaning, only July (2.7 °C higher then norm) was hot and dry. But summer of the next year, especial June and July, was cold (0.9 °C in June and 0.3 °C in July below norm) and very wet.

The optimum amount of precipitations for potato during growth season makes 300 mm (Agro-ecologic passport of potato variety, 2000); in 2004 there was about 231.7 mm of precipitation in Falenki breeding station. The drought period was from the third decade of June till second decade of August. Therefore the basic limiting factor during the growing season of potato in 2004 was insufficient moisture for normal tuber formation. But in June 2005 there was 33.6 mm rain above the norm, and in July the precipitation dropped out in regular intervals during all month on the norm level. Such conditions have appeared favourable for development of phytophthora. To the beginning of August potato herbage has appeared strongly struck by pathogen, therefore experiment was stopped.

Results and discussion

Close relation of earliness and phenologic phases was not found, however, it does not mean that such relationship can not be there. The early varieties Udacha and 211-00 germinated three days later then mid-season varieties, but for Udacha and 211-00 it was marked a strong affection of seedlings by rhizoctonia.

The dynamic test of varieties has shown, that the tuber growth was very strongly linked with presence or absence of precipitations ($r = 0.913$-0.923). The gain went in steps, being sometimes reduced up to zero during a long drought (Figure 1). The influence of temperature on rates of a gain in tuber weight was expressed slightly, as high day temperatures (about 32 °C) was compensated by lower night temperatures, when there was a basic outflow of nutritious substances into tubers.

On a gain in tuber weight at first year of study all samples were divided in groups by their reaction on fluctuations of weather conditions. The early varieties Vesna and Ognivo differ by a similar type of development at the first twenty days of the research period, but after long-time drought the gain in tubers of Vesna stopped owing to drying of the foliage, and Ognivo was sharply increased in tuber weight ($r = 0.059$), therefore they were referred to as groups I and II.

In the group of mid-early varieties the correlation on dynamics of a tuber gain between varieties Nevsky, 282-97 and 136-99 makes $r = 0.887$-0.896, and between Nevsky and Elisaveta $r = 0.279$, so separate group of mid-early varieties Elisaveta and Visa ($r = 0.817$) was selected. In group V there were hybrids which have occupied an intermediate position between mid-early and mid-season varieties ($r = 0.969$) and in group VI - mid-season varieties ($r = 0.910$) (Figure 1).

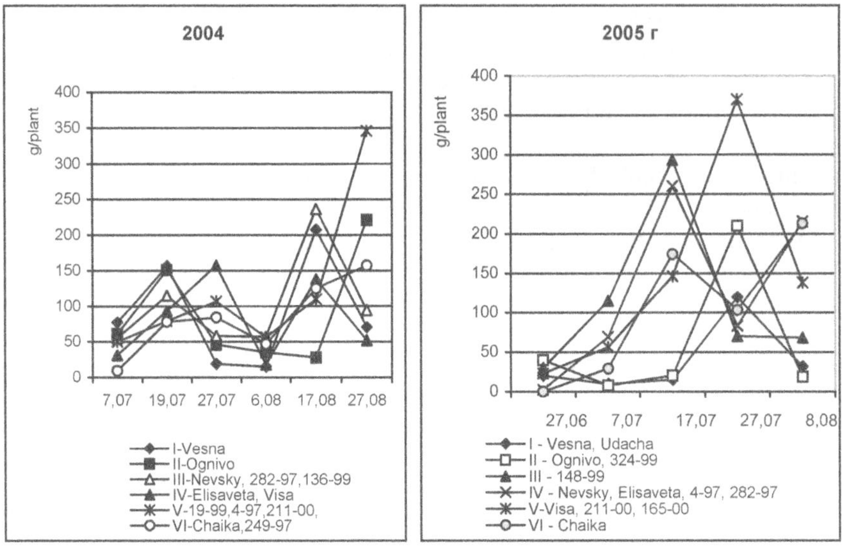

Figure 1. Dynamics of gain of productivity by decades.

In 2005 the similarity between varieties was marked despite distinctions of a mode of wetting. The separate group of early varieties was made with a hybrid 148-99 which had a sharp increase of productivity at the beginning of growth and subsequent decrease of gain in tuber weight. Variety Elisaveta has occupied a sliding position between groups, which at the first year was similar to variety Visa, and under conditions of sufficient moisture - to variety Nevsky. Unstable group V selected in 2004 obviously has shown the similarity with mid-early group.

The productivity is composed by the number of tubers and average weight of one tuber, what is pointed out by close correlation between amounts of marketable tubers and weight of one tuber. The relation of productivity and total number of tubers is not so obviously expressed and more closely correlates for late samples.

The dependence of productivity on its elements undergoes changes during growth season. At the initial period of tuber formation early varieties shown more expressed relation of productivity and total number of tubers, then this relation weakens, but simultaneously relation of productivity and number of tubers increases, and by the end of growth season weight of one tuber (Table 1) takes the first place. But such a picture was observed only under conditions of sufficient moisture.

The intensive gain in total number of tubers begins from the moment of forming of rudiments at the first days after germination. For early varieties simultaneous formation of the maximal tuber number goes. In further their gain stops, the productivity increases at the expense of increase of separate tubers, reaching marketable weight. At the first year of study during one and half a month total number of tubers of early varieties remained at the same level. For variety Vesna the number of tubers changed from 7 up to 6.8 tubers/bush, varieties Ognivo,

Table 1. Correlation between productivity and its elements, 2005.

Data of test digging	Coefficient of correlation between productivity and					
	total number of tubers		number of marketable tubers		weight of one marketable tuber	
	2004	2005	2004	2005	2004	2005
27.06		0.532		0.336		0.336
7.07	0.786	0.594		0.840		-0.090
17.07	0.531	0.512	0.794	0.939	0.533	0.682
27.07	-0.095	0.256	0.632	0.855	-0.214	0.538
8.08	0.139	0.284	0.378	0.855	0.509	0.734

Elisaveta, 282-97, 136-99 had similar changes. Mid-season variety Chaika showed continuous process of formation of new tubers and increase of their weight. The similar tendency was observed for variety Visa, which, despite of early accumulation of yield, can be physiologically referred to group of late varieties. In the drought period one common feature has united all samples, irrespective of ripening level – formation of tubers has stopped completely, and for early varieties the reduction of tuber rudiments began. This factor must be taking into account at selection of the early varieties. Even the small number of tubers in a nest, but large and levelled can serve as a reason for selection of plant on earliness.

The continuity of a gain in total number of tubers was precisely shown as well at the next year with sufficient moisture. The gain in above-ground part has doubtless interrelation with productivity. However direct correlation is too weak, is subject of casual, discounted factors and is badly caught by the direct correlation analysis. It is possible to note only, that height of a stem in the year with sufficient moisture tends to correlate better with productivity, e. g. more powerful plants increase tuber weight.

Conclusions

At the analysis of influence of weather factors on productivity of potatoes the primary influence of moisture on the degree of gain in tuber weight down to stopping of normal tuber growth in adverse dry periods was revealed. Additionally we now accept on the Falenki breeding station the technique of the definition of ripening groups of potato hybrids by trial diggings, we propose to begin digging of potato on the 50-th day after planting, i.e. 10 days earlier.

At selection of hybrids at early stages of selection it is necessary to pay attention to the size of tubers.

References

Agroecological password of a potato variety (hybrid). – M. - 2000. – 16 p. [in Russian]
Kol'adko I.I., Makhan'ko V.L., Vologdina L.N. Breeding of potato on earliness In: Proc. Intern. Conf. Deal with 75 anniversary of Institute of Potato Breeding of Belarus National Academy of Sciences. – Minsk, 2003. - V. 1– P. 34-40.[in Russian]
Kustarjov A.I.. И. Origin, evolution, ecology and breeding of potato In: Thesis of Dr. In Agriculture. - Brjansk, 2002. - 25 p. [in Russian]
Loginov S.I. Increasing of efficiency of selection for earliness within hybrid combinations of nematode-resistance parent forms In: PhD Thesis – M., 1999 - 25 p. [in Russian]
The Technique of research in culture of a potato – M., 1967 – 264 p. [in Russian]

New Potato Cultivars of Penza Research Institute of Agriculture

A.A. Kabunin
Penza Research Institute of Agriculture, Lunino, Penza region, 442731, Russia

Abstract

The Penza Science-Research Institute of Agriculture has been carrying out selection of potato for 20 years in cooperation with All-Russia Research Institute of Potato-growing (VNIIKH). On the basis of the starting material received from VNIIKH, in conditions of Middle-Volga region new highly productive and effective cultivars are created. Cultivars 'Utyonok', 'Ramzay', 'Russkiy souvenir', 'Daryonka' are included in the State register of selection achievement allowed to use. These varieties are effectively cultivated in various regions of Russia. New cultivars 'Baritone', 'Tyoscha', 'Batya' are under state testing. Brief characteristics of new and perspective cultivars are stated. Selection work proceeds, new cultivars are prepared to be transferred to the state testing.

Keywords: selection, cooperation, cultivars, adaptability

Introduction

Potatoes in Russia have traditionally been second in significance food product. Experience and practical work of farmers engaged in potato- growing in many developed European countries show that potato-growing is an effective production capable of both surviving in market economy and being profitable in agricultural business.

Selection of new potato cultivars occupies the leading position in the system of factors providing high productivity and effectiveness of potato growing. In the process of selection potato cultivars are subjected to a negative influence of environmental factors and gradually loose their productivity and quality. The development of industries producing prepared foods from potatoes demands such cultivars that possess specific qualities which were less important before, such as high content of dry substances and proteins, resistance to pulp darkening, low content of sugars etc.

In Russia potatoes are being grown in all areas. Most of it (97%) is done in small private farms. Vast spreading of potatoes and the diversity of natural climatic conditions demand many new cultivars, which are well adaptive to specific local conditions and agricultural technologies in different regions. Moreover, it is necessary to take into consideration habitual local preferences of potato consumers (in some regions people prefer potatoes with red tubers, in others, on the contrary with yellow tubers) and many other minor factors. Selection of new potato cultivars is undoubtedly necessary and most effective in the conditions of the suggested area of cultivation.

Now we have achieved practical results due to sufficient cooperation of the Russian Scientific Research Institute of potato-growing (VNIIKH) of the Russian Academy of Sciences and regional agricultural scientific institutions.

Methods

For more than 20 years the Penza Research Institute of Agriculture has been carrying out potato selection using hybrid-starting seeds received from VNIIKH. Cooperation of scientific institutions of different regions has certain administrative and economic advantages. Being a vegetative reproducing culture, potato makes it possible to use a shortened scheme of selection work in the regions. Selection effect of potatoes in any region depends mainly on the number and genetic diversity of the starting material. The choice of parental couples, where and in what conditions parental collections are sustained and hybridized is absolutely of no importance. Genetically stable tubers of potato hybrids are made in one centre and spread over all regions for which the new cultivars are necessary.

We have been selecting potatoes using the scheme since 1980 in some scientific centres of different regions of Russia. Penza Research Institute of agriculture has been carrying on potato selection using the starting material (potato hybrids)of the VNIIKH selection centre. Identical sets of hybrids are tested in different regions. The difference between them is caused by specific conditions of the region, which demands either phytophthora and drought resistant cultivars or overwet resistant ones. Legally cooperative work of all regional scientific institutions is based on written contracts because each of them makes part of state company Russian Agricultural Academy. The contracts specify the duties of the partners and their share in the authorship of the new cultivar. This way of cooperation between regional institutions and the centre has made it possible to spread the agro-ecological boundaries of testing hybrids, increase the effect of the work, arrange the selection work in the places where the number of qualified specialists is limited and what is most important double the number of new cultivars.

VNIIKH reports demonstrate this way of cooperation has allowed to create more than 20 cultivars for different regions. Positive results of the cooperation may make sufficient ground for the fereral conception of the selection work. Here we demonstrate the results of our work.

Our selection work is done in the fields in the traditional scheme of individual clone selection. Hybrids are being tested in the selection fields during 8-11 years (the time of the experiment depends on the quality of the hybrids). Geographical position, soil, and climatic conditions of Penza region (boundary position between dry and wet regions, annual fluctuations of weather conditions) make it possible to create new cultivares good for both central Russia and its southern regions. Our institute receives from 2 to 10 thousand potato hybrids from VNIIKH. Hybrids are created in the genetic laboratory of VNIIKH headed by Dr Yashina. They are selected on the basis of widely known varieties of home and foreign selection and artificial hybrids specially made of disease-resisting natural potatoes. Dr Yashina pays special attention to the selection hybrids resistant to Phytophthora infestans and virus diseases. Agro-technology of the selection work (the use of mineral and chemical fertilizers) is intermediate. The varieties created in such a way are well resistant to our conditions and perceptive to better agro-technology though they are not particular for a highly organized agro-technology.

The results of the cooperative work

At present about 50 thousand potato hybrids made in VNIIKH have been tested and marked in Penza RIA. About 500 selection patterns are being tested in the selection field. Four new cultivars of potatoes are included into the State register and allowed for use in the Middle-Volga region as the result of mutual cooperation. They are the two early-ripening cultivars 'Utyonok' and 'Daryonka' and the two middle-ripening cultivars 'Ramzay' and 'Russkiy Souvenir' Three more new cultivars are being tested.

Potato cultivars selected by Penza Research Institute Agriculture have proved fine in the Middle-Volga and some other regions. They are highly adaptive to the local climatic and soil conditions and characterized by high quality tubers.

New early-ripening 'Utyonok' has been included into the State register for use in the Middle-Volga region since 1998. It provides high yield (22.0-31.0 tons per hectare) without intensive agro-technology and minimum fertilizers. In case of better agro-technology the yield is 42.0 tons. Yield stability in changing weather is high (80-82%) The yield of young potatoes is rich and early (15-20 tons) and the cultivar is good for storage in autumn. 'Utyonok' is resistant to *Synchytrium endobioticum, Globodera rostochiensis, Rhizoctonia solani* and bacterial diseases and tolerant to viruses. Leaves are medium resistant to *Phytophthora infestans*, tubers are highly resistant to it. Tubers are highly resistant to larvae of *Agriotes sputator.*

Tubers are yellow, longish-oval, big (up to 250g) soft peel, small and rare dims. Starch content 12-20%. Taste qualities are high and very high. Pulp is light yellow, resistant to darkening when cut or cooked, soft when boiled. Protein content about 2%.. Longish tubers with the pulp resistant to darkening are good for cooking, storage qualities are high. The variety can be used both for early harvesting and winter consuming. It can be used for freezing and conservation

Both producers and consumers appreciate the cultivar for its exterior and taste. Even those who have been prejudiced against yellow potatoes now actively grow and buy 'Utyonok'. Though it is rather difficult for a small regional centre to promote to the market any new cultivar 'Utyonok' is being grown in the Volga region, in Chernosemye, in central Nechernosemye region, in Krasnodar region and in the North Caucasus republics. Elite seeds are sent annually to these regions. 'Utyonok' is highly appreciated by the farmers of Siberia, Far East and Kamchatka but they are too far that is why it is not so widely spread there.

'Ramzay' is a new universal mid-ripening cultivar (used in the Middle-Volga region since 2002). It is heat and drought resistant and characterized by early and intensive yield. Sometimes the yield of young potatoes can be taken earlier than from early-ripening cultivars. The yield (with minimum fertilizers) is 24.0-35.0 tons. Yield stability is 81-84%. 'Ramzay' is resistant to *Synchytrium endobioticum*, viruses, slightly subjected to *Phytophthora infestans* moderately to *Macrosporium solani.* Tubers are round and oval, big (up to 200g), evened, highly resistant to larvae *Agriotes sputator.* Taste qualities are good and excellent. Pulp is yellow, resistant to darkening when cut or cooked. Starch content is 13.0-20.0%; protein content is more than 2%. Storage is good. Tubers can be used for any kind of cooking, production of chips, freezing and conservation.

'Daryonka' is an early-ripening food cultivar. It was included into the State register in 2006. Its yield (without fertilizers) in 1996-2000 was 19. 1-28. 7 tons, young potato yield is 20tons. The cultivar is heat and drought resistant, its productivity is stable annually. It is sensitive even to a small quantity of fertilizers.

'Daryonka' is resistant to *Synchitrum endobioticum* and *Globodera rostochiensis*, slightly resistant to virus diseases, highly resistant to *Phytophthora infestans*. Tubers are resistant to *Strepomyces scabies* and other *Actinoyices* and *Rhizoctonia solani*. Tubers grow compactly, stolons are short. The cultivar can be well cultivated and harvested by machines. Tubers are yellow, big (60-150g) round-oval and narrowed to the bottom. Peel is soft, dims are small and rare. Pulp is yellow; resistant to darkening when cut or cooked, soft when boiled. Tubers are tasty and presentable. Starch content is 16. 8-20. 0%. Storage is good. The cultivar can be processed into potato products.

'Russkiy Souvenir' is a universal mid-early ripening cultivar (winter consuming and production of potato products). It was included into the State register for the Middle Volga and North Caucasus regions in 2005. It is heat and drought resistant, characterized by constant high productivity- 19. 8-25. 8 tons (in non-fertilized soils). During state testing the yield of 'Russkiy Souvenir' on some fields was about 70 tons. As was stated by the specialists from the Russian Research Institute of Phytopathology 'Russkiy Souvenir' leaves are most resistant to *Phytophthora infestans*, tubers are satisfactory resistant. The cultivar is resistant to *Synchytrium endobioticum*. Leaves are highly resistant to *Leptinotarsa decemlineata* virus diseases are insignificant. Tubers are resistant to *Streptomyces scabies* and other *Actinomyces*.

Tubers are yellow, big (150-200g) round-oval, resistant to mechanical pressure. Peel is slightly netted; dims are small. Pulp is light yellow, resistant to darkening when cut or cooked soft when boiled. Tubers are tasty, may be recommended for frying (chips, etc) Starch content is 14. 0-20. 0%; high protein content in tubers. Good for storage. The cultivar is good for industrial processing into prepared foods and chips.

At present new cultivars 'Bariton' (early-ripening), 'Tyoscha' (mid-early ripening) and 'Batya' (mid-ripening) are under state testing. They are selected by Penza SRIA in cooperation with BNIIKH. 'Baritone' is a new cultivate meant for production of young potatoes. It is an early ripening and high-qualitative variety. Though the testing period is not yet too long, 'Baritone' has got significant results of early production in the southern regions of Russia.

New cultivar 'Tyoscha' regularly gives high yield (37-40 tons and more) of big evened tubers. It is resistant to *Phytophthora*, adaptive to extreme weather conditions (from drought to over wetting). The cultivar is remarkable for a long period of physiological repose of tubers in storage. The cultivar is good for being processed into potato products.

The new cultivar 'Batya' is highly productive (up to 40 tons and more), resistant to *Phytophthora*. Tubers are characterized by high starch content. Moreover the cultivar is highly resistant to heat and droughts and is characterized by intensive growing of tubers. State testing has proved its resistance to *Globodera rostochiensis*. Tubers of this variety are traditionally white and their pulp is soft and very tasty.

Conclusion

Productivity and quality of new potato cultivars created in the Penza Research Institute of Agriculture make it possible to increase the effect of potato business in Russia.

Potato selection in the Penza Scientific Research Institute of Agriculture is carried out effectively. The nearest perspectives include the creation of new cultivars for the production of potato products. About 500 patterns are being tested in selection fields. State testing of some perspective cultivars is meant to develop Russian potato production further.

New cultivars strengthen the Russian potato production and agricultural business. Our specialists look forward to fruitful cooperation with colleagues from other scientific establishments and agricultural business.

Curriculum vitae – Alexander A. Kabunin

Alexander A. Kabunin – born in a family of farmers- studied agronomy in Penza State Agricultural Academy. In 1988 he received the scientific degree of the Candidate of Agriculture in Moscow Research Institute of Potatoes. In 1989 he started potato selection work in Penza Research Institute of Agriculture where in 1993 he organized and headed a scientific laboratory of potato selection. He is the author of 7 new cultivars of potatoes. He published more than 35 scientific papers and hundreds of columns en conference papers. He develops methods of effective potato selection and active in promotion of new potato cultivars and technologies into potato business. He is an Assistant Director of the Penza Research Institute of Agriculture.

Profile – Penza Research Institute of Agriculture

State scientific institution Penza Research Institute of Agriculture is one of the leading scientific agricultural institutions in the Volga region in Russia. It dates back to 1909. The principal task of the institute is to provide scientific ground for agricultural business of Penza and nearly regions. The main aim is working out and mastering of low-energy, low-recourse and ecologically safe technologies of biological agriculture. The Institute place premium on top quality research. The Institute combines scientific research with innovation and an acute eye for business. We work on more than 30 problems among which selection and seed growing of grain-crops, oil-producing crops, lentil, potato, hemps with low drug content, perennial and fodder grasses. Penza Research Institute of Agriculture studies and develops problems of soil fertility and biological technologies of crop growth. The institute provides post graduate courses in crop and potato growth and cooperates with many scientific institutions and businesses.

Advances in biotechnology

DNA markers for potato introgression breeding

M.P. Beketova[1,2], P.E. Drobyazina[1], E.A. Sokolova[1], E.A. Tsar'kova[1], V.A. Vorobiev[1], I.M. Yashina[2] and E.E. Khavkin[1]
[1]Institute of Agricultural Biotechnology, Russian Academy of Agricultural Sciences, 42 Timiryazevskaya ul., Moscow 127550, Russia, emil@iab. ac. ru;
[2]Institute of Potato Growing, Russian Academy of Agricultural Sciences, Korenevo, Moscow region, 140052, Russia

Abstract

To develop breeder-friendly sequence characterized amplified region (SCAR) markers characteristic of particular *Solanum* species and workable as tools for introgression breeding, we compared several polymorphic regions of *Solanum* genomes, such as repetitive sequences identical in the genomes of wild *Solanum* species and derived potato cultivars, the characteristic fragments of the genes for potato late blight resistance (R genes) and the hot spot of variation in intron 2 of the developmental gene *FLORICAULA/LEAFY.* Multiple alignments of these sequences became the basis for designing the species-specific PCR primers. The specificity of SCAR markers amplified with these primers was verified against the accessions of several *Solanum* species. The markers reliably discerned the germplasm of wild *Solanum* species in potato cultivars and were feasible for monitoring the transfer of such germplasm into potato cultivars. The presence of SCAR markers derived from the R genes was associated with the foliage and tuber resistance of potato cultivars to late blight; it follows that such markers are potential predictors of these phenotypic traits in new potato crosses.

Keywords: SCAR markers, dispersed repeats, *FLORICAULA/LEAFY,* R genes, late blight resistance

Introduction

Numerous, highly reproducible and easy-to-handle DNA markers open new vistas in *Solanum* taxonomy and potato breeding (Spooner *et al.*, 2005). An essential task is to extend the range of easily scored markers, so that they cover genome regions of diverse functions and evolution rates. Wild *Solanum* species are extensively used in potato breeding as the source of many economically important traits. For that reason reliable markers of species-specific germplasm and of the particular genes of agronomic importance would greatly facilitate the introgression breeding and help verify the complicated and often insufficiently documented pedigrees of potato cultivars, which frequently comprise the germplasm of several wild *Solanum* species. Among DNA markers, sequence characterized amplified region (SCAR) germplasm markers are most breeder-friendly: they are inexpensive, return rapid results and are easy to interpret.

Development of SCAR markers

To develop SCAR markers characteristic of particular *Solanum* species, we compared several polymorphic regions of *Solanum* genomes, including (1) repetitive sequences identical in the genomes of progenitor *Solanum* species and derived potato cultivars, (2) the characteristic fragments of the genes for potato late blight resistance (R genes) and (3) intron 2 of the developmental gene *FLORICAULA/LEAFY*.

Among the repetitive genome sequences, the fragments amplified with PawS primers were used as a novel source of species-specific SCAR markers. Initially these primers were designed by Rogowsky *et al.* (1992) to synthesize the rye-specific hybridization probes for chromosome analysis; however, later we found that PawS primers could be also employed as a universal tool for genotyping diverse species of higher plants, including those of *Solanum* (Zaitsev and Khavkin 2001). The genome fragments amplified with PawS primers (interPawS sequences) have not been sufficiently characterized in *Solanum* species. One of a few exceptions, a genome fragment about 800 bp in length was amplified with the PawS 5 primer from several *Solanum* species, cloned and sequenced, and the sequence of the *S. bulbocastanum* fragment was submitted to the Genbank as an accession number DQ010412. The comparative study of similar fragments cloned from *S. chacoense, S. kurtzianum* and *S. stoloniferum* and also obtained by the BLAST search (www.ncbi.nlm.nih.gov/BLAST) for DQ010412 homologs in the *S. demissum, S. lycopersicum* and *S. tuberosum* genomes produced several 85-90% identical sequences. These homologs seem to reside in the hot spot of pathogen resistance genes on chromosome 5 of *S. demissum* and *S. tuberosum* and on chromosome 4 of *S. lycopersicum*. The analysis of BAC sequences deposited in NCBI showed that interPawS sequences on chromosome 5 in *S. demissum* and *S. tuberosum* are dispersed repeats.

To develop a SCAR marker specific of *S. bulbocastanum*, we first compared the *S. bulbocastanum* fragment DQ010412 to its partial homologs from other *Solanum* species. From the multiple alignment of these sequences, a specific forward primer was designed for the SCAR marker Sblb, whereas the reverse primer corresponded to the conserved stretch of the fragment DQ010412 (Table 1). The resulting 509-bp SCAR marker is apparently specific of *S. bulbocastanum* germplasm (Figure 1a). In the case of *S. demissum*, by aligning over 20 sequences from *S. demissum* (Kuang *et al.*, 2005), *S. tuberosum* (Rickert *et al.*, 2003) and *S. caripense* (Trognitz and Trognitz 2005), we selected three conserved regions, each 10-20 Kb in length, which were subjected to the second round of multiple alignment. Several short (about 10 bp) variable stretches were checked with BLAST for species-specificity. Two of these sequences were used to design primers for the SCAR marker Sdem (Table 1). In addition to *S. demissum*, this 415-bp marker was also present in *S. stololoniferum* and absent from several other *Solanum* species (Figure 1b).

Another strategy was employed to develop SCAR markers of *Solanum* species from the already characterized genes. Eleven R genes coding for NBS-LRR kinases that recognize various races of late blight pathogen *Phytophthora infestans* were transferred into potato from *S. demissum*. Two of them, *R1* and *R3a*, have been characterized, and the genetic databases comprise many structural homologs of these genes of as yet unknown functions. PCR primers 76-F and 76-R (Table 1) designed on the basis of the variable LRR region in the *R1* gene (Genbank accession AF447489, Ballvora *et al.*, 2002) produce an easily discerned 1400-bp fragment and also several weaker bands. The SCAR marker R1 has been already verified against the differential sets and used to screen potato commercial cultivars (Gebhardt *et al.*, 2004; Beketova *et al.*, 2006; Trognitz and Trognitz 2007). It is noteworthy that, in addition to R1 (Figure 2a), this marker recognizes R6 and R9 differentials as well (Trognitz and Trognitz 2007). In our laboratory,

Table 1. Forward (F) and reverse (R) primers used to amplify SCAR markers.

Marker	Primer	5′ - 3′ sequence	Annealing temperature, °C
Sblb	SblbF	TCTTACTATCGGTCTTGTCTACAGCCCAGGT	66
	SblbR	AGATTCTCTCCTTTCAATATCGGATAAACATTCGG	
Sdem	SdemF	CTTTCCTAAGAACATCCTATTATTATA	61
	SdemR	CGGAAACAGAAACAGAAACAGGA	
R1	76-F*	CACTCGTGACATATCCTCACTA	55
	76-R*	CAACCCTGGCATGCCACG	
R3	R3a-F	CGA CAT GTA TTG ATC TCC CTG	63
	R3a-R	CTT CAC AGC TTC CCT TYT ATC	
RB	RGA2-F	GGCAAGCTCAGAATCAATTATCCACCCCAACTTTTAAAT	68
	RGA2-R	CAAGTATTGGGAGGACTGAAAGGTGTTGGGCATC	

*From Gebhardt *et al.* (2004). Other primers are designed by the authors.

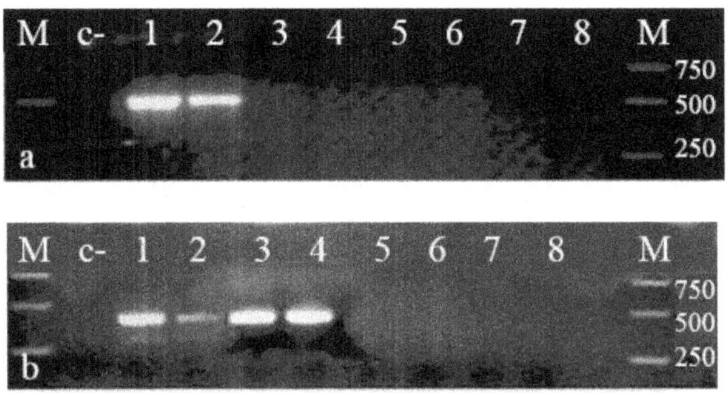

Figure 1. Verification of SCAR markers based on dispersed repeats.
(a) Sbc marker of S. bulbocastanum germplasm. M - Gene Ruler 1 kb DNA Ladder; c- – negative control (without DNA); 1, 2 – S. bulbocastanum*; 3 –* S. tuberosum *ssp.* tuberosum*; 4 –* S. phureja*; 5, 6 –* S. tuberosum *ssp.* andigena*; 7, 8 –* S. demissum.
(b) Sdem marker of S. demissum *germplasm. M - Gene Ruler 1 kb DNA Ladder; c- – negative control; 1, 2 –* S. demissum*; 3, 4 –* S. stoloniferum*; 5, 6 –* S. tuberosum *ssp.* andigena*; 7 –* S. phureja*; 8 –* S. bulbocastanum.

a SCAR marker was developed on the basis of the *R3a* gene (Genbank accession AY849382, Huang *et al.*, 2005) by aligning it to all structural homologs available in the NCBI database and designing a specific forward primer (Table 1). The 496–bp fragment corresponding to this marker was found in R3 and R7 differentials (Figure 2b). The SCAR markers R1 and R3 are located on chromosomes 5 and 11, respectively, and provide convenient tools to follow the introgression of *S. demissum* germplasm into potato cultivars. Genes R3 and R7 are known to reside in one cluster on chromosome 11 (Huang *et al.*, 2005); our data presume that these genes are structural homologs.

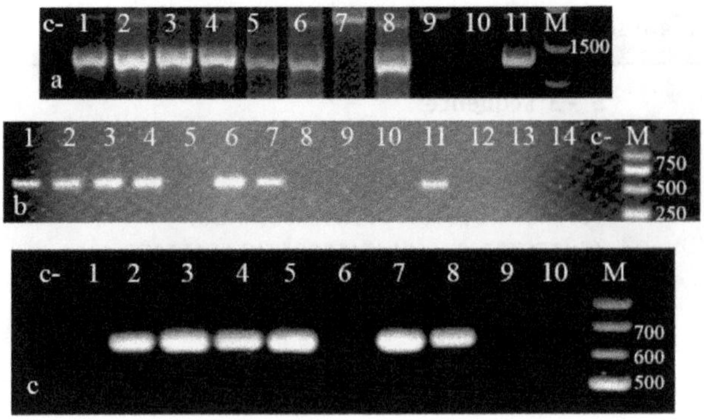

Figure 2. Development of SCAR markers based on R gene sequences using (a, b) the CIP differential sets and (c) the accessions of Solanum species.
(a) Marker R1: c- – negative control; 1, 2-9 - R1; 10 – r; 11 – R7; 12 – cv. Temp; M – 100 bp DNA Ladder Plus.
(b) marker R3: 1 – R2R3; 2 – R2R3R4; 3 – R1R2R3R4; 4 – R1R3R4; 5 – R1R2R4; 6 – R1R2R3; 7 – R3R4; 8 – R1R4; 9 – R1R2; 10 – R10; 11 – R7; 12 – R4; 13 – R1; 14 – r; M – GeneRuler 1 kb DNA Ladder.
(c) marker RB: 1 – negative control; 2, 7, 10, 11 – S. tuberosum*; 3-6, 8, 9 –* S. bulbocastanum*; M– 100 bp DNA Ladder Plus.*

To recognize the *RB* gene for late blight resistance in *S. bulbocastanum*, specific primers (Table 1) were designed on the basis of the RGA2 sequence (Genbank accession AY336128, Song *et al.*, 2003). A 638-bp fragment of *RB* was amplified from genomic DNA of *S. bulbocastanum*, cloned and sequenced. The sequence was 99. 4% identical to the prototype gene. This SCAR marker (Figure 2c) resides on chromosome 8 of *S. bulbocastanum*.

The second intron of the *FLORICAULA/LEAFY* gene of meristem identity (Flint2) is polymorphic in many plant species and therefore widely used in molecular systematic studies (Grob *et al.*, 2004). However, Flint2 has not been previously characterized in *Solanum* species, except in *S. lycopersicum* (Molinero-Rosales *et al.*, 1999). Hence we designed the primers recognizing the conserved regions in the exons 2 and 3 and amplified several gene fragments comprising FLint2, 900-1000 bp in length, from genomic DNA of potato cultivars and *S. demissum* accessions. These fragments were cloned, and sequences thus obtained were used to design the primers for the most variable (hot spot) region within FLint2. Among numerous fragments, 200-500 bp in length, corresponding to the FLint2 hot spot in several *Solanum* species, some were apparently characteristic of particular species. To further clarify FLint2 specificity as potential markers, we screened the series of available *S. demissum* and *S. stoloniferum* accessions. As a result, invariant species-specific patterns were recognized by Drobyazina and Khavkin (2007). Currently we clone and characterize these fragments in order to develop SCAR markers that would reliably discern the *demissum* and *stoloniferum* germplasm.

SCAR markers as breeding tools

To validate the SCAR markers of *Solanum* species as an instrument of introgression breeding, two conditions must be met. First, it is crucial to verify that the markers characterize the particular species. In the case of many potato cultivars and breeding lines, we deal with insufficiently documented pedigrees comprising several wild *Solanum* progenitors, and the particular parental forms of the wild *Solanum* species that had been used for hybridization many years ago are unknown or no longer available. We therefore must start from screening the range of available accessions of each particular *Solanum* species and then proceed by discerning the invariant species-specific genome fragments in the potato cultivars that allegedly comprise the particular germplasm. Such approach would fail if the far-away hybridization with wild *Solanum* species involved non-typical accessions. To illustrate the case, the most typical (invariant) Flint2 markers were selected in *S. demissum* and *S. stoloniferum;* all these invariants were also found in potato cultivars comprising the germplasm of these two species (Figure 3).

Second, we must establish that the markers stay in potato lines after several rounds of crosses and are not bred out of potato together with unfavourable traits of the wild species. This case is exemplified by the Sdem marker. The marker successfully recognized the germplasm of both *S. demissum* and *S. stoloniferum;* however, the marker was rapidly lost in 2-3 breeding cycles following the cross to the wild species (Figure 4). These data also proved that the presence of the Sdem marker was by no means related to cultivar resistance to late blight and the occurrence of the R1 marker.

On the contrary, the SCAR markers developed from the *R1* and *R3* gene sequences were stable in the progeny of *S. demissum* after many backcrosses when potato cultivars were bred for

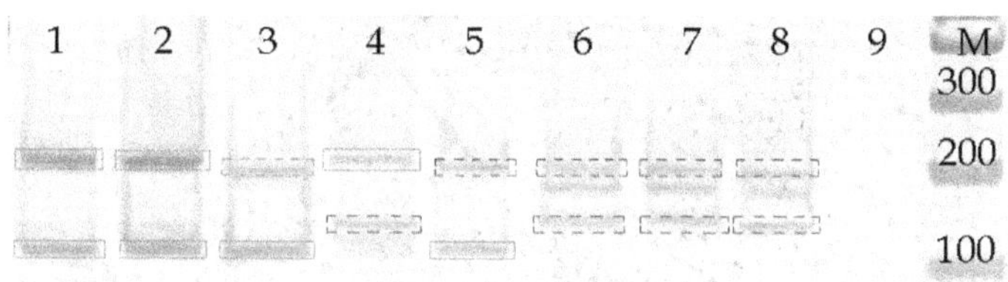

Figure 3. Introgression of FLint2 markers from wild Solanum species into potato cultivars.
1-2 - S. stoloniferum, *3-5 - potato cultivars Golubizna, Skoroplodnyi and Bryanskii Ranny, 6-8 -* S. demissum, *9 - negative control, M - Gene Ruler 100 bp DNA Ladder Plus.*

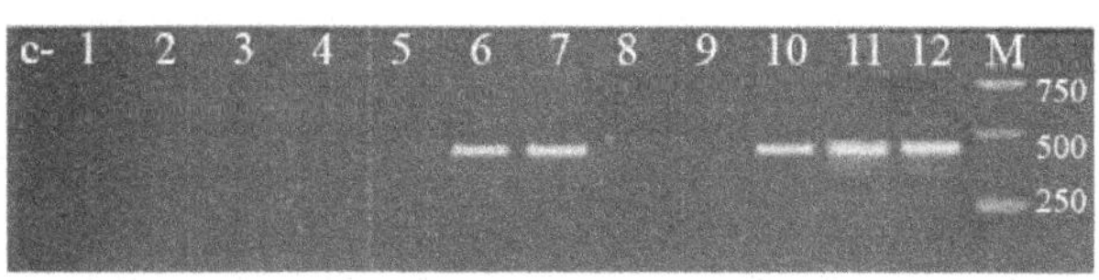

Figure 4. SCAR marker Sdem in potato cultivars comprising S. demissum germplasm.
M – Gene Ruler 1 kb DNA Ladder; c- – negative control; 1-12 – potato cultivars (wherever possible, the number of generations following the cross with S. demissum *is shown in parentheses): 1 – Nena (>4), 2 – Isna (>4), 3 – Clivia (>4), 4 – Olev (2), 5 – Temp (3), 6 – Krasnaya Rosa (3), 7 – Vesna (2), 8 – Effekt (3), 9 – Ramenskii (2,3), 10 – Udacha (3), 11 – Anoka (2), 12 – Skoroplodnyi (3).*

late blight resistance (Figure 5a and 5b, respectively). A significant relationship was observed between the frequency of the R1 marker occurrence in potato cultivars and both foliage and tuber resistance to this disease (Table 2). It follows that the SCAR markers based on the R gene fragments can directly prove the presence of the corresponding resistance genes; therefore these markers would immediately help monitor the trait of late blight resistance through numerous crosses and predict the resistance of the newly bred potato lines to late blight.

Conclusions

SCAR markers are very attractive because they are both specific and easy to handle and read, hence highly reliable. The already characterized fragments of *Solanum* genomes, whatever scarce, provided enough opportunity to develop five new SCAR markers. Together with the R1 marker developed earlier in another laboratory, these markers discern the particular *Solanum* germplasm and help discover it in potato cultivars. For this reason, SCAR markers of *Solanum* species involved in potato breeding present a promising tool for re-establishing cultivar pedigrees, verifying cultivar identities and monitoring germplasm transfer in the introgression breeding.

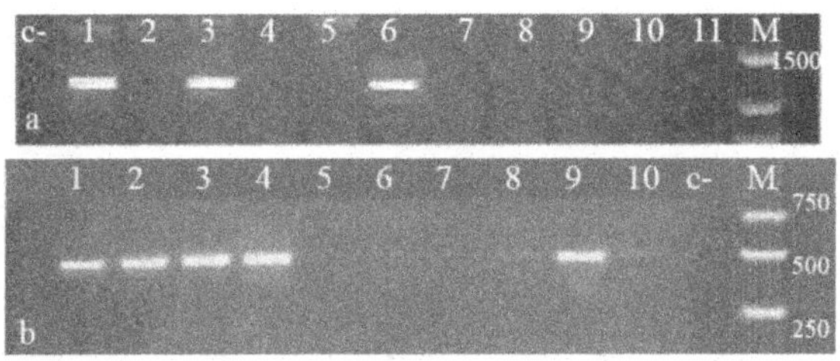

Figure 5. SCAR markers R1 and R3 in demissoid potato cultivars with high late blight resistance.
(a) Marker R1. c- - negative control; 1-11 – potato cultivars: 1 – Temp, 2 – Desiree, 3 – Wauseon; 4 – Vestik, 5 – Delfin, 6 - Kalinka, 7 – Nevsky, 8 – Vytok, 9 – Udacha, 10- Belorussky Ranny, 11 – Lorkh; M - 100 bp DNA Ladder Plus. (b) Marker R3. 1 – Effect; 2 – Resurs; 3 – Darenka; 4 – Bezhitskii; 5 – Temp; 6 – Bryanskii; 7 – Detskoselskij; 8 – Nevsky; 9 – Udacha; 10 – Anoka; c- - control; M – GeneRuler 1 kb DNA Ladder.

Table 2. The R1 marker is an index of late blight resistance.

Presence of R1	Total number of screened cultivars	The number of cultivars manifesting high (7-9 points) late blight resistance (percentage is given in parentheses)	
		foliage	tubers
+	49	22 (45)	27 (55)
-	98	16 (16)	29 (29)

Further validation of the SCAR markers of *Solanum* species is required to avoid covert pitfalls of false PCR positives resulting from putatively incomplete specificity of repetitive sequences against the vast span of insufficiently characterized *Solanum* genomes or from numerous pseudogenes (Kuang *et al.*, 2005) and paralogs of R genes.

Acknowledgements

The authors thank all colleagues from N. I. Vavilov Institute of Plant Industry (St. Petersburg, Russia), Institute of Phytopathology (Golitsyno, Russia), Centre for Potato, Fruit and Vegetable Growing (Samokhvalovichi, Belarus), the US potato gene bank (Sturgeon Bay, WI), the Gross Lüsewitz Kartoffel-Sortiment (Germany) and The Centre for Genetic Resources (Wageningen, the Netherlands) who provided the plant material. The study was supported in part by the ISTC-USDA-ARS project #3714p.

References

Ballvora, A., M.R. Ercolano, J. Weiss, K. Meksem, C.A. Bormann, P. Oberhagemann, F. Salamini, and C. Gebhardt (2002) The R1 gene for potato resistance to late blight (*Phytophthora infestans*) belongs to the leucine zipper/NBS/LRR class of plant resistance genes. Plant Journal 30: 361-371.

Beketova, M.P., P.E. Drobyazina and E.E. Khavkin (2006) The *R1* gene for late blight resistance in early and late maturing potato cultivars. Russian Journal of Plant Physiology 55: 384-389.

Drobyazina, P.E. and E.E. Khavkin (2007) Structural homologs of *CONSTANS* and *LEAFY* in potato and its wild relatives. Acta Horticulturae (in press).

Gebhardt, C., A. Ballvora, B. Walkemeier, P. Oberhagemann and K. Schüler (2004) Assessing genetic potential in germplasm collections of crop plants by marker-trait association: a case study for potatoes with quantitative variation of resistance to late blight and maturity type. Molecular Breeding 13: 93-102.

Grob, G.B.J., B. Gravendeel and M.C.M Eurlings (2004) Potential phylogenetic utility of the nuclear *FLORICAULA/LEAFY* second intron: comparison with three chloroplast DNA regions in *Amorphophallus* (Araceae). Molecular Phylogenetics and Evolution 30: 13-23.

Huang, S., E.A.G. van der Vossen, H. Kuang, V.G.A.A. Vleeshouwers, N. Zhang, T.J.A. Borm, H.J. van Eck, B. Baker, E. Jacobsen and R.G.F. Visser (2005) Comparative genomics enabled the isolation of the *R3a* late blight resistance gene in potato. Plant Journal, 42: 251-261.

Kuang, H., F. Wei, M. Marano, U. Wirtz, X. Wang, J. Liu, W. Shum, J. Zaborsky, L. J. Tallon, W. Rensink, S. Lobst, P. Zhang, C. Tornqvist, A. Tek, J. Bamberg, J. P. Helgeson, W. Fry, F. You, M. Luo, J. Jiang, C. Buell and B.J. Baker (2005) The *R1* Resistance Cluster Contains Three Groups of Independently Evolving, Type I *R1* Homologues and Shows Substantial Structural Variation Among Haplotypes of *Solanum demissum*. Plant Journal 44: 37–51.

Molinero-Rosales, N., M. Jamilena, S. Zurita, P. Gomez, J. Capel and R. Lozano (1999) *FALSIFLORA*, the tomato orthologue of *FLORICAULA* and *LEAFY*, controls flowering time and floral meristem identity. Plant Journal 20: 685-693.

Paal, J.M., H. Henselewski, J. Muth, K. Meksem, C.M. Menendez, F. Salamini, A. Ballvora and C. Gebhardt (2004) Molecular cloning of the potato *Gro1-4* gene conferring resistance to pathotype Ro1 of the root cyst nematode *Globodera rostochiensis*, based on a candidate gene approach. Plant Journal 38: 285-297.

Rickert, A.M., J.H. Kim, S. Meyer, A. Nagel, A. Ballvora, P.J. Oefner and C. Gebhardt (2003) First-generation SNP/InDel markers tagging loci for pathogen resistance in the potato genome. Plant Biotechnology Journal 1: 399-410.

Rogowsky, P.M., K.W. Shepherd and P. Langridge (1992) Polymerase chain reaction based mapping of rye involving repeated DNA sequences. Genome 35: 621-626.

Song, J., J.M. Bradeen, S.K. Naess, J.A. Raasch, S.M. Wielgus, G.T. Haberlach, J. Liu, H. Kuang, S. Austin-Phillips, C.R. Buell, J.P. Helgeson and J. Jiang (2003) Gene RB cloned from *Solanum bulbocastanum* confers broad spectrum resistance to potato late blight. Proceedings of the National Academy of Sciences 100: 9128–9133.

Spooner, D., R. van Treuren and M.C. de Vicente (2005) Molecular markers for genebank management. IPGRI Technical Bulletin no. 10, 126 p.

Trognitz, F.C. and B.R. Trognitz (2005) Survey of resistance gene analogs in *Solanum caripense*, a relative of potato and tomato, and update on *R* gene genealogy. Molecular and General Genomics 274: 595-605.

Trognitz, B.R. and F.C. Trognitz (2007) Occurrence of the *R1* allele conferring resistance to late blight in potato R-gene differentials and commercial cultivars. Plant Pathology 56: 150-155.

Zaitsev, V.S. and E.E. Khavkin (2001). Identification of plant genotypes by PCR analysis of the dispersed repeating sequences R173. Russian Agricultural Sciences 3: 2-5.

Curriculum vitae – Emil E. Khavkin

Emil E. Khavkin graduated from the Moscow Agricultural Academy in 1958, and it was this university wherefrom he received his MS and PhD in Agricultural and Biological Chemistry in 1958 and 1964. For two decades (1960-1981) spent with the Siberian Institute of Plant Physiology and Biochemistry in Irkutsk, he was engaged in the biochemical studies of plant growth and development; these studies became the basis of his DSci dissertation in Plant Physiology defended in 1974. Back to Moscow, Emil Khavkin studied plant nitrogen nutrition as related to fertilization technologies in the All-Russian Institute of Applied Agricultural Chemistry (1981-1987) until he moved to the Institute of Agricultural Biotechnology (1990 until presently). He is Professor in Biological Chemistry, 1993.

His major fields of research and teaching have been Plant Physiology and Biochemistry, Functional Genomics, and Molecular Breeding. Currently his major areas of research interest include plant developmental genes, with particular emphasis on the genes that control potato earliness, plant genes for pathogen resistance, in particular resistance genes for potato late blight, DNA markers of economically important genes and traits, and also various technologies for plant fingerprinting and molecular systematics. Emil Khavkin published two books and over 200 scholarly articles in refereed journals and the volumes of collected papers.

Profile – All-Russian Research Institute of Agricultural Biotechnology (ARRIAB)

The institute was founded in 1974 as the Institute of Applied Molecular Biology and Genetics and later changed to its present name. ARRIAB is engaged in a wide-scope research on plant genes of agronomic importance, with an emphasis of key genes of abiotic and biotic stress, genetic engineering of plants, animals, and viruses, plant cell and tissue culture, molecular breeding technologies, and molecular and immunological diagnostics of plant and animal diseases. Many of these issues are explored as international projects.

Generation of potato transgenic lines resistant to potato leafroll virus and potato spindle tuber viroid

L.M. Khromova[1], M.P. Beketova[1], I.V. Shmyglya[1], S.N. Kirsanova[1], B.A. Biryukova[1], T.E. Koshkina[2], O.A. Shulga[2] and S.K. Zavriev[3]
[1]All-Russian Potato Research Institute, 23, Lorch Street, Kraskovo, 140051, Moscow region, Russia
[2]All-RSIAB, 127550, Moscow, ul. Timiryazevskaya, 42, Russia
[3]Shemyakin and Ovchinnikov Institute of Bioorganic RAS, 117877 Moscow, ul. Miklukho-Maklaya, 16/10, Russia

Abstract

Resistance of transgenic potato lines of the cultivar Zhukovskiy ranniy to Potato Leafroll Virus (PLRV) and to Potato Spindle Tuber Viroid (PSTV) was studied. Genetic construct used for plant transformation carries an expression cassette of hybrid RNA containing the antisense RNA fragment of PLRV replicase (852-990a. a.) and the fragment of PSTV RNA (58-214n.). One from the four generated transgenic lines has demonstrated resistance to both pathogens in laboratory tests with *in vitro* plants. Leaves of transgenic potato plants were mechanically inoculated by sap from a viroid-infected plant. Infection development was tested by RT-PCR. Transgenic potato plants were inoculated with purified PLRV through cut roots. PLRV accumulation in plants was detected by ELISA of samples from the lateral shoot of the virus-infected plants.

Keywords: potato, transgenic lines, potato leafroll virus, potato spindle tuber viroid, resistance

Introduction

Application of gene transfer technologies for improvement of agricultural cultivar characteristics, controllable by transferred genes, offers great potentials for practice. Modern selection of a potato already creates quit perfect blocks of genes borrowed from many wild and primitive species on the basis of interspecific hybridization, and frequently there is a task for revision of separate parameters but not radical reorganization of variety genome. At the same time the potato has vegetative type of propagation and introduced genes are stable inherited.

On the other hand, procedures of potato transformation, regeneration and cloning are well developed and do not represent problems at a stage of works with *in vitro* culture. In this study we generated potato transgenic lines carrying the expression cassette of antisense RNA of a replicase gene of PLRV and the RNA fragment of PSTV. Here, we analyzed the resistance of the transgenic potato plants to PLRV and PSTV.

Materials and methods

To develop resistance to PLRV and to PSTV in transgenic potato plants we used the repression action of an antisense RNA. With this purpose we cloned the cDNA fragment of PSTV (58-214n.) and the cDNA fragment of PLRV replicase (852-990 a. a.). Hybrid RNA consisting from the fragment of the PLRV replicase antisense RNA fused to the RNA fragment of the PSTV will be expressed in plant cells under the control of the enhanced promoter of the CaMV 35S RNA and the nopalinsynthase terminator. The hybrid RNA expression cassette has been cloned in a binary plasmid vector containing also a kanamycin-resistance gene (*NPTII*) for selection of transgenic plants. This vector named pVL1 has been transferred in *Agrobacterium tumefaciens* AGLO and used for transformation of potato plants.

In vitro potato plants of the cultivar Zhukovskiy ranniy (All-RSIP selection) unresisting to virus disease have been transformed in this study. *Agrobacterium*-mediated transformation was done according to early-elaborated in All-RSIP procedure (1, 2, 3). Four individual plants for each generated transgenic line have been tested on resistance to PSTV, 6 to 10 individual plants for each transgenic line have been tested on resistance to PLRV.

For testing of transgenic lines on PSTV resistance we used 1:1 dilution of inocula from *in vitro* plant of hybrid h-946-3 containing PSTV. Tested transgenic *in vitro* plants were released from an agar and placed in glasses with liquid MS-media (4) diluted in 2 times without sugar and hormones. Media was changed on fresh each 3-4 days. A middle leaf of each plant was dusted with carborundum and rubbed with a drop of 20µl of PSTV inocula. Each infected leaf was marked with a thread. In parallel for each transgenic line one plant was processed with carborundum and with the buffer without PSTV. The infected plants for each line were united and exhibited at separate glasses in the common plastic chamber under 100% by humidity in a phytotron (16 hours day, 4000 lux, 25°C). Plants were tested during 30-60 days after inoculation by RT-PCR analysis according procedure for PSTV-testing kit (firm 'Agrodiagnostic'). Samples for analysis were prepared from whole plant without primary infected leaf.

For testing of transgenic lines on PLRV resistance we used a PLRV-isolate from the infected *in vitro* plant (cultivar *Flor Blanka*, All-RSIP collection). Plants were inoculated through the cut roots of *in vitro* plants (8). We took plants from cultural tubes, washed roots from an agar, cut roots for 1/3 lengths, placed plants on MS-solution with 2µg/ml purified PLRV (5). After incubation within 6 hours on natural light at 20°C roots were washed from unsorbed virus, and plants were transferred on liquid MS-media without sucrose and hormones. Plants were cultivated in phytotron. Media was changed on fresh each 3-4 days. After a week plant apex was deleted for provocation of growth of lateral shoots. Lateral shoots were tested by ELISA to detect PLRV according to instruction for diagnostic kit, All-RSIP production (7).

Purified PLRV sample and sap of the infected plants was used as a positive control, sap of non-infected plants was used as a negative control. Obtained data were processed by a method of the dispersive analysis according to (6).

Results and discussion

Potato plants regenerated after transformation on selective medium with kanamycin were tested by the PCR-analysis to demonstrate the genomic integration of the hybrid RNA expression cassette (Figure 1).

According to the PCR results transgenic potato lines Zhukovskiy ranniy R1-4, R1-5a, R1-9, and R1-13/1 contain the insert of a target gene. The transgenic potato lines have been multiplied by sterile cuttings and tested on resistance to PLRV and PSTV infection.

After micropropagation cuttings of each transgenic line and a control non-transformed plant (a control-2) were inoculated mechanically with PSTV, mock-inoculated transgenic potato plants were used as a control-1. The PSTV presence was detected within 60 days by RT-PCR (Figure2). All plants without the transgene insert (lines R1-5, R1-3, R1-13/2) were infected (Table1). Among lines of a potato for which the transgene integration was proved, line R1-9 is allocated. This line was not infected, while others transgenic potato lines (R1-13/1, R1-5a and R1-4) were infected.

The same lines were tested for resistance to PLRV. This virus is not mechanically transmissible, but only by aphid vector. So, we had elaborated the procedure of the early diagnostic test for estimation of plant resistance to PLRV with use of cut roots (8). This method was used to test transgenic potato lines with integrated transgene (Table 2).

Threshold levels to determine infection were established according to (7), where samples were considered infected (P) if the optical density at 450nm (OD_{450}) were above the average of the non-infected controls plus three times the standard deviation:
$P = 0.219.\ (_{P0.219} = _{X0.\ 041} + 3E_{\ (3 \times 0.059 = 0.177)})$.

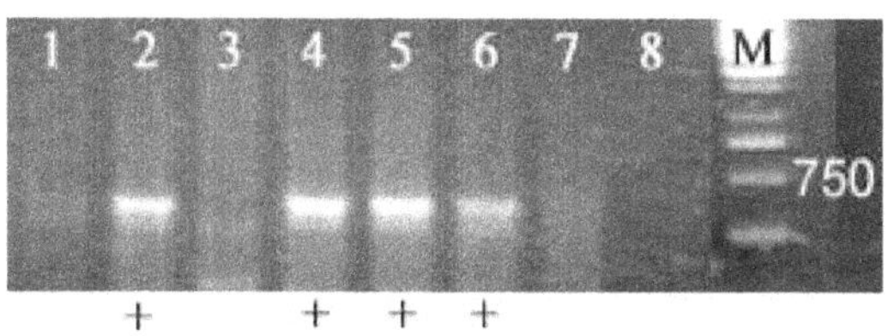

Figure 1. Agarose gel electrophoresis of PCR products amplified with VL1A1/VL1B1 primers from total DNA from potato kanamycin resistant plants.
1 line Zhukovskiy ranniy R1-3,
2 line Zhukovskiy ranniy R1-4,
3 line Zhukovskiy ranniy R1-5,
4 line Zhukovskiy ranniy R1-5a,
5 line Zhukovskiy ranniy R1-9,
6 line Zhukovskiy ranniy R1-13/1,
7 line Zhukovskiy ranniy R1-13/2,
8 negative control,
9 O'GeneRuler 1kb DNA Ladder ('Fermentas'). Expected fragment size is 655 b. p.

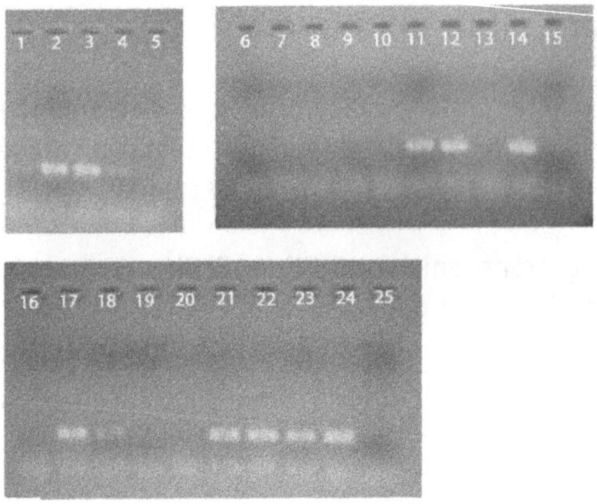

Figure 2. Agarose gel electrophoresis of RT-PCR products amplified from extractsof viroid-infected plants. Lanes 1-4 - line Zhukovskiy ranniy R1-13/1, infected with PSTV, lane 5- line Zhukovskiy ranniy R1-13/1, control-1. Lanes 6-9 - line Zhukovskiy ranniy R1-9, infected with PSTV, lane 10 - line Zhukovskiy ranniy R1-9, control-1. Lanes 11-12 - line Zhukovskiy ranniy R1-5a, infected with PSTV, lane 13 - line Zhukovskiy ranniy R1-5a, control-1. Lane 14 – PSTV sample, positive control, lane 15 – negative control.
Lanes 16-19 - line Zhukovskiy ranniy R1-4, infected with PSTV, lane 20 - line Zhukovskiy ranniy R1-4, control-1. Lanes 21-24 – non-transgenic Zhukovskiy ranniy, infected with PSTV, control-2, lane 25- non-transgenic Zhukovskiy ranniy, control-1.

Table 1. Analysis of transgenic potato lines on the PSTV resistance.

Plants	**PCR detection of transgene insert**	**PSTV presence in the inocula**	**RT-PCR detection of PSTV presence**
Zhukovskiy ranniy (control-2)	–	+	+
R1-5	–	+	+
R1-3	–	+	+
R1-13/2	–	+	+
Zhukovskiy ranniy (control-1)	–	–	–
R1-4	+	+	+
R1-4 (control-1)	+	–	–
R1-9	+	+	–
R1-9 (control-1)	+	–	–
R1-5a	+	+	+
R1-5a (control-1)	+	–	–
R1-13/1	+	+	+
R1-13/1 (control-1)	+	–	–

Table 2. ELISA values at various days post-inoculation with PLRV of infected and non-infected transgenic potato plants.

Plants	**ELISA value (OD_{450}) at**		**Standard deviation**	**(+)Positive infected (–) Negative infected**
	18 dpi	**52 dpi**		
Zhukovskiy ranniy, non-transgenic	0.100	0.950	±0.279	+
Zhukovskiy ranniy R1-4	0.030	0.268	±0.131	–/+
Zhukovskiy ranniy R1-5a	0.080	0.620	±0.192	+
Zhukovskiy ranniy R1-9	0.050	0.175	±0.096	–
Zhukovskiy ranniy R1-13/1	0.105	0.440	±0.207	+
Positive control	0.100	0.449		
background (-)	0.100	0.041		
E- the maximal deviation from an average among values of a background		0.059		
P- threshold of reliability		0.219		

As can be noticed in Table 2 it was not possible to infect authentically with PLRV the line Zhukovskiy ranniy R1-9, other lines have shown the values exceeding a threshold of reliability that testifies to their ability to infection. Summary data on resistance of transgenic potato lines to both pathogens as the ability to resist to PLRV and to PSTV infections at testing conditions are shown in the Table 3.

Our results indicate that the line Zhukovskiy ranniy R1-9 has complex resistance to both pathogens. It is important to mention that these results should be considered as preliminary as they are received on *in vitro* plants with application of the tests developed for juvenile plants. The used tests have for an object to reject pathogen sensitive lines and to select perspective lines for the further tests on pathogen resistance.

Table 3. The summarized results of the laboratory testing of transgenic potato lines on resistance to PLRV and to PSTV.

Variety/line	**Resistance to**	
	PSTV	**PLRV**
Zhukovskiy ranniy St	–	–
Zhukovskiy ranniy R1-4	–	–
Zhukovskiy ranniy R1-9	+	+
Zhukovskiy ranniy R1-5a	–	–
Zhukovskiy ranniy R1-13/1	–	–

Line Zhukovskiy ranniy R1-9 was micropropagated. The mini tubers were produced in a greenhouse for further field investigations.

References

Rassadina, G.V., N.O. Yur'eva, N.N. Efremova (1995). Methodical instructions for reception of the transformed plants of a potato, M.

Butenko, R.G, L.M. Khromova, G.V. Sednina (1984). Methodical instructions for reception of alternative cellular lines and plants of different cultivars of a potato, M.

Simakov, E.A., A.I. Uskov, Yu.A. Varitsev, V.P. Knyazeva, S.M. Musin, G.I. Antonova, V.V. Boyko (2000). New technology of the improved initial material in elite seed-growing of a potato, M.

Murashige, T., F. Scoog (1962). A revised medium for rapid growth and bioassays with tobacco tissue cultures In: Physiol plantarum, v. 15, pp. 473-497.

Varitsev, Yu.A., V.K. Novikov, L.V. Chugunova and G.P. Varitseva (1991). Some methodical aspects of reception of an ELISA diagnostic kit to a Potato Leafroll Virus', Works of All-RSIP, pp. 25-33.

Dospekhov, B.A. (1973). Technique of field experience, M. Kolos, p. 336.

The instruction on application of an ELISA diagnostic kit for definition of potato virues (2006). SSI All-RSIP, M., Korenevo.

Khromova, L.M., Yu.A. Varitsev, I.V. Shmyglya, S.N. Kirsanova, M.P. Beketova, G.P. Varitseva (2005). Development of the early diagnostic test for an estimation of efficiency of the genes supervising resistance to a Potato Leafroll Virus In: Materials of the 2nd All-Russia congress on protection of plants 'Fytosanitory improvement of ecosystems', S.-Petersburg, T. 1, pp. 574-577.

Means of biochemical identification of potato varieties

A.S. Sologub and P.A. Melnyck
Ukrainian scientific-research station of plants' quarantine,UAAS, v. Boyani, Novoselitza district, Chernivtsy region, Ukraine, ndskr@myrealbox. com

Abstract

The protein of tubers of investigated varieties of potato by a method of isoelectrofocusing in PAAG, have individual protein spectrum, which does not repeat; that permits to characterize objectively a breed by its originality. We consider, that all varieties of potato, which are included in the state register, should pass biochemical passportization. Together with this the given method is irreplaceable at breed expertise of seed material, and also permits a quarantine service to control import of varieties at a zone of distribution of potato wart. The method of biochemical identification of potato varieties gives a possibility to determine more exactly the unknown variety by biochemical parameters in two hours.

Keywords: isoelectrofocusing, modificational, genetic markers, protein spectra, polyacrylamide gel, ampholytes, supernatant, identification

Introduction

The majority of farm-valuable plants is exposed to significant modificational variability under influence of the environment. It complicates selection of pairs at crossing and selection of desirable genotypes from hybrid populations. It is necessary to have such methods of estimation, which help to judge authentically a genotype of a plant. The selection of valuable genotypes presently is conducted intuitively, and for a reliable estimation it is necessary to have the exact information on a genotype on the basis of genetic markers.

There is a known way of identification of potato varieties to their morphological features – form and stain of tubers, flowers, stalks and leaves. Such (this) variant of identification of varieties and hybrids of a potato is long and labour-consuming, as it occupies all the vegetative period of plant development. From the applied point of view the usage of generically determined protein polimorphysm is especially important at potato varieties passportization, as the success of selection work depends on their correct identification. The proteins are markers of concrete structural genes. Allelic variants are inherited codominantly, their structure does not depend on conditions of life, they are rather easily determined with the help of electrophoresis.

Methods

The identification of varieties of potato by protein spectra at the usage of electrophoretic method is simple, does not require any special expenses and time, due to what the estimation of breed features is possible as the protein structures.

In our work the version of electrophoresis - isoelectrofocusing in polyacrylamide gel was used, where the proteins are divided in an electric field simultaneously both in a gradient of a voltage, and in a gradient pH, which are created in gel by ampholyte-carriers. Ampholytes form a linear gradient pH, and each protein is concentrated in a zone pH according to its isopoint, i.e. the protein is focused owing to loss of a charge. The material for research was the varieties of potato of the register of Ukraine given by Institute of Potato Production, UAAS.

For biochemical identification of varieties a tuber tissue was pounded at a cold in a porcelain mortar. The supernatant after extraction and centrifugation of sap was used for the analysis. Proteins isoelectrofocusing conducted in polyacrilamide gel (T-5 of %, C - 3%) with ampholyte of a range pH 3.0-10.0. The gel from ampholytes was washed by a solution Tno:alcohol:water in the ratio (10:25:65). Colouring of protein spectra was made with the help of a stain Cumassie blue P-250 in a solution acetic acid: alcohol:water (10:25:65). Surplus of a stain from the gel was washed by the same solution (Figure 1). Account of protein tracks was conducted with the help of the program developed in UkrSRSPQ.

The offered way of identification permits to determine a breed of potato in 24 hours.

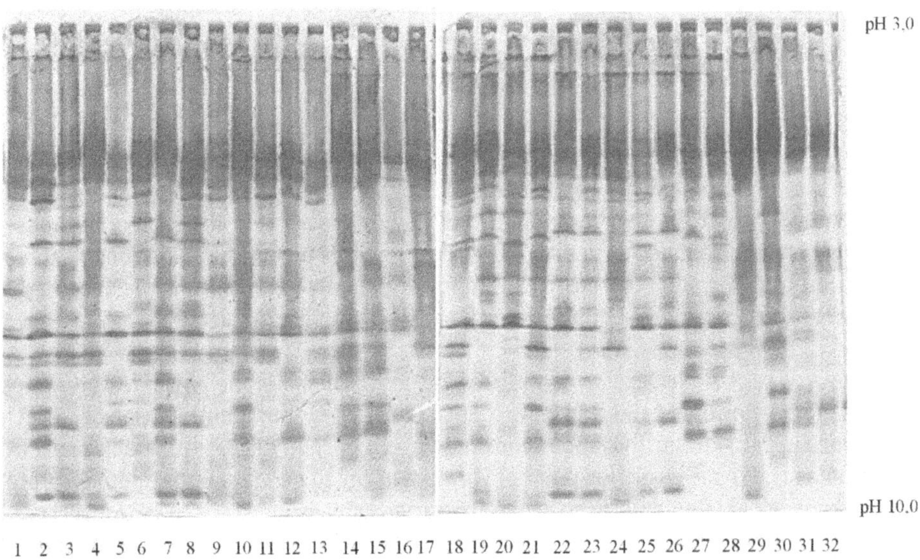

Figure 1. Protein spectrums (passports) of potato varieties of Ukraine register.

1	*Bagryana*	*9*	*Kosin*	*17*	*Olvia*	*25*	*Serpanok*
2	*Borodyans'ka pink*	*10*	*Leleka*	*18*	*Rakurs*	*26*	*Malych*
3	*Bozhedar*	*11*	*Nezabudka*	*19*	*Fantasia*	*27*	*Kobza*
4	*Buyan*	*12*	*Nemeshayivska*	*20*	*Lugovska*	*28*	*Kupava*
5	*Vodogray*	*13*	*Slovyanka*	*21*	*Lybid'*	*29*	*Zov*
6	*Dobrochin*	*14*	*Svitanok Kyyivsky*	*22*	*Polliska pink*	*30*	*Zarevo*
7	*Zlagoda*	*15*	*Povin'*	*23*	*Virynea*	*31*	*Horos*
8	*Gart*	*16*	*Poran*	*24*	*Obriy*	*32*	*Yavir*

Conclusions

We consider, that all varieties of potato, which are included in the state register, should pass biochemical passportization. Together with this the given method is irreplaceable at breed expertise of a seed material, and also permits to a quarantine service to control import of varieties at a zone of distribution of potato wart. The given method is to be used by selective and research organizations of Ukraine for biochemical identification of potato varieties.

Acknowledgements

The Proceedings were justified by the scientific Board of UkrSRSPQ in December 7, 2006, Protocol No.5.

References

Nazar V.M. Instruction on potato approbation. K., 1989. – 7p. (Назар В. М. Руководство по апробации картофеля. К. 1989. - 7 с.).

Kuchko A.A. Our potato varieties – your success' guarantee Nemishayevo. 1998. – 18p. (Кучко А.А. Наши сорта картофеля - залог вашего успеха. Немішаєво. 1998. - 18 с.).

Osypchuk A.A. Main characteristics of potato varieties. Nemishayevo . 1999. – 82p. (Осипчук А.А. Основные характеристики сортов картофеля. Немішаєво. 1999. - 82 с.).

Righetti T. Isoelectric focusing. Theory, methods, implementation. G., Mir, 1986. – 300p. (Ригетти Т. Изоэлектрическое фокусирование. Теория, методы, применение. Г.: Мир, 1986. - 300 с.).

Troyitskiy I.G. Isoelectric focusing of proteins in self-organizing and artificial pH gradients. – K. Nauchnaya mysl, 1984. – 280p. (Троицкий И. Г. Изоэлектрическое фокусирование белков в самоорганизующихся и искусственных рн градиентах. - К. Научная мысль, 1984. - 280 с.).

Sologub A.S., Zelya A.G., Melnyck P.O. Method of biochemical identification of potato varieties. Patent UA No. 64565 A from 16-02-2004. (Сологуб А. С., Зеля А. Г., Мельник П. А. 'Способ биохимической идентификации сортов картофеля' - Патент UA No.64565 A от 16-02-2004 г.).

Morphological characters and appearing resistance in potato somaclones, transgenic plants and somatic hybrids

G.A. Yakovleva, I.A. Rodkina, T.V. Semanyuk, V.L. Dubinich, S.V. Manarkhovich and N.I. Podobed
Research and Practical Centre of National Academy of Sciences of Belarus for Potato, Fruit and Vegetable Growing, Potato Biotechnology Laboratory, 2-a Kovaleva St., Samokhvalovichi, 223013 Minsk, Belarus

Abstract

Somaclonal variability, transgenic plants and somatic hybridis are studied in the laboratory of biotechnology from the point of view of their significance for potato breeding. Potato somaclones were selected for resistance to common scab caused by 5 species of *Streptomyces* on infected backgrounds *in vitro* and *in vivo*. The resistance to common scab for 77.9% of all preliminary selected *in vitro* somaclones was in the range of 7-9 scores during three years of field experiments. Appearance of three new species of *Streptomyces* in natural population of pathogen in Belarus has changed the situation and decreased the number of resistant somaclones.

Transgenic potato lines of Belarusian cultivar Belorusky 3 (B3) with coat protein (CP) gene of potato virus Y (PVY) were obtained from the Moscow Bioengeneering Centre. Transgenic lines were tested in the field for 6 years after inoculation by PVY. Healthy lines according to phenotype and ELISA-test were revealed among transgenic plants. A three year scheme was proposed for evaluation and selection of PVY-resistant transgenic potato with stable display of target trait. It includes two years inoculation by PVY and two years of field tests. PVY-resistant lines were also revealed among sexual progeny of transgenic potato cv. B3. Bloomed somatic hybrids have been obtained in following combinations: 1) $78563\text{-}76_{cld}$ (tbr, 4x) + *S. bulbocastanum*; 2) LDH (tbr, 2x) + *S. bulbocastanum*; 3) $78563\text{-}76_{cld}$ + *S. polyadenium*; 4) $78563\text{-}76_{cld}$ + (*S. brevidens* × *S. etuberosum*); 5) 86-6 [(LDH × *S. chacoense*), 2x] + *S . cardiophyllum*; 6) 86-6 + *S. etuberosum*; 7) 86-6 + (*S. etuberosum* × *S. brevidens*). The berries with seeds and seedlings have been obtained in combinations No 1, 3, 6, 7. The somatic hybrids had intermediate phenotype and formed the tubers by the type of cultivated parter.

Keywords: somaclones, common scab, transgenic plants, PVY, somatic hybrids

Introduction

Potato biotechnology is based on the capacity to regenerate shoots *in vitro*. Potato plants were successfully regenerated from protoplasts, tissue culture of leaf, stem, root and tuber. Regeneration ability depends on particular genotype. Callus appearance is possible source of useful changes for potato breeding and it is undesirable in genetic transformation because it presents potential danger for commercial cultivar loss. Development of cell and tissue culture methods provide potato breeding with non-traditional possibilities which are determined by utilization of somaclonal variability, transgenic plants and somatic hybridization.

Selection of potato somaclones on resistance to common scab

Somaclones of cvs. Yavar, Altair, Lazurit, Resy, Revelino and 78563-76 were obtained from primary callus culture and selected for resistance to common scab on infection backgrounds *in vitro*, in the field and also under natural conditions (Table 1).

Suspension of 5 species of *Streptomyces: S. scabies, S. griseus, S. violaceus, S. candidus, S. globisporus* (3.6-4.6×10^6 spores/ml) was used for inoculation of microtubers *in vitro*. Infection in the field was conducted by introduction under each tuber the mixture of 3-5 species of *Streptomyces* and skin from heavy infected tubers. The presence of selection pressure in the

Table 1. Selection of potato somaclones resistant to common scab.

Years of evaluation	**1991-1997**	**1995-1999** **4 series: 1995-1997; 1996-1998; 1997-1999; 1998-1999**	**1998-2000**
Initial material	Microtubers of 623 somaclones	Tubers of 96 somaclones	Tubers of 78 somaclones
Conditions	*In vitro* pH 5.6-5.7, 28°C, dark	*The field* pH: 1995 – 6.0, 1996 – 4.8, 1997 – 5.2, 1998 – 4.9, 1999 – 5.4	*The field* pH: 1998 – 6.1, 1999 – 6.4, 2000 – 4.6
Infection background	Artificial 5 species of *Streptomyces*: *S. scabies, S. violaceus, S. globisporus, S. candidus, S. griseus*	Artificial 3-5 species of *Streptomyces*: *S. scabies, S. violaceus, S. globisporus, S. candidus, S. griseus* The skin from heavy infected tubers of previous year	Natural 1999 – new aggressive species: *S. chromofuscus, S. violaceoruber, S. melanosporofaciens*
Selection pressure – comparison of resistance in scores between resistant and susceptible genotypes with Student criterion (*t*).	-	*t*: 1995 – 12.8 1996 – 5.2 1997 – 9.5 1998 – 17.3 1999 – 7.3	*t*: 1998 – 7.1 1999 – 10.5 2000 – 10.8
Resistant genotypes, %	15.4	81.3	0.8

field was confirmed by authentic significance of Student criterion, when comparing resistance in scores between resistant and susceptible genotypes. The resistance to common scab for 77.9% of all preliminary selected *in vitro* somaclones was in the range of 7-9 scores during three years.

The situation has changed in dry and hot 1999 year when nearly all commercial cultivars and early selected resistant somaclones were affected by common scab lesions in natural conditions. Disease development didn't nearly depend on pH. One of the reasons is change in natural population of pathogen and discovery in Belarus of 3 new aggressive species of *Streptomyces*: *S. chromofuscus, S. violaceoruber, S. melanosporofaciens* (Ivanyuk *et al.*, 2005). Somaclones which kept on medium resistance to common scab under changeable environment were introduced into the culture *in vitro*.

The evaluation of somaclones on resistance to common scab was made anew in five years by introduction of suspension of eight species of *Streptomyces* into area of microtuber formation *in vitro*. These somaclones revealed the preservation of character 'resistance to common scab'. The greenhouse experiment didn't confirm their level of resistance to disease.

Tests for distinctess, homogeneity and stability revealed somaclones which differed from initial genotype by morphological characters and exceeded it - 20-40% by content of total protein and dry matter. All somaclones were identical to initial cultivar by electrophoretic patterns of tuber soluble proteins.

Selection of transgenic potato resistant to PVY in vegetative and sexual generations of cv. Belorusky 3 plants with gene of coat protein Y-virus

Transgenic potato lines of cv. Belorusky 3 (B3) with coat protein (CP) gene of potato virus Y (PVY) strain N were obtained in Moscow Bioengeneering Centre by agrobacterial transformation of internodal stem segments with 4 cassettes of expression: A – p35SPLYCP (8 lines); B - p35SPLYCPdelATG (8 lines); C – p35SX28YCP (7 lines); D - p35SX28YCPdelATG (10 lines). They expressed foreign RNA and (or) protein (Bagyan *et al.,* 1997). The shoot regeneration was induced under the conditions of minimal development of callus on the selective medium with kanamycin.

We studied the accumulation of PVY in systemic leaves of transgenic B3 after artificial two-year inoculation and following growing on infection background in the field by analysis of visual symptoms of disease and ELISA-test. We tested transgenic plants during long-term vegetative propagation (6-7 years) and their sexual progeny.

Plants of non-transformed cv. B3 and transgenic line without target gene were heavy affected by Y-virus at the end of the experiment. It is noted that the tubers of initial B3 lost the ability to germination after mechanical inoculation for two years running. Healthy lines according to phenotype and ELISA-test were revealed among transgenic CP PVY plants. These lines were healthy or weakly affected by Y-virus in the year of inoculation (1996) and healthy after repeated inoculation (1997) and during the following years of propagation of one and two year of inoculated plants. They also didn't contaminate the indicator plants of tobacco cv. Samsun. Inoculation of tobacco plants by juice of field transgenic plants is necessary for determination in analyzed samples the intact virus particles which able to infect the potato. Seven transgenic sources of resistance to PVY: B – 2, C – 4, D – 1 have been selected between 33 transgenic lines.

As a result we propose the scheme of evaluation and selection of PVY-resistant transgenic potato with the sequence of Y-virus genome for three years:

- First year – first inoculation by PVY of 3-6 plants for every line in greenhouse;
- Second year – second inoculation in the field where 3 tubers of all healthy and weakly Y-virus affected plants of every line are planted;
- Third year – three tubers of every healthy plants are planted in the field between heavy affected plants; tests with indicator plants of tobacco cv. Samsun and graft on Y-virus affected (Y^N+Y^o, Y^{NT}) tomato in the greenhouse for determination of type of resistance and revealing PVY-immune transgenic lines.

During second and third years we tested the morphological characters according to guidelines of UPOV (*Solanum tuberosum* L.) for determination the safety of principal parameters of commercial cultivar. Such scheme permits to select transgenic line with stable display of target trait in the third year of the experiment.
Transgenic lines of B3 were identical to the initial cultivar by morphological characters, yield, content of total protein, dry matter and electrophoretic patterns of tuber soluble proteins.

We used proposed 3-year scheme for analysis of kanamycin-resistant seedlings of 3 crosses: (1) c1 × cv. Kapris (17 hybrids, 91 plants); (2) c3 × cv. Kapris (51 hybrid, 293 plants); (3) cv. Arkhideya × c9 (32 hybrids, 185 plants). The splitting of hybrids by homogeneity of display the target trait was found for all three years of the experiment. The kanamycin-resistant hybrids were divided into three groups: (1) all plants of the hybrid were healthy; (2) different plants of one and the same hybrid had negative (healthy plant) or positive (affected plant) significance by the ELISA-test; (3) all plants of hybrid were PVY-affected according to ELISA (Table 2).

The plants of cv. Arkhideya for two years tests were in second group in contrast to cv. Kapris where all plants were affected by PVY during first and second years of the experiment. All plants of both cultivars accumulated Y-virus in the third year according to ELISA data. The plants of cv. Arkhideya had heavy symptoms of PVY affecting.

Tendency of non-stability in expression of target trait was found in vegetative generations of F_1 of transgenic lines. Nevertheless, in all combination there were selected hybrids which were homogeneous and stable in displaying target trait 'resistance to PVY' during three years: c1 × Kapris – 1 hybrid, c3 × Kapris – 14 hybrids, Arkhideya × c9 – 10 hybrids.

The productivity of kanamycin-resistant hybrids and high resistance to PVY were not correlated in all the cases. We have been selected 25 hybrids which were stable in displaying target trait 'resistance to PVY' and free from viruses PVX, PVM, PVA. Among them 11 had yield equal or more than control variants (transgenic lines, cvs. Kapris, Arkhideya, Belorusky 3).

Table 2. Distribution of kanamycin-resistant hybrids from 3 crosses of transgenic lines with CP PVY gene and early cultivars by homogeneity in displaying trait 'resistance to PVY'.

Combination	Year of evaluation	Hybrids in all	Hybrids of 1-st group		Hybrids of 2-d group		Hybrids of 3-d group	
			n	%	n	%	n	%
c1 × Kapris	first	17	16	94.12	1	5.88	-	-
	second	17	4	23.53	13	76.47	-	-
	third	12	1	8.33	9	75.0	2	16.67
c3 × Kapris	first	51	39	76.47	12	23.53	-	-
	second	49	19	38.78	30	61.22	-	-
	third	44	14	31.82	23	52.27	7	15.97
Arkhideya × c9	first	32	32	100	-	-	-	-
	second	31	22	70.97	9	29.03	-	-
	third	28	10	35.71	12	42.86	6	21.43

Increasing the potato gene pool accessible for breeding by somatic hybridization

Protoplasts of cultivated and wild genoty[pes were isolated from leaves of *in vitro* plants and fused with PEG according to protocols N3 and N6 described by Sidorov *et al.* (1985). Following culture and regeneration of shoots were conducted on media, which are presented in author's certificate of SU (Sidorov *et al.,* 1990). Identification of hybrids was conducted by analysis of peroxidase isoenzymes from plants *in vitro* and (or) electrophoretic patterns of tuber soluble proteins and morphological characters of plants and tubers. Initial cultivated parters in somatic hybridization: chlorophyll deficient mutant of *S. tuberosum,* 4x (78563-76_{cld}); *S. tuberosum,* 2x (LDH – dihaploid cv. Lasunak); interspecific sexual hybrid *S. tuberosum* × *S. chacoense* (86-6, 2x). As wild parters were used tuberous diploid species and non-tuberous diploid species and sexual hybrids. Tuberous wild species: *S. bulbocastanum* (Sb), *S. polyadenium* (L39-2), *S. cardiophyllum* (L8). Non-tuberous parters: *S. etuberosum* (E55-1), *S. brevidens* × *S. etuberosum* (L48-2), *S. etuberosum* × *S. brevidens* (L49-2).

Somatic hybrids were regenerated in 7 combinations (Table 3). Hybrid nature of shoots regenerated in combinations with non-tuberous wild parters 48, 2D, 4D was confirmed by isoperoxidase; with tuberous wild parter F – only by profiles of soluble proteins and SB, DL, 1D by both markers. Peroxidase isoenzymes of regenerates F were identical to cultivated parent 78563-76.

Somatic hybrids of six combinations excluding F showed the different extension of anthocyanin coloration for stem of plant *in vitro*. Wild parent of combination F did not show anthocyanin

Table 3. Verification of hybridity of plants regenerated in 7 combinations of protoplast fusion by peroxidase isoenzymes or profiles of soluble proteins.

Combination	Parters for somatic hybridization		Verification of hybridity of regenerants. Isoperoxidase* or profiles of soluble proteins**	
	Cultivated	Wild	Analyzed shoots	Hybrds
SB	78563-76$_{cld}$	*S. bulbocastanum*	34	34*
DL	LDH	*S. bulbocastanum*	17	17*
F	78563-76$_{cld}$	*S. polyadenium*	14	14**
1D	86-6	*S. cardiophyllum*	148	6*
48	78563-76$_{cld}$	*S. brevidens×S. etuberosum*	12	12*
2D	86-6	*S. etuberosum*	987	59*
4D	86-6	*S. etuberosum×S. brevidens*	47	47*

coloration in any part of the plant. All other wild parters revealed anthocyanin coloration of stem of different intensity and localization in contrast to cultivated parents, which were not stained by anthocyanin. It is possible that anthocyanin coloration of stem *in vitro* is bound with wild species in pedigree of interspecific hybrid.

Morphological characters of parents were displayed in plants of interspecific somatic hybrids of all combinations *in vivo* (greenhouse or field) in different degree by traits of leaf (shape), rarely of corolla and stem and presence of anthocyanin coloration in different parts of the plant. Parent characters evidently were shown in plants of somatic hybrids of SB, DL and 1D. Somatic hybrids of combinations SB and DL revealed traits specific to *S. bulbocastanum*: anthocyanin coloration of leaf and stem, and *S. tuberosum* - leaf blade with leaflets (dissected leaf), faceted stem. For somatic hybrids of combination 1D morphological character of 86-6 was presence of anthocyanin coloration in axil of leaf, *S. cardiophyllum* – ovoid shape of terminal leaflet and cordate shape of base of terminal leaflet. Lines with morphological characters of both parents and with intermediate characters were presented in populations of somatic hybrids of 48, 2D and 4D. Morphological characters of parents in combination F were negligible.

The cultivated parents of combinations F, SB, 48 is one and the same - 78563-76 which formed yield of ware tubers on short stolons, trait specific to commercial potato cultivars. Mexican species of *S. bulbocastanum* and *S. polyadenium* – wild parents of combinations SB and F formed in conditions of Belarus small tubers with diameter of 1 - 2 cm on long stolons at the end of vegetative season. Essential character inherent to somatic hybrids of F and SB is ability to set medium and large tubers typical for tetraploid *S. tuberosum* L. Initial wild parter *S. brevidens* × *S. etuberosum* of combination 48 is unable to set tubers. The appearance of tubers in greenhouse was noted for 10 lines from 11 ones in population of somatic hybrids of combination 48.

Cultivated parents of combinations of DL, 1D, 2D, 4D are presented by diploid potato: dihaploid of cv. Lasunak (LDH) for DL and interspecific sexual hybrid *S. tuberosum* × *S. chacoense* (86-6) for 1D, 2D, 4D. The formation of tubers of DL was like to cultivated parent LDH with its characteristic defects – tubers of seed fraction with the diameter not more than 3 cm on long stolons. The tubers both parents of combination 1D: 86-6 (+) *S. cardiophyllum* are on long stolons. But the tubers of 86-6 were larger than *S. cardiophyllum*, 3 and 1 cm accordingly. Somatic hybrid 1D-37-4 had the tubers on long stolons as both parents but their size corresponded to cultivated parent 86-6. Wild parters of combination of 2D and 4D are non-tuberous. The formation of stolons, tuber-like structures and tubers was noted in populations of somatic hybrids 2D and 4D. Presence of the trait 'ability to set tubers' was noted for 17 hybrids of combination 2D and 10 – 4D.

Thus, the trait 'formation of tubers' for somatic interspecific hybrids of 7 combinations is determined by cultivated parent when displaying this character.

The somatic hybrids of all combinations were able to bloom. But intensity of florescence and also inflorescence size, frequency of flowers were different between separate somatic hybrids of one and the same combination and various combinations. Somatic hybrids of combination 48 were unable to set berries, while berries of hybrids of 1D and DL as a rule were without seeds. Successful backcrosses with *S. tuberosum* were conducted with somatic hybrids of combinations F, 2D, 4D. First sexual generation was obtained for somatic hybrids of F, SB, 2D, 4D and second – for SB, 2D and 4D. The formation of single plump seeds and regeneration of normal seedlings is possible for somatic hybrids of combination DL.

Conclusion

Somaclonal variability can be used to obtain new forms of potato which differ from initial genotype by 1 – 2 traits. The use of somaclonal variability is limited by considerable expenses to conduct numerous tests because the desirable trait is accidentally changed and the proportion of phenotypes which are useful for breeding is very small. The situation is simplified if there is the possibility to select the desirable trait *in vitro* when this trait can be modeled at the present stage of development. It is very difficult to model the condition to show the trait 'resistance to common scab' because this disease can be caused by more than 20 species of *Streptomyces* and natural population of pathogene can change. It is necessary to carry out two-three years of vegetative propagation to stabilise the showing of one or other trait. We observed the predominance of a tendency for recovery of morphological characters of somaclones from anther culture to phenotype of initial genotype during tuber propagation (Rodkina and Yakovleva, 2001).

Revealing the target trait is changed in transgenic potato plants of cv. Belorusky 3 with CP PVY gene in direction as its improvement and worsening in comparison with initial cultivar. The determination of character change of a target trait is possible for three years on the condition of creation of a strong selection pressure. The special approach is required to make the conditions of adequate selection to resistance. We used two-year inoculation by PVY in contrast to generally accepted one-year inoculation.

The sources of resistance to PVY have been selected between transgenic lines cv. Belorusky 3. Most of clones of seven selected transgenic sources of resistance to PVY were healthy according to ELISA test during long-term vegetative propagation in conditions of artificial infection background after two-year Y-virus inoculation. PVY-resistant hybrids have been selected between kanamycin-resistant hybrids of 3 crosses with breeding lines by proposed scheme which includes three years of evaluation in conditions of hard selective pressure.

Interspecific somatic hybrids have been obtained in 7 combinations of protoplast fusion between cultivated and wild potato by protocol of chemical fusion with PEG. Tuberous and non-tuberous forms of wild potato were used. Somatic hybrids were characterized by intermediate phenotype and ability to form tubers by type of cultivated parent. Somatic hybrids have been successfully crossed with breeding lines. Therefore these interspecific somatic hybrids can be used to increase the potato gene pool accessible for breeding.

References

Bagyan, I.L., Gulina, I.V., Kraev, A.S., Mironov, V.N., Padegimas, L.V., Poogin, M.M., Revenkova, E.V., Shhhchennikova, A.V., Shulga, O.A., Socolova, M.A., Vecente-Corbajosa, Yu., Yakovleva, G.A., Skryabin, K.G. (2005). Plant Gene Technology. In: C. Nicolini (Ed.) Genome Structure and Function. From Chromosomes Characterization to Genes Technology. Kluwer Academic Publishers, Dordrecht/Boston/London, NATO ASI Series 31: 279-318.

Ivanyuk, V.G., Banadicev, S.A., Zhuromsky, G.K. Proteection of potato from diseases, pests and weeds (2005). Belprint, Minsk, 696 p. (In Russ.).

Rodkina I.A. and Yakovleva G.A. Variability of morphological characters of potato androclones during vegetative reproduction (2001). Biotechnology 2: 31-39. (In Russ.).

Sidorov, V.A., Piven, N.M., Gleba, Y.Y., Sytik, K.M. (1985) Somatic hybridization of Solanaceae. Naukova Dumka, Kiev (In Russ.).

Sidorov, V.A., Samoilov, V.M., Gleba, Y.Y. Method of obtaining plant-regenerants from potato protoplasts (1990). Author's Certificate SU 1604275 A1, A 01 H 4/00. Published 07-11-90. Bulletin N41.

Curriculum vitae – Galina A. Yakovleva

Galina A. Yakovleva studied plant biochemistry at the Kazan State University for 1970 – 1975. Then, she took the post-graduate course at the Institute of Plant Physiology USSR Academy of Sciences, Moscow. Galina A. Yakovleva received PhD degree from this Institute in 1984 based upon her work on transport of ion through thylakoid membranes of chloroplasts. She began to work with potato culture in 1984. Now she is the head of potato biotechnology laboratory at Research and Practical Centre of National Academy of Sciences of Belarus for Potato, Fruit and Vegetable Growing. Galina A. Yakovleva leads the research projects in the state programs of Belarus on genetic resources, potato breeding, genetic engeneering. Her main scientific interests in potato breeding and biotechnology are: use of tissue culture approaches and molecular markers in plant breeding; use of wild species, in particular by somatic hybridization; genetic engineering of plants. She is the author of over 100 scientific publications, including conferences. Galina A. Yakovleva is a leader of the group in professional organisations: Belarusian Society of Geneticists and Breeders.

Profile – Research and Practical Centre of National Academy of Sciences of Belarus for Potato, Fruit and Vegetable Growing

Potato is one of the leading agricultural crops in Belarus. Research and Practical Centre of National Academy of Sciences of Belarus for Potato, Fruit and Vegetable Growing includes Potato Department, which presents all directions of scientific and practical research on potato in Belarus. The basic directions in work of the Potato Department are the following: collection and study of world potato gene pool; theory elaboration of physiological-biochemical, genetic, immunological, biotechnological methods of breeding; release of new potato varieties with high consumer qualities for different economic purposes, resistant to main pests and diseases; working up the methods and production technologies, accelerated propagation and growing of seed potato varieties of high quality; creation and production of modern resource-saving technologies for potato cultivars. Elaborations, research potential of our Institute are world interest. The priority of Belarusian selection for breeding highly-starch varieties is generally accepted. We support and develop relations with many countries, firms and we are interested in expansion of cooperation with all the parters.

www.ingramcontent.com/pod-product-compliance
Lightning Source LLC
Chambersburg PA
CBHW081124180926
51611CB00004B/44
* 9 7 8 9 0 8 6 8 6 0 4 2 5 *